T5-CVD-964

CRYSTALLOGRAPHIC GROUPS

CRYSTALLOGRAPHIC GROUPS

T. JANSSEN

Katholieke Universiteit Nijmegen,
Nijmegen, The Netherlands

1973

NORTH-HOLLAND PUBLISHING COMPANY – AMSTERDAM • LONDON
AMERICAN ELSEVIER PUBLISHING COMPANY, INC. – NEW YORK

Library of Congress Catalog Card Number: 72–96140

North-Holland ISBN: 0 7204 0194 1
American Elsevier ISBN: 0 444 10436 4

Publishers:
NORTH-HOLLAND PUBLISHING COMPANY – AMSTERDAM
NORTH-HOLLAND PUBLISHING COMPANY, LTD. – LONDON

Sole distributors for the U.S.A. and Canada:
AMERICAN ELSEVIER PUBLISHING COMPANY, INC.
52 VANDERBILT AVENUE, NEW YORK, N.Y. 10017

PRINTED IN THE NETHERLANDS

PREFACE

The present work grew out of a series of lectures given in 1968 and 1969 by the author at the Universities of Nijmegen and Leuven. The intention of these lectures was to give a working knowledge of group theoretical methods used in solid state physics. There are many books on physical applications of group theory, but very few of them are devoted to the symmetry of the solid state. Therefore, when mimeographed notes appeared, the author was pressed from several sides to publish them as a book. In the present book the aim is to establish a bridge between the formal language of mathematical group theory and the more down-to-the-earth way of speaking about symmetries by solid state physicists. This means that on one side the fundamental mathematics is discussed in a rather rigorous way, but that often full proofs will be omitted. On the other hand rather general problems in solid state physics are discussed, but without going into all the details of the various specific problems. The emphasis is on the properties of the symmetry groups of crystal physics and on the specialization of the general group theoretical methods to formulations as used in solid state physics. Reference will often be made to the standard mathematical literature for proofs of propositions, and to current papers for more specific problems. It is hoped that, afterwards, the reader will be able to read the current literature by himself.

Although most of the material is not new, many results scattered throughout the literature are brought together. For the form of the presentation, I owe very much to many people. In the first place I am indebted to Prof. A. Janner who introduced me into the field and with whom, during many years, I had very stimulating discussions. I thank Prof. J. Burckhardt for his advice concerning the publication of this book and for his critical remarks to a preliminary version of the manuscript. I also wish to thank Dr. J. Beltman for many critical remarks and for composing many of the exercises. I am greatly indebted to Dr. M. Boon for reading the manuscript and for his many very useful suggestions. Finally I thank all those students who by their questions and remarks contributed considerably to the final form of this book.

CONTENTS

NOTATION

$a = \{\mathbb{1} \mid \boldsymbol{a}\} \in U$	
$\boldsymbol{a}$	primitive translation
$\boldsymbol{a}_i$	basis translation
a_{qj}	absorption operator
$a_{qj}^{\dagger}$	creation operator
$\boldsymbol{a}(\boldsymbol{q})$	column of a_{qj}
$\boldsymbol{A}(\boldsymbol{r}, t)$	vector potential
$A(\boldsymbol{r}, t) = (\phi, \boldsymbol{A})$	four-potential
$A(3)$	affine group
$\boldsymbol{b}_i$	basis reciprocal lattice
c	velocity of light
C_i	conjugacy class
C_n	cyclic group of order n
$\mathcal{C}_i$	class sum
$\mathbb{C}$	complex number field
d, d_α	dimension
$D(G)$	representation of G
$D_\alpha(G)$	irreducible representation
$D_g(H)$	conjugated representation
$D \downarrow H$	subduced representation
$D \uparrow G$	induced representation
$D^{(l)}$	irreducible representation $SO(3)$
$D(\boldsymbol{q})$	dynamical matrix
D_n	dihedral group
e	unit element, elementary charge
$\boldsymbol{e}_i$	basis vector
$\boldsymbol{e}(\boldsymbol{q}, j)$	polarisation vector
$e(\boldsymbol{q})$	polarisation matrix
$E(3)$	Euclidean group
E	energy
$\boldsymbol{E}(\boldsymbol{r}, t)$	electric field
$f(g, h)$	factor system
F	field
$g \in G$	
g_{ij}	metric tensor
G	group, space group
G^{d}	double group of G
$G_{\boldsymbol{k}}$	group of $\boldsymbol{k}$
$GL(3, \mathbb{R})$	linear group
$h \in H$	
H	group, subgroup
$\boldsymbol{H}(\boldsymbol{r}, t)$	magnetic field
H	Hamiltonian
$\mathcal{H}, \mathcal{H}_{\boldsymbol{k}}$	Hilbert space
i	central inversion
Im_φ	image of mapping φ
J_n	extension of R by $S(4)$
$\boldsymbol{k}$	vector in reciprocal space
$\boldsymbol{K}$	vector reciprocal lattice
K	point group
K^{d}	double point group
$K_{\boldsymbol{k}}$	point group of $G_{\boldsymbol{k}}$
Ker_φ	kernel of homomorphism φ
L	little group
$\boldsymbol{L}$	angular momentum
m	mass
$m(g, h)$	factor system
m_α	multiplicity of $D_\alpha(G)$
n	order of a group
$\boldsymbol{n} = n_1\boldsymbol{a}_1 + n_2\boldsymbol{a}_2 + n_3\boldsymbol{a}_3$	

$n(g,h)$	factor system J_n
n_i	order class C_i
N	order of G
O	octahedral group
O_i	tensor operator
$O(n)$	orthogonal group
$\boldsymbol{p}$	momentum operator
$p_\alpha\binom{\boldsymbol{n}}{i}$	momentum particle at $\boldsymbol{x}\binom{\boldsymbol{n}}{i}$
$P(G)$	projective representation
P_g	substitution operator
$\boldsymbol{q}$	reciprocal space vector (phonons)
$\mathbb{Q}$	rational number field
Q_n	symmetry group of potential
r	number of classes
$\boldsymbol{r}$	position
$R \in O(n)$	
$\mathbb{R}$	real number field
s	index, number of points in a star
$s(g,h)$	spin factor system
S	set
$\boldsymbol{S}$	spin operator
$S(4)$	Shubnikov group
$SO(3)$	rotation group
S_n	group of permutations of n elements
$\boldsymbol{t}$	translation
$\boldsymbol{t}_R$	nonprimitive translation
T	tetrahedral group
T	time reversal
$T(3)$	translation group
T_g	linear operator in a representation
$u(\alpha)$	phase factor for $P(\alpha)$
$u(R) \in SU(2)$	mapped on $R \in SO(3)$
$\boldsymbol{u}\binom{\boldsymbol{n}}{i}$	displacement vector

$U_\chi = \exp\left(\frac{-\mathrm{ie}}{\hbar c}\chi\right)$	
U	translation subgroup of space group
V	vector space
$V(\boldsymbol{r})$	potential
$W(\boldsymbol{r})$	potential
$\boldsymbol{x}\binom{\boldsymbol{n}}{i}$	position vector in a crystal
ZG	integral group ring of G
$\mathbb{Z}$	ring of integers
α	label of irreducible representations
$\alpha, \beta, \ldots$	elements abstract point group
Γ_i	irreducible representations point group
$\Gamma(\boldsymbol{q},g)$	matrix (see eq. (6.30))
$\Delta(\boldsymbol{q},g)$	matrix (see eq. (6.32))
$\delta^*(\boldsymbol{k}) = \Sigma_{\boldsymbol{K}\in\Lambda^*}\delta(\boldsymbol{k}-\boldsymbol{K})$	
$\epsilon = \pm 1$	
θ_0	complex conjugation operator
θ	time reversal operator
Λ	direct lattice
Λ^*	reciprocal lattice
ρ_{ij}^α	operator (see eq. (2.5))
σ_i	Pauli spin matrix ($i = 1, 2, 3$)
$\boldsymbol{\tau}_j$	position in unit cell
ξ_i	coordinate
χ	character, gauge function
χ_g	compensating gauge function for g
$\psi(\boldsymbol{r})$	wave function
$\psi \in \mathcal{H}$	
φ	angle
$\varphi(K)$	arithmetic point group
$\Phi(\boldsymbol{r},t)$	potential
$\Phi_{\alpha\beta}\binom{\boldsymbol{n}}{i\ j}$	harmonic term phonon Hamiltonian

$\Phi_i(\alpha_1, ..., \alpha_\nu) = \epsilon$	defining relation
$\omega, \omega_j(\boldsymbol{q})$	frequency
$\omega(g,h)$	factor system
$\omega_s(g,h)$	spin factor system
Ω	volume
Ω_0	volume unit cell
$\Omega(\boldsymbol{q})$	diagonal matrix $\omega_j^2(\boldsymbol{q})$
$\mathbb{1}$	unit matrix, identity mapping
$\mathbb{1}_d$	d-dimensional unit matrix
$\sim$	equivalent to
$\cong$	isomorphic
$[\,]$	equivalence class
$\boxtimes$	semidirect product
$\equiv$	equal modulo $\boldsymbol{K} \in \Lambda^*$
$\in$	element of
$\tilde{S}$	transpose of S
$S^\dagger$	Hermitian conjugate to S
$*$	complex conjugation

$\{s \in S \mid \text{property } A\}$ = set of all elements of S with the property A

$\{\varphi(s) \mid \text{all } s \in S\}$ = set of all elements which are in the image under φ of the set S

INTRODUCTION

More and more it is being realized that group theory is of fundamental importance for physics. This has not always been the case. More than one eminent physicist considered group theory as an unnecessary burden for the physicists. Indeed, many of the simpler results can also be obtained without a knowledge of group theory; but, especially after the development of quantum mechanics, many consequences were derived from symmetry which were by no means trivial. One has even seen that many of the fundamental notions of physics can be formulated in group theoretical terms. Nowadays, group theory has become an indispensable part of the physicist's luggage. It has been applied in atomic, molecular, nuclear and elementary physics, and, perhaps with the most succes, in solid state physics. The solid state shows a great variety of very rich symmetries. At the same time, everything can be formulated in the frame of quantum mechanics, where no problems of interpretation occur as in elementary particle physics.

Everybody has an intuitive notion of symmetry. A figure has symmetry, if there are transformations which leave the distance of the points of the figure the same (solid motions) and which move the figure into itself. When one considers a crystal as a regular pattern of point particles, all solid motions which transform this pattern into itself are symmetry transformations. The set of all symmetry transformations of the pattern can be given a structure, which is known as a group structure. The group of all symmetries is the space group of the crystal. The study of symmetries of crystals is nearly as old as mathematical group theory. In the 19-th century the major properties and the derivation of all space groups were given by Schoenflies, Fedorov and others. The study of space groups belonged for a long time to mathematics and crystallography.

However, the space group symmetry has also important consequences for physics. Consider an electron in such a crystal. One can assume that a space group transformation of the electron coordinates will not change the physics of the problem. Now the electron is described by a wave function and transformation of the electron coordinates means transformation of this wave function. One has to determine which operator gives the transformation of the wave function for a given space group transformation. The relation between the space group and the corresponding operators is an example of a representation. This indicates why the representation theory is so important. Some very elementary properties of representations have far reaching consequences for physical problems. They lead to classification of eigenstates, to selection rules, to simplification of perturbation calculations and so on. The classification of energy eigenstates of a particle in a centrally symmetric potential by their eigenvalues of angular momentum and of its z-component is an example of a consequence of the rotation symmetry. For a particle in a one-dimensional symmetric potential $V(x) = V(-x)$ the eigenfunctions can be chosen to be even or odd. The fact that the matrix element of x between two odd functions vanishes is a trivial example of a selection rule.

The fact that we considered solid motions as symmetry transformations is related to the fact that the laws of physics are generally assumed to be unchanged by solid motions and by inhomogeneous Lorentz transformations. On the other hand, postulating the transformations leaving the laws unchanged restricts the possible forms of these laws (equations of motion). Therefore, there is a close relationship between the mechanics considered (relativistic or nonrelativistic, with and without spin) and the symmetry transformations allowed (Galilean transformations, Lorentz transformations). Here we shall only consider nonrelativistic mechanics and allow only solid motions and time reversal. Moreover, as solid state problems are many-body problems and the particles (e.g. electrons) are indistinguishable, we will also consider transformations which interchange particles. This permutation symmetry will add another aspect to our symmetry considerations.

In the present book we want to derive the fundamental consequences of the symmetry of the solid state. To do so we first have to master a basic knowledge of group theory. This is done in Ch. 1. In Ch. 2 we give a more precise definition of symmetry and we see which consequences one can derive in general from symmetries in quantum mechanical problems. In the two following chapters we study the consequences of point group and space group symmetries. To obtain these we study the properties and the representations of these groups. In Ch. 5 the group of possible transformations is extended to include time reversal. Moreover, in that chapter we discuss the difference between problems of particles with spin and those without. Finally in Ch. 6 we will become more specific and give some examples of concrete problems in solid state physics which can be attacked using group theoretical methods. Applications are found in the theory of electron bands, of lattice vibrations, in the problems of an electron in a crystal in an electromagnetic field or in a crystal with magnetic moments. Although these examples are very important ones, it represents only a choice.

CHAPTER I

MATHEMATICAL INTRODUCTION

In this first chapter we will introduce some elementary mathematical notions we need for our symmetry considerations in the physics of the solid state. We give here a brief treatment of elementary group theory, a short review of linear algebra, and the most important properties of group representations. This last topic will turn out to be of paramount importance for our physical conclusions. For a deeper understanding we refer to other books on group theory, like Hall [1959], Hamermesh [1962], Boerner [1967], and Jansen and Boon [1967].

1.1. Elementary group theory

1.1.1. Fundamental notions

The most fundamental concept in group theory is the concept of group. A *group* is a set G with a composition law, called product, which satisfies the following postulates.

1) For each pair of elements a, b of G the composition law determines an element $c = ab$ of the set G, called the *product* of a and b.

2) This product is *associative*, which means $(ab)c = a(bc)$.

3) To G belongs an element e, called the *right unit element*, such that for any element a of G $(a \in G)$ one has $ae = a$.

4) For each $a \in G$ there exists an element $a^{-1} \in G$, called the *right inverse* of a, such that $aa^{-1} = e$.

Some simple examples of groups are the following.

1) The set of real numbers $\neq 0$ forms a group, if we take for the composition law the usual product. The right unit element is the number one, and the right inverse of a number is its usual inverse.

2) The set of real numbers can be given another structure of group, if we choose for the composition law the addition of elements. In this case the

right unit element is the number zero, and the right inverse of a number is its negative.

3) Consider the set of all rotations which transform a plane rectangle in three-dimensional space into itself (fig. 1.1). It has four elements: the identity, and three 180° rotations a, b, and c. If we define the product of two elements ab as the rotation obtained by executing a after b, this set forms a group.

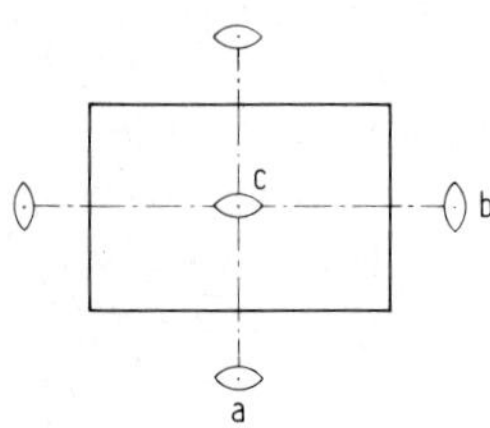

Fig. 1.1. Symmetries of a plane rectangle.

With respect to the four group postulates we have to make the following remarks. From the second postulate (the associativity) it follows that from an arbitrary product the parentheses may be omitted. The last two postulates are equivalent with the following two (see Hall [1959], p. 4).

3′) G contains an element e, such that for any $a \in G$ one has $ae = ea = a$.

4′) For each $a \in G$ there exists an element a^{-1}, such that $aa^{-1} = a^{-1}a = e$.

Of course for a given group the postulates 3 and 4 are easier to verify. However, once 3′) and 4′) are established one can speak about *the* unit element, and *the* inverse of an element, omitting the adjective "right", because it can easily be shown that the elements e and a^{-1} are unique.

A group is *Abelian*, if its product is commutative, which means that for any pair of elements a, b of G one has $ab = ba$. A group is *finite*, if the number of its elements is finite. This number is the *order of the group*. A group which is not finite is *infinite*. For a finite group the product may be given explicitly by its *multiplication* or *Cayley table*. In this table to each element correspond a row and a column. At the intersection of the row corresponding to the element a, and the column corresponding to the element b the product ab is written. It is easily proved that each element of G occurs exactly once in each row and each column. This statement is sometimes referred to as the rearrangement theorem. For the group of example 3 the

Cayley table looks like

	e	a	b	c
e	e	a	b	c
a	a	e	c	b
b	b	c	e	a
c	c	b	a	e

Consider in a group G an clement a, and its powers $a^2 = aa$, $a^3 = aaa$, etc. If for a certain integer m one has $a^m = e$, the element is of finite order. The smallest positive integer m such that $a^m = e$ is called the *order of the element* a. If no such integer exists the element is of infinite order. It can happen that the powers of an element a form already the whole group. In this case the group is called *cyclic*, and the element a is called a *generator* of the group. For a finite cyclic group of order N the elements are $a, a^2, ..., a^N = e$. In general, a set of elements of a group G is called a *set of generators*, if any element of the group can be written as a product of powers of the generators and their inverses. In general this set is not unique. In our example 3 one can choose the elements a and b as generators. The two other elements c and e are products of these two. One can as well choose a and c, or b and c.

When we know that a group is generated by an element a, and that N is the smallest natural number such that $a^N = e$, it is clear that it is a cyclic group of order N. The expression $a^N = e$ is called a relation in the group. As in this case the group is completely determined by this relation it is called a *defining relation*. On the other hand we can consider all the elements of a group as a set of generators. Then all products $ab = c$ are relations in the group. In general, given any set of generators of a group, one can consider relations between these generators. If these relations determine the group completely one calls these relations *defining relations* for the group. For the set of generators a, b in our example 3 one can choose as defining relations $a^2 = e$, $b^2 = e$, $ab = ba$.

A subset H of G which forms a group under the product rule of G is a *subgroup* of G. Trivial subgroups of G are G itself, and the unit element e, which forms a group of order one. A *proper subgroup* of G is a subgroup which differs from G and has order greater than one. To verify that H is a subgroup of G it is sufficient to check that 1) for each pair a, b in H the product is also in H, and 2) that for any $h \in H$ also $h^{-1} \in H$.

When G and H are groups one can give a group structure to the set consisting of all pairs (g, h) of elements $g \in G$, and $h \in H$, when one introduces as multiplication law

$$(g_1,h_1)(g_2,h_2) = (g_1g_2,h_1h_2) \quad \text{for any} \quad g_1,g_2 \in G, \text{ and } h_1,h_2 \in H\,.$$

The group obtained in this way is called the *direct product* $G \times H$ of the groups G and H. Suppose that a group G contains two subgroups, H_1 and H_2, satisfying the following conditions:

1) any element of H_1 commutes with any element of H_2 : $h_1h_2 = h_2h_1$,
2) the only element belonging both to H_1 and to H_2 is the unit element,
3) any element of G can be written as the product of an element $h_1 \in H_1$ with an element $h_2 \in H_2$.

From condition 2) it follows that the decomposition of an element of G is unique. To see this, assume that g can be written in two ways as a product $g = h_1h_2 = h_1'h_2'$ with $h_1, h_1' \in H_1$, and $h_2, h_2' \in H_2$. Then the element $(h_1')^{-1}h_1 = h_2'h_2^{-1}$ belongs both to H_1 and to H_2. From condition 2) it follows that this is the unit element and consequently one has $h_1 = h_1'$, and $h_2 = h_2'$, which shows that the decomposition is indeed unique. If we denote the element $g = h_1h_2$ by the pair (h_1,h_2) the product in G coincides with the product in the direct product $H_1 \times H_2$. Therefore, one can identify G with the direct product $H_1 \times H_2$. In fact the identification can be better described as an isomorphism as we will see in § 1.5.

1.1.2. Equivalence relations

In the following we will frequently have to identify certain objects. In order to do this in a proper way it is convenient to introduce the notion of equivalence relation. An *equivalence relation* between elements of a set S (element s is equivalent to element t is denoted by $s \sim t$) is a relation which is

1) *reflexive*, i.e. $s \sim s$ for any $s \in S$,
2) *symmetric*, i.e. if $s \sim t$, then also $t \sim s$,
3) *transitive*, i.e. if $s \sim t$, and $t \sim u$, then also $s \sim u$.

An example is given by the equality relation (=). It is easily verified, that this relation satisfies the three conditions.

An equivalence relation can be used to give a decomposition of the set S. To obtain this one takes together in one equivalence class all elements which are equivalent to each other. If we denote the *equivalence class* to which s belongs by $[s]$, it is defined as the subset of S consisting of all elements which are equivalent to s. We can write this definition as

$$[s] = \{t \in S \mid t \sim s\}\,.$$

From this definition it follows that $[s] = [t]$ if and only if $s \sim t$. Furthermore, $[s]$ and $[t]$ have no elements in common unless $s \sim t$. From each equivalence class one can choose one representative element. If $s_1, s_2, ...,$ are these representatives, the set S can be written as the union of the subsets $[s_1], [s_2], ...$ It is easily seen that this partitioning $S = [s_1] + [s_2] + ...$ does not depend on the choice of the representatives.

1.1.3. Equivalence relations in groups

In a group G one can define several kinds of equivalence relations. First we introduce a relation for which the equivalence classes are called the *left cosets* of a given subgroup H. The equivalence relation is defined by: $a \sim b$ if and only if there is an element $h \in H$ such that $a = bh$. Hence a left coset of H is a set $[g] = gH = \{gh \mid \text{all } h \in H\}$ for a fixed $g \in G$. One coset is the subgroup H itself. When $g_1 = e, g_2, g_3, ...$ are coset representatives the group G can be decomposed into

$$G = H + g_2H + g_3H + ...$$

Each coset has as many elements as the order n of H (n is also called the order of the coset). If G is a finite group of order N, the number of different left cosets is $s = N/n$, called the *index* of H in G.

In a similar way one defines a right coset of H in G as an equivalence class for the relation: $a \sim b$ if and only if there is an element $h \in H$ such that $a = hb$. A right coset of H to which the element g belongs is a class $[g] = Hg = \{hg \mid \text{all } h \in H\}$ and G can be decomposed into

$$G = H + Hg_2' + Hg_3' + ... ,$$

where in general the right coset representatives $g_1', g_2', ...$ are different from the left coset representatives $g_1, g_2, ...$ Also the left and right cosets of H are in general not the same. However, the number of left cosets is equal to the number of right cosets, and equal to the index of H in G.

A third equivalence relation we want to introduce in a group is the conjugation by an element: $a \sim b$ (one says a and b are conjugate) if and only if there is an element $g \in G$ such that $a = gbg^{-1}$. Again it is easily verified that this is indeed an equivalence relation. The equivalence classes are called the *conjugacy classes* (or simply the classes). The class to which an element a belongs is defined by

$$[a] = \{gag^{-1} \mid \text{all } g \in G\} .$$

This is the subset of G consisting of all elements which can be written as gag^{-1} for some $g \in G$. Notice that for different elements g_1 and g_2 in G the elements $g_1 a g_1^{-1}$ and $g_2 a g_2^{-1}$ are not necessarily different. Contrary to the cosets the classes have in general not the same number of elements. The unit element is always a class by itself, but only in Abelian groups each class contains only a single element.

1.1.4. Invariant subgroups

For a subgroup H of G the set $gHg^{-1} = \{ghg^{-1} \mid \text{all } h \in H\}$ for fixed $g \in G$ forms a subgroup, called a *conjugate subgroup*. On the other hand two subgroups H_1 and H_2 are conjugate if there is an element $g \in G$ such that $H_2 = gH_1g^{-1}$. A subgroup which is only conjugate to itself, i.e. for which $gHg^{-1} = H$ for any $g \in G$, is an *invariant subgroup* of G. In the literature one uses also the terms normal subgroup, and normal divisor. The formula $gHg^{-1} = H$ means that for any $g \in G$, and $h \in H$ one has $ghg^{-1} \in H$.

The right and left cosets of an invariant subgroup are identical, since $gH = Hg$. The set of cosets of an invariant subgroup can be given the structure of a group by the following definition of product of two cosets. When g_1H is the coset to which g_1 belongs, and g_2H the one to which g_2 belongs, the product of the two cosets is the coset to which g_1g_2 belongs: $(g_1H)(g_2H) = (g_1g_2)H$. The coset to which g_1g_2 belongs is well determined. Moreover, this definition of product does not depend on the choice of the coset representatives g_1 and g_2 as one can see as follows. Suppose g_1' is another element of g_1H. Then there is an element $h_1 \in H$ such that $g_1' = g_1h_1$. Similarly one has an element $h_2 \in H$ with $g_2' = g_2h_2$ another element of g_2H. The product of the two cosets $g_1H = g_1'H$ and $g_2H = g_2'H$ is given by $(g_1'H)(g_2'H) = (g_1'g_2')H = (g_1h_1g_2h_2)H = (g_1g_2g_2^{-1}h_1g_2h_2)H = g_1g_2H = (g_1H)(g_2H)$ because $g_2^{-1}h_1g_2 \in H$ and consequently $g_2^{-1}h_1g_2h_2H = H$. It is now easy to verify the group postulates

1) the product is well defined,
2) the product is associative: $[(g_1H)(g_2H)](g_3H) = (g_1g_2g_3)H$,
3) the coset H is the unit element: $(gH)H = (gH)$,
4) for the coset gH the coset $g^{-1}H$ is its inverse.

In this way the cosets form a group, called the quotient or *factor group*. It is denoted by G/H. The order of the factor group is the number of different cosets of H in G, and therefore equal to the index of the invariant subgroup H in G.

1.1.5. Mappings

A *mapping* ϕ of a set S to a set T is a rule which assigns to each element $s \in S$ a unique element $t = \phi(s)$ of T. The subset of T consisting of all elements t which can be written as $\phi(s)$ for some $s \in S$ is the *image* of ϕ, denoted by Im_ϕ. When $Im_\phi = T$ the mapping is called onto, otherwise, if T contains elements which are not in Im_ϕ, it is called into.

When G and H are groups, they have a structure which can be used to put additional restrictions on mappings from G to H. If $\phi : G \to H$ is such that it conserves the multiplication law, i.e. if $\phi(a)\phi(b) = \phi(ab)$ for any pair a, b in G, it is called a *homomorphism*. Notice that the unit element of G is mapped onto the unit element of H, because $\phi(e)\phi(a) = \phi(a)\phi(e) = \phi(a)$ for any $a \in G$. In general several elements of G are mapped onto the unit element of H. These elements of G form the *kernel* of the homomorphism ϕ, denoted by Ker_ϕ. This Ker_ϕ is an invariant subgroup of G, as one sees as follows. In the first place it is a subgroup, because 1) for $a, b \in G$ such that $\phi(a) = \phi(b) = e \in H$ one has $\phi(a)\phi(b) = \phi(ab) = e$, and consequently $ab \in Ker_\phi$, and 2) for an element $a \in Ker_\phi$ also $a^{-1} \in Ker_\phi$ because $\phi(a^{-1}) = \phi(a)^{-1} = e$. Moreover, it is invariant because for $a \in Ker_\phi$, and $b \in G$ one has $\phi(bab^{-1}) = \phi(b)\phi(b^{-1}) = e$ and therefore $bab^{-1} \in Ker_\phi$, which proves that Ker_ϕ is really an invariant subgroup. Also the image Im_ϕ forms a subgroup (of H), but in general it is not an invariant subgroup.

An example of a homomorphism ϕ: $G \to H$ is the mapping which assigns the unit element of H to each element of G. In this case $Im_\phi = e \in H$, and $Ker_\phi = G$. A second example is given by the mapping of a group G with an invariant subgroup H to the factor group G/H when we define the mapping $\phi : G \to G/H$ by : to each element of G one assigns the coset to which it belongs. The unit element of G/H is H. Therefore, in this case $Im_\phi = G/H$, and $Ker_\phi = H$. This homomorphism is called the *canonical epimorphism* of G onto the factor group G/H.

A homomorphism ϕ of G onto H with $Ker_\phi = e \in G$ is an *isomorphism*. Here one has a one-to-one correspondence between the elements of G and those of H. If ϕ is an isomorphism, there exists an isomorphism $\psi : H \to G$ such that $\psi(\phi(g)) = g$ for any $g \in G$, and $\phi(\psi(h)) = h$ for any $h \in H$. This isomorphism is denoted by ϕ^{-1}. The groups G and H are called *isomorphic groups*.

Furthermore, we will use the following terminology. A homomorphism of G onto H is an *epimorphism*. An isomorphism onto a subgroup of H is a *monomorphism*. A homomorphism of G into itself is an *endomorphism*. An isomorphism of G onto itself is an *automorphism*.

Table 1.1
Various kinds of homomorphisms

Homomorphism $\varphi : G \to H$	Im_φ	Ker_φ
Epimorphism	H	
Monomorphism		$e \in G$
Isomorphism	H	$e \in G$

An example of an automorphism is given by conjugation with a fixed element $g \in G$: one has $\phi_g(a) = gag^{-1}$ for any $a \in G$. It is easily verified that this defines indeed an automorphism. Such an automorphism, obtained by conjugation with an element, is called an *inner automorphism*.

Two automorphisms ϕ and ψ of G applied successively result again in an automorphism. Defining the product of ϕ and ψ by $(\phi\psi)(g) = \phi(\psi(g))$ for any $g \in G$, the set of all automorphisms of G forms a group. The unit element is the identity automorphism which maps each element of G onto itself. The group of automorphisms of G is denoted by $Aut(G)$.

1.1.6. Other algebraic structures

Although the concept of group plays the most fundamental role in our considerations here, there are some other structures which are also important. We will give definitions of the notions of ring, field, vector space, and algebra.

A *ring with unit element* is an Abelian group for which a second composition law is defined such that the following relations are verified. We denote the first composition law under which the set is an Abelian group by a + sign, and the second composition by the usual product notation. The elements of the ring are denoted by Greek letters. The unit element for the additive group is denoted by 0, that of the multiplication by ϵ.

1) The second law is associative: $(\alpha\beta)\gamma = \alpha(\beta\gamma)$.

2) There exists an element ϵ such that $\alpha\epsilon = \epsilon\alpha = \alpha$ for any ring element α.

3) $\alpha(\beta+\gamma) = \alpha\beta + \alpha\gamma$ and $(\alpha+\beta)\gamma = \alpha\gamma + \beta\gamma$.

In the following we omit the addition "with unit element", because we only consider here rings of this type. An example of a ring is the set of integers, which forms an Abelian group under addition (with unit element 0), and for which also multiplication is defined (with unit element 1).

A *field F* is a ring which has in addition to the relations 1), ..., 3) the following properties

4) Commutativity under the second law: $\alpha\beta = \beta\alpha$.

5) For any element $\alpha \neq 0$ there exists an inverse α^{-1}.
Examples of fields are the sets of real numbers ($F = \mathbb{R}$), of rational numbers ($F = \mathbb{Q}$), and of complex numbers ($F = \mathbb{C}$). In this book we will mean by field always one of these three fields. However, if we do not want to specify the field we denote it by F.

A *linear vector space* over a field F is an Abelian group V for which a scalar left multiplication by elements of the field F is defined such that

i) $\boldsymbol{x} + \boldsymbol{y}$ and $\alpha\boldsymbol{x}$ are well defined elements of V for any $\boldsymbol{x},\boldsymbol{y} \in V$, and any $\alpha \in F$,

ii) $(\boldsymbol{x}+\boldsymbol{y}) + \boldsymbol{z} = \boldsymbol{x} + (\boldsymbol{y}+\boldsymbol{z})$ and $(\alpha\beta)\boldsymbol{x} = \alpha(\beta\boldsymbol{x})$,

iii) There is an element $\boldsymbol{O} \in V$ such that $\boldsymbol{x} + \boldsymbol{O} = \boldsymbol{O} + \boldsymbol{x}$ for any $\boldsymbol{x} \in V$,

iv) For any $\boldsymbol{x} \in V$ there is an element $-\boldsymbol{x} \in V$ such that $\boldsymbol{x} + (-\boldsymbol{x}) = \boldsymbol{O}$,

v) $\boldsymbol{x} + \boldsymbol{y} = \boldsymbol{y} + \boldsymbol{x}$,

vi) $\alpha(\boldsymbol{x}+\boldsymbol{y}) = \alpha\boldsymbol{x} + \alpha\boldsymbol{y}$,

vii) $(\alpha+\beta)\boldsymbol{x} = \alpha\boldsymbol{x} + \beta\boldsymbol{x}$,

viii) $\epsilon\boldsymbol{x} = \boldsymbol{x}$.

We have here once more included the properties of Abelian groups in order to state the definition of a linear vector space more clearly. The elements of V are called *vectors*, and the elements of the field F *scalars*.

An *associative algebra* over a field F is a linear vector space V over this field F where V is at the same time a ring. This means that, apart from the sum of two vectors, also their product is defined, and the multiplication law satisfies the relations

ix) $\boldsymbol{x}\boldsymbol{y}$ is well defined and $(\boldsymbol{x}\boldsymbol{y})\boldsymbol{z} = \boldsymbol{x}(\boldsymbol{y}\boldsymbol{z})$ for any $\boldsymbol{x},\boldsymbol{y},\boldsymbol{z} \in V$,

x) $\boldsymbol{x}(\boldsymbol{y}+\boldsymbol{z}) = \boldsymbol{x}\boldsymbol{y}+\boldsymbol{x}\boldsymbol{z}$ and $(\boldsymbol{x}+\boldsymbol{y})\boldsymbol{z} = \boldsymbol{x}\boldsymbol{z}+\boldsymbol{y}\boldsymbol{z}$ for any $\boldsymbol{x},\boldsymbol{y},\boldsymbol{z} \in V$,

xi) $\alpha(\boldsymbol{x}\boldsymbol{y}) = (\alpha\boldsymbol{x})\boldsymbol{y} = \boldsymbol{x}(\alpha\boldsymbol{y})$ for any $\boldsymbol{x},\boldsymbol{y} \in V$,

which appear as additional to the 8 relations for a linear vector space. An example of an associative algebra is the set of real $n \times n$ matrices. Clearly sum and product of two matrices are well defined, as is the product of a real number with a matrix. Moreover the set satisfies all conditions i) to xi). The field F is here the field of real numbers. Instead of "algebra over the field $\mathbb{R}$" one usually calls it a real algebra.

1.1.7. Group algebra

For a set S an *S-valued function* on a group G is a mapping $\mathfrak{f} : G \to S$. This means that to each element $g \in G$ is assigned an element $\mathfrak{f}(g) \in S$. It is a generalization of the notion of function with real or complex numbers as argument. In particular we will consider cases where S is the set of integers $\mathbb{Z}$, of real numbers $\mathbb{R}$, or of complex numbers $\mathbb{C}$. A special kind of function on a group, which will play an important role, is a class function, defined as a function $\mathfrak{f}$ such that $\mathfrak{f}(a) = \mathfrak{f}(b)$ whenever a and b belong to the same class: $\mathfrak{f}(a) = \mathfrak{f}(gag^{-1})$ for any $g \in G$.

Addition of two functions and multiplication of a function by an element of S can be defined in a natural way if S is a ring or a field, by

$$(\mathfrak{f}_1 + \mathfrak{f}_2)(g) = \mathfrak{f}_1(g) + \mathfrak{f}_2(g)$$

$$(\alpha\mathfrak{f})(g) = \alpha\mathfrak{f}(g) \qquad\qquad (\alpha \in S)\,.$$

We consider a special type of these functions defined as follows. For any element $a \in G$ one defines a function $\mathfrak{a}(g)$ by

$$\mathfrak{a}(g) = \begin{cases} 1 & \text{if} \quad g = a \\ 0 & \text{otherwise}\,. \end{cases}$$

Now we restrict ourselves to finite groups. Then an arbitrary function can be written as

$$\mathfrak{f}(g) = \sum_{a \in G} \mathfrak{f}(a)\mathfrak{a}(g)$$

or

$$\mathfrak{f} = \sum_{a \in G} \mathfrak{f}(a)\,\mathfrak{a}\,.$$

One can define multiplication of two functions $\mathfrak{f}_1$ and $\mathfrak{f}_2$ by

$$(\mathfrak{f}_1 * \mathfrak{f}_2)(g) = \sum_{a \in G} \mathfrak{f}_1(a)\mathfrak{f}_2(a^{-1}g)\,.$$

The product denoted by $*$ is called the *convolution product*. When we consider functions with values in a field F, the functions on a finite group form an algebra over F with this convolution product and the addition and scalar multiplication defined above. The algebra is called the *group algebra*. For any pair of elements $a, b \in G$ the convolution product $\mathfrak{a} * \mathfrak{b}$ is given by

$$(\mathfrak{a} * \mathfrak{b})(g) = \sum_{h \in G} \mathfrak{a}(h)\mathfrak{b}(h^{-1}g) = (\mathfrak{ab})(g)\,.$$

Moreover, one has for the convolution product of two arbitrary functions

$$[\sum_{a\in G} \mathfrak{f}_1(a)\mathfrak{a}] * [\sum_{b\in G} \mathfrak{f}_2(b)\mathfrak{b}] = \sum_{a,b\in G} \mathfrak{f}_1(a)\mathfrak{f}_2(b)\,\mathfrak{a} * \mathfrak{b}\,.$$

This means that the elements of the group algebra can formally be considered as linear combinations of group elements with coefficients from F.

When F is the ring of integers, the functions $\mathfrak{n}\colon G \to \mathbb{Z}$ can be written as

$$\mathfrak{n} = \sum_{g\in G} \mathfrak{n}(g)\mathfrak{g}\,.$$

It is easily verified that these functions form a ring, the *integral group ring ZG* of G. As $\mathfrak{n}(g)$ is an integer, the elements of the group ring are formal sums over group elements with integral coefficients. To emphasize this fact we denote the product in ZG by fg instead of $\mathfrak{f} * \mathfrak{g}$. To this integral group ring ZG belong the elements $\mathfrak{C}_i = \Sigma_{a\in C_i} a$, where C_i $(i = 1, ..., r)$ is a class of the group G. Such an element $\mathfrak{C}_i$ of ZG is called a *class sum*. It is a function which has the value one on all the elements of the class C_i, and zero elsewhere. Any integral class function $\mathfrak{c}$ can be written as a sum of class functions

$$\mathfrak{c} = \sum_{i=1}^{r} \alpha_i \mathfrak{C}_i\,.$$

The coefficient $\alpha_i \in \mathbb{Z}$ is the value which $\mathfrak{c}$ takes on the class C_i.

PROPOSITION 1.1: A class sum $\mathfrak{C}_i$ has the property $\mathfrak{g} * \mathfrak{C}_i * \mathfrak{g}^{-1} = \mathfrak{C}_i$ for any $g \in G$.

Proof. If $a \in C_i$, also $gag^{-1} \in C_i$. Moreover $gag^{-1} = gbg^{-1}$ if and only if $a = b$. Hence

$$\sum_{a\in C_i} a = \sum_{a\in C_i} gag^{-1}$$

which is the stated property. From this proposition it follows that for any class function $\mathfrak{c}$ one has $\mathfrak{g} * \mathfrak{c} * \mathfrak{g}^{-1} = \mathfrak{c}$. On the other hand one has

PROPOSITION 1.2: If an element $\mathfrak{f}$ of the group algebra of G over F has the property $\mathfrak{g}\mathfrak{f}\mathfrak{g}^{-1} = \mathfrak{f}$ for any $g \in G$, then $\mathfrak{f}$ is a class function.

Proof: Consider elements $b, c \in C_i$. There is an element $g \in G$ such that $b = gcg^{-1}$. Because

$$\sum_{a\in G} \mathfrak{f}(a)\mathfrak{a} = \mathfrak{f} = \mathfrak{g} * \mathfrak{f} * \mathfrak{g}^{-1} = \sum_{a\in G} \mathfrak{f}(a)\, gag^{-1} = \sum_{b\in G} \mathfrak{f}(g^{-1}bg)\,\mathfrak{b}$$

one has

$$\mathfrak{f}(b) = \mathfrak{f}(g^{-1}bg) = \mathfrak{f}(c)\,.$$

Hence $\mathfrak{f}$ is a class function and can be written as $\mathfrak{f} = \Sigma_{i=1}^{r} \alpha_i \mathcal{C}_i$.

Consider now the product of two class sums. It has the property

$$\mathcal{C}_i \mathcal{C}_j = (g \mathcal{C}_i g^{-1})(g \mathcal{C}_j g^{-1}) = g \mathcal{C}_i \mathcal{C}_j g^{-1} .$$

Therefore $\mathcal{C}_i \mathcal{C}_j$ is an integral class function. This means that there are integers c_{ijk} called *class multiplication constants* such that

$$\mathcal{C}_i \mathcal{C}_j = \sum_{k=1}^{r} c_{ijk} \mathcal{C}_k \qquad (i, j = 1, ..., r) .$$

1.1.8. Example

In this section we will demonstrate the various concepts we introduced for groups in the preceding sections on the example of the group of permutations of three elements. A permutation which replaces element 1 by i_1 (= 1, 2, or 3), element 2 by i_2, and element 3 by i_3, where i_1, i_2, i_3 is a permutation of 1, 2, 3, is denoted by $\binom{1\ \ 2\ \ 3}{i_1\ i_2\ i_3}$. The six permutations of three elements are

$$e = \binom{1\,2\,3}{1\,2\,3},\ a = \binom{1\,2\,3}{2\,3\,1},\ b = \binom{1\,2\,3}{3\,1\,2},\ c = \binom{1\,2\,3}{2\,1\,3},\ d = \binom{1\,2\,3}{3\,2\,1},\ f = \binom{1\,2\,3}{1\,3\,2} .$$

The product of two permutations is defined as the permutation obtained by successive application. Here we write to the right the permutation to be performed first. As an example $ac = \binom{123}{231}\binom{123}{213} = \binom{123}{321}$, because by permutation c element 1 goes to element 2, and this element goes by a to 3. The element 3 goes by c to 3, and by a to 1, whereas element 2 goes by c to 1, and by a to 2, and is in this way left unchanged by the product ac. The Cayley table is given by

	e	a	b	c	d	f
e	e	a	b	c	d	f
a	a	b	e	d	f	c
b	b	e	a	f	c	d
c	c	f	d	e	b	a
d	d	c	f	a	e	b
f	f	d	c	b	a	e

The order of the group is six. It is a non-Abelian group. Because $e = a^3$,

$b = a^2, d = ac, f = ad = a^2c$ the elements a and c form a set of generators of the group. Relations between the generators a and c are $a^3 = c^2 = e$, and $acac = e$. From these relations it follows that $ca = a^2c$. Therefore all elements can be written in the form $a^m c^n$ with $m = 0$, 1, or 2, and $n = 0$, or 1. As these are exactly the elements of the group the given three relations are really defining relations. All other relations, in particular those in the Cayley table may be deduced from these relations. The order of element e is 1, of the elements a and a^2 is 3, that of the elements c, ac, and a^2c is 2. Subgroups are G itself, $H_1 = \{e, a, a^2\}$, $H_2 = \{e, c\}$, $H_3 = \{e, ac\}$, $H_4 = \{e, a^2c\}$, and $\{e\}$. The subgroup H_1 is an invariant subgroup of index two. The subgroups H_2, H_3, and H_4 have index three, but are not invariant. Cosets of H_1 are $H_1 = \{e, a, a^2\}$, and $cH_1 = \{c, ac, a^2c\}$. Left cosets of H_2 are $H_2 = \{e, c\}$, $aH_2 = \{a, ac\}$, and $a^2H_2 = \{a^2, a^2c\}$. Right cosets of H_2 are $H_2 = \{e, c\}$, $H_2a = \{a, a^2c\}$ and $H_2a^2 = \{a^2, ac\}$. The conjugation classes are $C_1 = \{e\}$, $C_2 = \{a, a^2\}$, $C_3 = \{c, ac, a^2c\}$. H_2, H_3, and H_4 are conjugate subgroups. The Cayley table for the factor group G/H_1, which is a group of order 2 is given by

	H_1	cH_1
H_1	H_1	cH_1
cH_1	cH_1	H_1

The canonical epimorphism $\phi : G \to G/H_1$ is given by $\phi(e) = \phi(a) = \phi(a^2) = H_1$, and $\phi(c) = \phi(ac) = \phi(a^2c) = cH_1$. The group G is isomorphic to the group of rotations which leave a plane equilateral triangle invariant. This group consists of the rotations through 0°, 120°, and 240° around an axis orthogonal to the triangle (these are denoted by E, A, and B), and the three 180° rotations around an axis through the midpoint of the triangle and one of its vertices (these are denoted by C, D, and F: fig. 1.2). The isomorphism is given by the

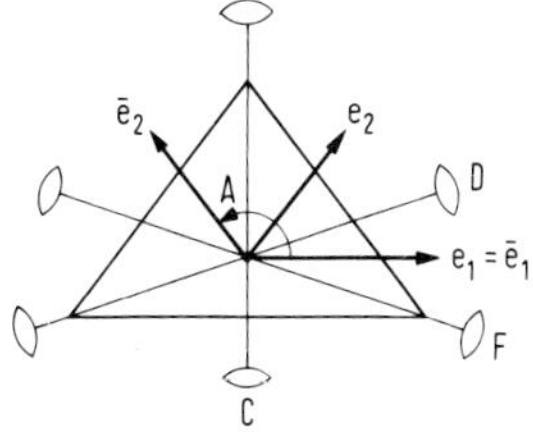

Fig. 1.2. Bases for representations of S_3.

mapping ψ such that $\psi(e) = E$, $\psi(a) = A$, and so on. Automorphisms of G:

	e	a	a^2	c	ac	a^2c
ϕ_0	e	a	a^2	c	ac	a^2c
ϕ_1	e	a	a^2	ac	a^2c	c
ϕ_2	e	a	a^2	a^2c	c	ac
ϕ_3	e	a^2	a	c	a^2c	ac
ϕ_4	e	a^2	a	ac	c	a^2c
ϕ_5	e	a^2	a	a^2c	ac	c

The group of automorphisms $Aut(G)$ is given by its Cayley table.

	ϕ_0	ϕ_1	ϕ_2	ϕ_3	ϕ_4	ϕ_5
ϕ_0	ϕ_0	ϕ_1	ϕ_2	ϕ_3	ϕ_4	ϕ_5
ϕ_1	ϕ_1	ϕ_2	ϕ_0	ϕ_4	ϕ_5	ϕ_3
ϕ_2	ϕ_2	ϕ_0	ϕ_1	ϕ_5	ϕ_3	ϕ_4
ϕ_3	ϕ_3	ϕ_5	ϕ_4	ϕ_0	ϕ_2	ϕ_1
ϕ_4	ϕ_4	ϕ_3	ϕ_5	ϕ_1	ϕ_0	ϕ_2
ϕ_5	ϕ_5	ϕ_4	ϕ_3	ϕ_2	ϕ_1	ϕ_0

Hence it follows that G and $Aut(G)$ are isomorphic groups. The isomorphism may be given by χ: $\chi(e) = \phi_0$, $\chi(a) = \phi_1$, $\chi(a^2) = \phi_2$, $\chi(c) = \phi_3$, $\chi(ac) = \phi_4$, $\chi(a^2c) = \phi_5$. Note that this isomorphism is not unique. If ϕ is an automorphism, also $\chi' = \chi\phi$ is an isomorphism. An element of the group ring is given by six integers: $n(e)$, $n(a)$, $n(a^2)$, $n(c)$, $n(ac)$, and $n(a^2c)$. Examples are the three class sums

$$\mathcal{C}_1 = e$$

$$\mathcal{C}_2 = a + a^2$$

$$\mathcal{C}_3 = c + ac + a^2c \, .$$

The class multiplication constants are found from

$$e_1e_1 = e_1 \qquad e_2e_1 = e_2 \qquad e_3e_1 = e_3$$

$$e_1e_2 = e_2 \qquad e_2e_2 = 2e_1 + e_2 \qquad e_3e_2 = 2e_3$$

$$e_1e_3 = e_3 \qquad e_2e_3 = 2e_3 \qquad e_3e_3 = 3e_1 + 3e_2 .$$

This means

$$c_{111} = c_{212} = c_{313} = c_{122} = c_{222} = c_{133} = 1$$

$$c_{221} = c_{323} = c_{233} = 2$$

$$c_{331} = c_{332} = 3$$

whereas the other class multiplication constants vanish.

1.2. Linear algebra

1.2.1. Finite-dimensional vector spaces

In section 1.6 of this chapter we defined a linear vector space over a field F. Here we take the field F to be the real or complex number field and we refer to the vector space V as a real or complex vector space. A set of vectors $\boldsymbol{x}_1, ..., \boldsymbol{x}_k$ is said to be *linearly independent* if from $\Sigma_{j=1}^k \alpha_j \boldsymbol{x}_j = \boldsymbol{0}$ it follows that $\alpha_1, ..., \alpha_k$ are all zero. A set of linearly independent vectors $\boldsymbol{e}_1, ..., \boldsymbol{e}_n$ is a *basis* of V, if each element of V can be written as

$$\boldsymbol{x} = \Sigma_{j=1}^n \xi_j \boldsymbol{e}_j , \tag{1.1}$$

wherc the real or complex numbers $\xi_1, ..., \xi_n$ are uniquely determined because the basis elements are linearly independent. The numbers $\xi_1, ..., \xi_n$ are the *components* of the vector $\boldsymbol{x}$ with respect to the basis $\boldsymbol{e}_1, ..., \boldsymbol{e}_n$. The number of elements of the basis is the *dimension* of V. It does not depend on the choice of the basis. We assume in the following that V is finite dimensional if we do not state otherwise. If $\xi_1, ..., \xi_n$ are the components of $\boldsymbol{x}$, and $\eta_1, ..., \eta_n$ are the components of a vector $\boldsymbol{y}$, the components of $\alpha\boldsymbol{x}$ (with $\alpha \in F$) are $\alpha\xi_1, ..., \alpha\xi_n$, and the components of $\boldsymbol{x} + \boldsymbol{y}$ are $(\xi_1 + \eta_1), ..., (\xi_n + \eta_n)$.

When another basis is chosen, this new basis $\bar{e}_1, ..., \bar{e}_n$ is related to the old one by a matrix S (which is nonsingular, meaning that its inverse exists) such that

$$\bar{e}_i = \sum_{j=1}^{n} S_{ji} e_j . \tag{1.2}$$

With respect to the new basis a vector x has components $\bar{\xi}_1, ..., \bar{\xi}_n$ such that

$$\xi_i = \sum_{j=1}^{n} S_{ij} \bar{\xi}_j . \tag{1.3}$$

We recall the definition of the product of a $p \times q$ matrix A by a $q \times m$ matrix B. It is a $p \times m$ matrix with $(AB)_{ij} = \Sigma_{k=1}^{q} A_{ik} B_{kj}$ ($i = 1, ..., p$ and $j = 1, ..., m$). If we write $e_1, ..., e_n$ as a row vector e (i.e. a $1 \times n$ matrix) and $\xi_1, ..., \xi_n$ as a column vector ($n \times 1$ matrix) the eqs. (1.1)–(1.3) can be written as

$$x = e\xi \qquad \bar{e} = eS \qquad \xi = S\bar{\xi} .$$

1.2.2. *Linear transformations*

If V_1 and V_2 are linear vector spaces, and if A is a mapping from V_1 to V_2 such that $A(\alpha x + \beta y) = \alpha A(x) + \beta A(y)$ for any $x, y \in V_1$ and $\alpha, \beta \in F$, then A is a *linear mapping*. When we choose bases $e_1^{(1)}, ..., e_n^{(1)}$ in V_1, and $e_1^{(2)}, ..., e_m^{(2)}$ in V_2 this linear mapping can be described by a $m \times n$ matrix A given by $Ae_i^{(1)} = \Sigma_{j=1}^{m} A_{ji} e_j^{(2)}$ ($i = 1, ..., n$). A linear mapping of the n-dimensional vector space V into itself is a linear transformation, which can be described by a $n \times n$ matrix. For the components of a vector one has the following relation. If $x' = Ax$, $x = \Sigma_{j=1}^{n} \xi_j e_j$, and $x' = \Sigma_{j=1}^{n} \xi_j' e_j$ one has

$$Ax = \sum_{i,j=1}^{n} \xi_i A_{ji} e_j .$$

Hence

$$\xi_j' = \sum_{i=1}^{n} A_{ji} \xi_i .$$

In short-hand notation one has $e' = eA$, and $\xi' = A\xi$.

The product of two linear transformations A and B is defined by their successive application: $(AB)(x) = A(B(x))$. It is easily verified that with this product rule the nonsingular linear transformations (those for which the mapping is onto) form a group. After a choice of basis the transformations

A and B correspond to matrices, and their product corresponds to a matrix given by $ABe_i = \Sigma_{j=1}^n B_{ji}(Ae_j) = \Sigma_{k,j=1}^n A_{kj}B_{ji}e_k$. Hence

$$(AB)_{ki} = \sum_{j=1}^{n} A_{kj}B_{ji}$$

which is the usual product rule for the matrices. The identity mapping $x \to x$ is for any basis represented by the unit matrix. Both unit mapping and matrix are denoted by $\mathbb{1}$.

When another basis is chosen for V, the linear transformation A is described according to eq. (1.2) by another matrix $\bar{A}_{ij}$ such that $\bar{e}' = \bar{e}\bar{A}$ and $\bar{\xi}' = \bar{A}\,\bar{\xi}$. One has

$$\bar{A} = S^{-1}AS \tag{1.4}$$

because $\bar{e}' = e'S = eAS$ and on the other hand $\bar{e}' = \bar{e}\bar{A} = eS\bar{A}$. Two matrices A and $\bar{A}$ for which the relation (1.4) holds are called *equivalent* or *similar*. Under such a *similarity transformation* the characteristic polynomial $\det(A - \lambda\mathbb{1})$ is invariant. Because

$$\det(A - \lambda\mathbb{1}) = \sum_{j=0}^{n} (-1)^{n-j}\sigma_j\lambda^j\,,$$

in particular the determinant $\det A = (-1)^n\sigma_0$, and the trace $\operatorname{tr} A = \sigma_{n-1} = \Sigma_{j=1}^n A_{jj}$ are invariant. This means that $\det A = \det \bar{A}$, and $\operatorname{tr} A = \operatorname{tr} \bar{A}$.

To a given matrix A are related its transpose $\tilde{A}$, and its Hermitian conjugate $A^\dagger$. These are defined by $\tilde{A}_{ij} = A_{ji}$ and $A^\dagger_{ij} = A^*_{ji}$, where $*$ denotes complex conjugation. A matrix can be symmetric $(\tilde{A} = A)$, Hermitian $(A^\dagger = A)$, unitary $(AA^\dagger = A^\dagger A = \mathbb{1})$, orthogonal $(A\tilde{A} = \tilde{A}A = \mathbb{1})$, diagonal (all elements A_{ij} with $i \neq j$ are zero).

On a complex vector space V a *Hermitian function or Hermitian form* is a mapping $f(x,y)$ which assigns to each pair $x,y \in V$ a complex number such that

1) $f(x,y) = f(y,x)^*$, (any $x,y \in V$)

2) $f(x, \alpha y + \beta z) = \alpha f(x,y) + \beta f(x,z)$, (any $x,y,z \in V$ and $\alpha,\beta \in \mathbb{C}$).

With respect to a basis of V one can describe f by a matrix defined by $f_{ij} = f(e_i, e_j)$. Then one has for arbitrary $x = e\xi$ and $y = e\eta$ the relation $f(x,y) = \Sigma_{i,j=1}^n \xi_i^* \eta_j f_{ij} = \xi^\dagger f\eta$. Because $f_{ij}^* = f(e_i, e_j)^* = f(e_j, e_i) = f_{ji}$ the matrix f_{ij} is Hermitian. With respect to another basis $\bar{e} = eS$ the same Hermitian form can be given by a matrix $\bar{f}_{ij}$ such that

$$\bar{f}_{ij} = \sum_{k,l=1}^{n} S^*_{ki} S_{lj} f_{kl} \quad \text{or} \quad \bar{f} = S^\dagger f S \,. \tag{1.5}$$

An example is obtained by putting $f = \mathbb{1}$. Then $f(\boldsymbol{x},\boldsymbol{y}) = \xi^\dagger \eta = \Sigma_{i=1}^n \xi_i \eta_i$ which is the expression for a scalar product in a complex vector space. It is an example of a positive definite Hermitian form. A Hermitian form is *positive definite* if and only if $f(\boldsymbol{x},\boldsymbol{x}) \geqslant 0$ for any $\boldsymbol{x} \in V$, and if $f(\boldsymbol{x},\boldsymbol{x}) = 0$ is equivalent with $\boldsymbol{x} = \boldsymbol{0}$. Notice that $f(\boldsymbol{x},\boldsymbol{x})$ is real because $f(\boldsymbol{x},\boldsymbol{x})^* = f(\boldsymbol{x},\boldsymbol{x})$. Without proof we state that *for any positive definite Hermitian form there is a basis, with respect to which its matrix is the identity matrix* $\mathbb{1}$.

On a real vector space we define analogously a *symmetric bilinear form* as a mapping which assigns to each pair of elements $\boldsymbol{x},\boldsymbol{y} \in V$ a real number $f(\boldsymbol{x},\boldsymbol{y})$ such that

1) $f(\boldsymbol{x},\boldsymbol{y}) = f(\boldsymbol{y},\boldsymbol{x})$ for any $\boldsymbol{x},\boldsymbol{y} \in V$,

2) $f(\boldsymbol{x},\alpha\boldsymbol{y} + \beta\boldsymbol{z}) = \alpha f(\boldsymbol{x},\boldsymbol{y}) + \beta f(\boldsymbol{x},\boldsymbol{z})$ for any $\boldsymbol{x},\boldsymbol{y},\boldsymbol{z} \in V$ and $\alpha, \beta \in \mathbb{R}$.

Its matrix f_{ij} is defined by $f_{ij} = f(\boldsymbol{e}_i, \boldsymbol{e}_j)$ for a given basis. It is a symmetric matrix. With respect to another basis $\bar{\boldsymbol{e}} = \boldsymbol{e}S$ one obtains a matrix f with

$$\bar{f} = \tilde{S} f S \,. \tag{1.6}$$

Also in this case one can prove that for any positive definite symmetric bilinear form there is a basis such that its matrix is the identity matrix.

A Hermitian or symmetric bilinear form f is *invariant* under a linear transformation A if $f(\boldsymbol{x},\boldsymbol{y}) = f(A\boldsymbol{x},A\boldsymbol{y})$ for any $\boldsymbol{x},\boldsymbol{y} \in V$. This means that $f_{ij} = f(A\boldsymbol{e}_i, A\boldsymbol{e}_j)$ or

$$f = A^\dagger f A \quad (\text{in the real case } f = \tilde{A} f A) \,. \tag{1.7}$$

If there is a basis for which f_{ij} becomes the identity matrix eq. (1.7) says that with respect to this basis $A^\dagger A = \mathbb{1}$ ($\tilde{A}A = \mathbb{1}$ for a real vector space). In other words, with respect to a basis for which a Hermitian form is described by the identity matrix, a linear transformation which leaves this Hermitian form invariant is described by a unitary matrix.

1.2.3. Subspaces

If V is a linear vector space over a field F a subset W of V is a *subspace* if for each $\boldsymbol{x},\boldsymbol{y} \in W$ and $\alpha \in F$ also $\boldsymbol{x} + \boldsymbol{y} \in W$, and $\alpha\boldsymbol{x} \in W$. When $\boldsymbol{e}_1, ..., \boldsymbol{e}_m$ form a basis for W any element of W is of the form $\Sigma_{i=1}^m \xi_i \boldsymbol{e}_i$. One has m = dimension of $W \leqslant n$ = dimension of V.

A subspace W of V is *invariant* under a linear transformation A of V if $A(W) \subseteq W$. By choosing a basis of V consisting of $\boldsymbol{e}_1, ..., \boldsymbol{e}_m, ..., \boldsymbol{e}_n$ such that $\boldsymbol{e}_1, ..., \boldsymbol{e}_m$ is a basis of W, the matrix A is of the form

$$A = \begin{pmatrix} A^{(1)} & A^{(2)} \\ 0 & A^{(3)} \end{pmatrix} \tag{1.8}$$

where $A^{(1)}$ is a $m \times m$ matrix. Notice that the subspace with basis $\boldsymbol{e}_{m+1}, ..., \boldsymbol{e}_n$ is not necessarily an invariant subspace. This will be the case only if $A^{(2)} = 0$ in eq. (1.8). When V has an invariant subspace W with $W \neq V$ and $W \neq \boldsymbol{0}$, it is called *reducible* with respect to A. Otherwise V is *irreducible*.

Let V and W be linear vector spaces. The *direct sum* of V and W consists of all pairs $(\boldsymbol{x}, \boldsymbol{y})$ with $\boldsymbol{x} \in V$ and $\boldsymbol{y} \in W$. It has the structure of a linear vector space if one defines $(\boldsymbol{x}_1, \boldsymbol{y}_1) + (\boldsymbol{x}_2, \boldsymbol{y}_2) = (\boldsymbol{x}_1 + \boldsymbol{x}_2, \boldsymbol{y}_1 + \boldsymbol{y}_2)$, and $\alpha(\boldsymbol{x}, \boldsymbol{y}) = (\alpha\boldsymbol{x}, \alpha\boldsymbol{y})$. This vector space is denoted by $V \oplus W$. If V is n-dimensional, and W is m-dimensional the space $V \oplus W$ is $(n+m)$-dimensional. For bases $\boldsymbol{e}_1, ..., \boldsymbol{e}_n$ of V, and $\boldsymbol{e}_{n+1}, ..., \boldsymbol{e}_{n+m}$ of W a basis for $V \oplus W$ is given by $(\boldsymbol{e}_1, 0), ..., (\boldsymbol{e}_n, 0), (0, \boldsymbol{e}_{n+1}), ..., (0, \boldsymbol{e}_{n+m})$. For linear transformations A on V, and B on W the direct sum is the linear transformation

$$(A \oplus B)(\boldsymbol{x}, \boldsymbol{y}) = (A\boldsymbol{x}, B\boldsymbol{y}) .$$

On the basis of $V \oplus W$ mentioned above the linear transformation $A \oplus B$ has the form

$$A \oplus B = \begin{pmatrix} A & 0 \\ 0 & B \end{pmatrix} . \tag{1.9}$$

One has $\det(A \oplus B) = \det(A) \det(B)$, and $\mathrm{tr}(A \oplus B) = \mathrm{tr}(A) + \mathrm{tr}(B)$. It is evident from eq. (1.9) that $V \oplus W$ is reducible with respect to $A \oplus B$.

For given vector spaces V and W we can construct another vector space in the following way. Consider all pairs $(\boldsymbol{x}, \boldsymbol{y})$ with $\boldsymbol{x} \in V$, and $\boldsymbol{y} \in W$. We denote this pair by $\boldsymbol{x} \otimes \boldsymbol{y}$. They generate a linear vector space by the postulates

1) $(\alpha\boldsymbol{x}) \otimes \boldsymbol{y} = \alpha(\boldsymbol{x} \otimes \boldsymbol{y})$,
2) $(\boldsymbol{x}_1 + \boldsymbol{x}_2) \otimes \boldsymbol{y} = \boldsymbol{x}_1 \otimes \boldsymbol{y} + \boldsymbol{x}_2 \otimes \boldsymbol{y}$,
3) $\boldsymbol{x} \otimes (\alpha\boldsymbol{y}) = \alpha(\boldsymbol{x} \otimes \boldsymbol{y})$,
4) $\boldsymbol{x} \otimes (\boldsymbol{y}_1 + \boldsymbol{y}_2) = \boldsymbol{x} \otimes \boldsymbol{y}_1 + \boldsymbol{x} \otimes \boldsymbol{y}_2$.

The vector space $V \otimes W$ defined in this way is the *tensor product* of V and W.

It is a vector space of dimension nm. Because $\boldsymbol{x} \otimes \boldsymbol{y} = \Sigma_{i=1}^{n} \Sigma_{j=1}^{m} \xi_i \eta_j \boldsymbol{e}_i \otimes \boldsymbol{e}_j$ the elements $\boldsymbol{e}_i \otimes \boldsymbol{e}_j$ form a basis of $V \otimes W$. The tensor product of the linear transformations A on V, and B on W is a linear transformation $A \otimes B$ on $V \otimes W$ defined by

$$(A \otimes B)(\boldsymbol{x} \otimes \boldsymbol{y}) = (A\boldsymbol{x}) \otimes (B\boldsymbol{y}) . \tag{1.10}$$

With respect to the basis $\boldsymbol{e}_i \otimes \boldsymbol{e}_j$ the matrix of $A \otimes B$ is given by

$$(A \otimes B)(\boldsymbol{e}_i \otimes \boldsymbol{e}_j) = \Sigma_{k=1}^{n} \Sigma_{l=1}^{m} A_{ki} B_{lj} \boldsymbol{e}_k \otimes \boldsymbol{e}_l$$

or

$$(A \otimes B)_{kl,ij} = A_{ki} B_{lj} , \quad (i,k=1, ..., n \text{ and } l,j=1, ..., m) .$$

We note that $\det(A \otimes B) = \det(A) \det(B)$, and $\mathrm{tr}(A \otimes B) = (\mathrm{tr}\, A)(\mathrm{tr}\, B)$. Furthermore one has the property

$$(A \otimes B)(C \otimes D) = (AC) \otimes (BD)$$

because $(A \otimes B)(C \otimes D)(\boldsymbol{x} \otimes \boldsymbol{y}) = (A \otimes B)(C\boldsymbol{x} \otimes D\boldsymbol{y}) = (AC\boldsymbol{x}) \otimes (BD\boldsymbol{y}) = (AC \otimes BD)(\boldsymbol{x} \otimes \boldsymbol{y})$.

1.2.4. Sets of linear transformations

Consider a set of linear transformations on an n-dimensional vector space V. The number of elements is called the *order of the set*. Its dimension is n. By choosing a basis $\boldsymbol{e}_1, ..., \boldsymbol{e}_n$ in V a set of linear transformations corresponds to a set of $n \times n$ matrices $\mathcal{A} = \{A_1, ..., A_N\}$. For another basis $\bar{\boldsymbol{e}}_1, ..., \bar{\boldsymbol{e}}_n$ (cf. eq. (1.2)) one obtains another set of matrices $\bar{\mathcal{A}} = \{\bar{A}_1, ..., \bar{A}_N\}$, where $\bar{A}_i = S^{-1} A_i S$ $(i = 1, ..., N)$ for some nonsingular matrix S. The sets $\mathcal{A}$ and $\bar{\mathcal{A}}$ are called *equivalent*. Notice that the matrix S is the same for all elements of the set.

It is convenient to define sets of linear transformations formed from the sets $\mathcal{A} = \{A_1, ..., A_N\}$ on V and $\mathcal{B} = \{B_1, ..., B_M\}$ on W. For $N=M$ one defines the *direct sum* of $\mathcal{A}$ and $\mathcal{B}$ by $\mathcal{A} \oplus \mathcal{B} = \{A_1 \oplus B_1, ..., A_N \oplus B_N\}$. It is a set of transformations on $V \oplus W$, has order N and dimension $n+m$. Also for $N=M$ the *inner Kronecker product* is the set $\mathcal{A} \otimes \mathcal{B} = \{A_1 \otimes B_1, ..., A_N \otimes B_N\}$. It is a set of linear transformations on $V \otimes W$. It has order N and dimension nm. Notice that this definition of direct sum and inner Kronecker product assumes a fixed ordering of the elements of the two sets.

The *outer Kronecker product* is also defined for $N \neq M$ by $\mathcal{A} \times \mathcal{B} = \{A_1 \otimes B_1, ..., A_1 \otimes B_M, ..., A_N \otimes B_M\}$. It has order NM and dimension nm.

If V has a subspace W which is invariant under all elements of a set $\mathcal{A}$ of linear transformations, W is called $\mathcal{A}$*-invariant*. When $W \neq V$ and $W \neq \mathbf{0}$, all matrices can simultaneously be brought into the form of eq. (1.8) by a suitable choice of basis. Such a set is called *reducible*. In particular, a set which is equivalent to the direct sum of two sets is called *fully reducible*. A set which is not reducible is *irreducible*.

1.2.5. Schur's lemma

PROPOSITION 1.3 (Schur). Let $\mathcal{A}$ and $\mathcal{B}$ be two irreducible sets of matrices, of dimensions n and m respectively. Let S be an $m \times n$ matrix such that $S\mathcal{A} = \mathcal{B}S$. Then either $S = 0$, or S is nonsingular and $\mathcal{A}$ and $\mathcal{B}$ are equivalent.
Proof. Let $\mathcal{A}$ correspond to a set of linear transformations on an n-dimensional vector space V, $\mathcal{B}$ to one on an m-dimensional vector space W, and S to a linear mapping $S : V \to W$. Then consider $Im_S \subseteq W$. Take $\boldsymbol{x} \in V$. Then $S\boldsymbol{x} \in Im_S$. Moreover, $\mathcal{B}S\boldsymbol{x} = S\mathcal{A}\boldsymbol{x} \in Im_S$. Therefore, Im_S is a subspace invariant under $\mathcal{B}$. As $\mathcal{B}$ is irreducible one has either $Im_S = 0$, or $Im_S = W$. In the first case $S = 0$. Next we look at V. Here Ker_S is an invariant subspace, because for $\boldsymbol{x} \in Ker_S$ one has $S\boldsymbol{x} = \mathbf{0}$, and $S\mathcal{A}\boldsymbol{x} = \mathcal{B}S\boldsymbol{x} = \mathbf{0}$. Hence $\mathcal{A}\boldsymbol{x} \subsetneqq Ker_S$. As $\mathcal{A}$ is irreducible one concludes that either $Ker_S = \mathbf{0}$, or $Ker_S = V$. In the latter case $Im_S = \mathbf{0}$ and $S = 0$. For the case $Ker_S = \mathbf{0}$ one has $Im_S = W$ and therefore S is a one-to-one-mapping from V to W. Hence either $S = 0$ or S is a non-singular $n \times n$ matrix and in the latter case $\mathcal{A} = S^{-1}\mathcal{B}S$ is equivalent to $\mathcal{B}$.

Consequences of Schur's lemma.
1) If S links two nonequivalent irreducible sets, i.e. $S\mathcal{A} = \mathcal{B}S$, one has $S = 0$.
2) If S links two irreducible sets of the same dimension and $\det S = 0$, then $S = 0$.
3) If an $n \times n$ matrix commutes with all the elements of an irreducible set of matrices, then S is a multiple of the unit matrix $\mathbb{1}$. To prove this one con-

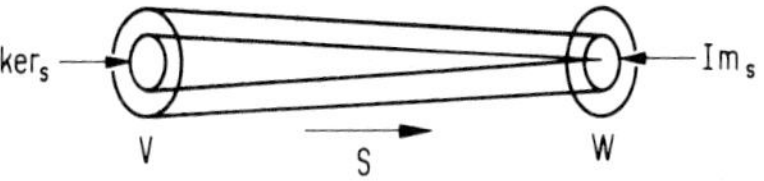

Fig. 1.3. Schur's lemma.

siders the two cases: S is singular or not. If S has determinant equal to zero, from 2) it follows that $S = 0$. If S is nonsingular, there is an eigenvalue $\lambda \neq 0$ of S. In this case the matrix $S - \lambda\mathbb{1}$ commutes with all the elements of the set and $\det(S - \lambda\mathbb{1}) = 0$. Hence $S = \lambda\mathbb{1}$. Contrary to the first two consequences, the third is only valid for complex matrices, because in general the eigenvalue λ is complex.

1.3. Representations of finite groups

1.3.1. Matrix representations

In section 2.2 we saw that the nonsingular transformations of a linear vector space V over a field F form a group. This group is denoted by $GL(V)$. After a choice of basis in V this group corresponds to a group of matrices, denoted by $GL(n,F)$, where n is the dimension of the vector space. We will now consider subgroups of $GL(n,F)$ which are homomorphic to a given group G. A *representation of a group* G in a vector space V is a homomorphism T from G into the group $GL(V)$. A *matrix representation* of G is a homomorphism D from G into $GL(n,F)$. Of course, choosing a basis in V a representation in V is associated to a matrix representation. On the other hand, for a matrix representation, the matrices correspond to linear transformations in an arbitrary n-dimensional vector space, if we choose a basis in this vector space. Therefore, there is a close connection between the two kinds of representations. However, sometimes it is convenient to distinguish between them. The *dimension* of a representation is the dimension of the vector space. Two representations D and D' are called *equivalent* if there exists a nonsingular matrix S such that $D(g) = S^{-1}D'(g)S$. Here we will only consider *representations in complex vector spaces of finite dimension*. Moreover, we will restrict ourselves to *finite groups*, although many of the properties and definitions treated here can be used as well for arbitrary groups.

An example of a representation of a group G is given by the homomorphism which assigns the number 1 to each element of G. This representation of dimension 1 is called the *trivial representation*. The kernel of the homomorphism, also called the *kernel of the representation*, is the complete group G. If the kernel of the representation is only the unit element, the representation is called *faithful*. In this case different elements of G are mapped on different matrices. An example is given by the following representation which plays a role in the next sections. Consider the complex functions on the finite group G. These form a linear vector space, even an algebra as we saw in section 1.7. We now define for each $a \in G$ a linear transformation T_a on this space by $(T_a\mathfrak{f})(g) = \mathfrak{f}(a^{-1}g)$. This is a homomorphism because $(T_b(T_a\mathfrak{f}))(g) = (T_a\mathfrak{f})(b^{-1}g) = \mathfrak{f}(a^{-1}b^{-1}g) = (T_{ba}\mathfrak{f})(g)$. To obtain corresponding matrices we notice that a basis in the vector space is given by $\mathfrak{g}_1, ..., \mathfrak{g}_N$, where $g_1, ..., g_N$ are the elements of G, and where $\mathfrak{g}$ is the function which takes the value 1 on the element g and the value 0 everywhere else. A matrix for T_a is obtained from $T_a\mathfrak{g}_i = \mathfrak{a}\mathfrak{g}_i = \mathfrak{g}_j$, because $T_a\mathfrak{g}_i(g) = \mathfrak{g}_i(a^{-1}g) = \mathfrak{a}\mathfrak{g}_i(g)$. The index j is uniquely determined from $g_j = ag_i$. The matrix corresponding to T_a is $D(a)_{ki} = \delta_{kj}$. Here δ_{kj} is the Kronecker δ-symbol. The matrix $D(a)$ has exactly one 1 in each row and each column and 0 on the remaining places. The representation obtained in this way is called the *regular representation*. It is a faithful representation. We remark that the regular representation can also be defined in a less sophisticated manner. To do this we choose again a fixed ordering $g_1, ..., g_N$ for the elements of G. Then one defines a $N \times N$ matrix $D(a)$ by

$$ag_i = \sum_{j=1}^{N} D(a)_{ji} g_j \; .$$

Since the group is mapped onto itself in a one-to-one way by left multiplication, this is a meaningful definition. One has still to show that we really obtain a representation in this way. This is seen from $(g_ig_j)g_k = \Sigma_m D(g_ig_j)_{mk} g_m = g_i \, \Sigma_l D(g_j)_{lk} g_l = \Sigma_l D(g_j)_{lk} \, \Sigma_m D(g_i)_{ml} g_m = \Sigma_m \, \{(D(g_i)D(g_j)\}_{mk} g_m$. In section 3.4 we will treat other examples of representations. There we will illustrate also the notions and properties to be discussed in this section.

As we will see, in physics we are especially interested in representations consisting of unitary matrices, called *unitary representations*.

PROPOSITION 1.4. Any representation of a finite group is equivalent to a unitary representation.

Proof. On the representation space V one considers a positive definite Hermitian form $f(\boldsymbol{x}, \boldsymbol{y})$, e.g. the function $f(\boldsymbol{x}, \boldsymbol{y}) = \xi^\dagger \eta$ (cf. §2.2). If T_g is the linear transformation representing the element $g \in G$, one defines a function $F(\boldsymbol{x}, \boldsymbol{y})$ on V by

$$F(\boldsymbol{x}, \boldsymbol{y}) = \sum_{g \in G} f(T_g \boldsymbol{x}, T_g \boldsymbol{y}) \, .$$

It is easily verified that $F(\boldsymbol{x}, \boldsymbol{y})$ is also a positive definite Hermitian form. Moreover, it is invariant under G, because for any $h \in G$ one has

$$F(T_h \boldsymbol{x}, T_h \boldsymbol{y}) = \sum_{g \in G} f(T_g T_h \boldsymbol{x}, T_g T_h \boldsymbol{y}) = \sum_{gh \in G} f(T_{gh} \boldsymbol{x}, T_{gh} \boldsymbol{y}) = F(\boldsymbol{x}, \boldsymbol{y})$$

as summation over gh for fixed h is the same as summation over g. In section 2.1 we mentioned the fact that for every positive definite Hermitian form there is a basis such that the corresponding matrix is the unit matrix. With respect to this basis the transformations T_g are unitary matrices, because of eq. (1.7).

Given two representations $D(G)$ and $D'(G)$, the direct sum of the two sets is again a representation denoted by $(D \oplus D')(G)$. It is called the sum representation. Its matrices can be written as

$$(D \oplus D')(g) = \begin{pmatrix} D(g) & 0 \\ 0 & D'(g) \end{pmatrix} \qquad (\text{any } g \in G) . \tag{1.11}$$

If the dimensions of D and D' are d and d', respectively, the dimension of the sum representation is $d + d'$. It is evident that this procedure can be generalized to the sum of several representations. Moreover, one can take $D = D'$, or in general several terms in the sum equal. One writes

$$\sum_{\alpha=1}^{s} m_\alpha D_\alpha(G) = (m_1 D_1 \oplus m_2 D_2 \oplus ... \oplus m_s D_s)(G) .$$

Here $mD(G) = D(G) \oplus D(G) \oplus ... \oplus D(G)$ (m terms). A representation is reducible if it is *reducible* as a set of transformations. It is *fully reducible* if it is fully reducible as a set of transformations. A representation which is not reducible is *irreducible*. One has

PROPOSITION 1.5. A reducible representation of a finite group is fully reducible.

Proof. A representation of a finite group is equivalent to a unitary representation. Therefore, on the representation space V there exists an invariant, positive definite Hermitian function F. As the representation is reducible, there is an invariant subspace W. Now the subset of all elements $\boldsymbol{x}$ of V for which $F(\boldsymbol{x},\boldsymbol{y}) = 0$ for any $\boldsymbol{y} \in W$ is again a subspace W'. Moreover, it is also an invariant subspace, because for any $g \in G$ and any $\boldsymbol{x} \in W'$ one has $F(T_g\boldsymbol{x},\boldsymbol{y}) = F(\boldsymbol{x},T_g^{-1}\boldsymbol{y}) = 0$ for any $\boldsymbol{y} \in W$. Finally $W \oplus W' = V$. This proves the statement.

From this proposition it follows that a representation of a finite group is equivalent to ($\sim$) a sum of irreducible representations.

$$D(G) \sim \sum_{\alpha=1}^{r} m_\alpha D_\alpha(G)$$

where $D_1(G), ..., D_r(G)$ are irreducible representations, the *irreducible components* of $D(G)$. The integer m_α is called the *multiplicity* of the component $D_\alpha(G)$. We will show later on that this decomposition of $D(G)$ is unique. This means the following. If $D(G) \sim \Sigma_{\alpha=1}^{r} m_\alpha D_\alpha(G)$ and $D(G) \sim \Sigma_{\beta=1}^{s} m'_\beta D'_\beta(G)$, there is a one-to-one correspondence between $(D_1(G), ...$ $..., D_r(G))$ and $(D'_1(G), ..., D'_s(G))$ such that $r = s$, $m_\alpha = m'_\alpha$, and $D_\alpha(G) \sim D'_\alpha(G)$. As a consequence of Schur's lemma every irreducible representation of an Abelian group is one-dimensional. Hence every representation of a finite, Abelian group has an equivalent representation, consisting of diagonal matrices. Note however, that this consequence of Schur's lemma is only valid because we consider complex representations. In the case of real representations this is no longer true. When D and D' are representations of dimensions d and d' respectively, the inner Kronecker product $D(G) \otimes D'(G)$ gives another representation of G because

$$[(D \otimes D')(g)]\,[(D \otimes D')(g')] = [D(g) \otimes D'(g)]\,[D(g') \otimes D'(g')]$$

$$= D(gg') \otimes D'(gg') = (D \otimes D')(gg')\,.$$

This representation is called the *product representation*. In general it is reducible, even when D and D' are irreducible representations.

For two representations $D(G)$ and $D'(H)$ of two groups G and H, the outer Kronecker product is also a representation, in this case of the direct product $G \times H : (D \times D')(g,h) = D(g) \otimes D'(h)$. It is a representation, because

$$[(D \times D')(g,h)]\,[(D \times D')(g',h')] = [D(g) \otimes D'(h)]\,[D(g') \otimes D'(h')]$$

$$= D(gg') \otimes D'(hh') = (D \times D')(gg', hh')\,. \tag{1.12}$$

1.3.2. Orthogonality relations

Suppose that $D_1(G)$ and $D_2(G)$ are two irreducible representations of the group G, with dimensions d_1 and d_2 respectively. We take an arbitrary $d_1 \times d_2$ matrix R, and define

$$S = \sum_{g \in G} D_1(g) R D_2^{-1}(g)\,.$$

This matrix S satisfies $SD_2(g) = D_1(g)S$ for any $g \in G$, because

$$SD_2(g) = \sum_{h\in G} D_1(h)RD_2^{-1}(h)D_2(g)$$

$$= D_1(g) \sum_{h\in G} D_1^{-1}(g)D_1(h)RD_2^{-1}(h)D_2(g)$$

$$= D_1(g) \sum_{h\in G} D_1(g^{-1}h)RD_2^{-1}(g^{-1}h) = D_1(g)S\,.$$

Because of the irreducibility of $D_1(G)$ and $D_2(G)$ it follows from Schur's lemma that either $D_1(G) \sim D_2(G)$ or $D_1(G) \nsim D_2(G)$, in which case one has

$$0 = S_{ij} = \sum_{g\in G} \sum_{k,l} D_1(g)_{ik}R_{kl}D_2(g^{-1})_{lj}\,.$$

Since R is arbitrary we find in this case

$$\sum_{g\in G} D_1(g)_{ik}D_2(g^{-1})_{lj} = 0\,.$$

One finds this relation by taking for R the matrix with 1 in the kl-place, and 0 everywhere else.

When $D_1(G) \sim D_2(G)$ we consider only the case that $D_1(G) = D_2(G)$, i.e. the case that the matrices corresponding to an element $g \in G$ are the same. Then S commutes with all elements of $D_1(G) = D_2(G)$. Hence $S = \lambda\mathbb{1}_d$, where d is the dimension both of D_1 and of D_2. The trace of S is given by

$$\operatorname{tr} S = \sum_{g\in G} \operatorname{tr}(D_1(g)RD_1(g^{-1})) = N \operatorname{tr} R$$

when N is the order of G. It follows that $\lambda d = N \operatorname{tr} R$, and consequently

$$S_{ij} = \frac{N}{d} \operatorname{tr} R\,\delta_{ij} = \sum_{g\in G} \sum_{k,l} D_1(g)_{ik}R_{kl}D_1(g^{-1})_{lj}\,.$$

We conclude that in this case

$$\sum_{g\in G} D_1(g)_{ik}D_1(g^{-1})_{lj} = \frac{N}{d}\delta_{ij}\delta_{kl}\,. \qquad (1.13)$$

Now we have proved the orthogonality relations for irreducible representations

$$\sum_{g\in G} D_\alpha(g)_{ik} D_\beta(g^{-1})_{lj} = \frac{N}{d}\delta_{ij}\delta_{kl}\delta'_{\alpha\beta} \tag{1.14}$$

with the symbol $\delta'_{\alpha\beta}$ defined by

$$\delta'_{\alpha\beta} = \begin{cases} 0 \quad \text{if} \quad D_\alpha(G) \not\sim D_\beta(G) \\ 1 \quad \text{if} \quad D_\alpha(G) = D_\beta(G) \\ \text{undefined if } D_\alpha(G) \sim D_\beta(G), \text{ but } D_\alpha(G) \neq D_\beta(G)\,. \end{cases}$$

For unitary representations the eq. (1.14) can also be written in the form

$$\sum_{g\in G} D_\alpha(g)_{ik} D^*_\beta(g)_{jl} = \frac{N}{d}\delta_{ij}\delta_{kl}\delta'_{\alpha\beta} \tag{1.15}$$

because then $D(g^{-1})_{ij} = (D(g)^{-1})_{ij} = (D(g))^\dagger_{ij} = D(g)^*_{ji}$.

1.3.3. Characters

A very important concept will be the notion of character of a representation. The *character of an element* $g \in G$ in a representation $D(G)$ is defined by

$$\chi(g) = \operatorname{tr} D(g)\,.$$

The character of the representation is the complex function on G which assigns $\chi(g)$ to $g \in G$. Some important properties are the following.

The character is a class function. It is a function on the group, because to each element of the group is assigned a complex number. Moreover, the characters of conjugate elements are the same since $\chi(gag^{-1}) = \operatorname{tr}(D(g)D(a)D(g)^{-1}) = \operatorname{tr} D(a) = \chi(a)$. From the invariance of the trace under similarity transformations follows also that equivalent representations have the same character: if $D(g) = SD'(g)S^{-1}$ one has $\operatorname{tr} D(g) = \operatorname{tr} D'(g)$. We shall even show that the characters completely characterize classes of equivalent representations. We note that the character of the unit element, i.e. $\operatorname{tr} D(e) = \operatorname{tr} \mathbb{1}_d$, is equal to the dimension of the representation.

For the characters one can derive orthogonality relations from those for the representations. One has from eq. (1.14) putting $i=k$, $l=j$, and summing over k and j, that

$$\sum_{g\in G} \chi_\alpha(g)\,\chi_\beta(g^{-1}) = N\delta_{\alpha\beta}\,.$$

For unitary representations $D(g^{-1}) = D(g)^\dagger$, and therefore $\chi(g^{-1}) = \chi^*(g)$. As every representation of a finite group is equivalent to a unitary one, and the characters of equivalent representations are the same, one has for all irreducible representations, and not only for unitary ones

$$\sum_{g\in G} \chi_\alpha(g)\chi_\beta^*(g) = N\delta_{\alpha\beta} . \tag{1.16}$$

Notice that here we have used the Kronecker δ, and not the special symbol δ' like in eq. (1.14).

When the group G has r conjugacy classes $C_1, ..., C_r$, we can construct an r-dimensional vector $\theta(D)$ for each representation $D(G)$ by

$$\theta(D) = \left(\left(\frac{n_1}{N}\right)^{1/2} \chi(C_1), ..., \left(\frac{n_r}{N}\right)^{1/2} \chi(C_r)\right) . \tag{1.17}$$

where n_i is the number of elements in class C_i and $\chi(C_i)$ the value of the character on the class C_i. For two nonequivalent irreducible representations these vectors are orthogonal, as one sees from

$$\theta(D_\alpha)\,\theta(D_\beta)^\dagger = \sum_{i=1}^{r} \frac{n_i}{N}\chi_\alpha(C_i)\,\chi_\beta^*(C_i) = \frac{1}{N}\sum_{g\in G} \chi_\alpha(g)\,\chi_\beta^*(g) = \delta_{\alpha\beta} .$$

Because there are at most r mutually orthogonal vectors in a r-dimensional space, the number of nonequivalent irreducible representations of G is at most equal to r. Later on we will show that it is exactly equal to r.

If $D_1(G)$ and $D_2(G)$ are representations of dimension d_1 and d_2 respectively, the character of the direct sum $D(G) = (D_1 \oplus D_2)(G)$ is given by $\chi(g) = \mathrm{tr}\,(D_1 \oplus D_2)(g) = \mathrm{tr}\,D_1(g) + \mathrm{tr}\,D_2(g) = \chi_1(g) + \chi_2(g)$. For a representation $D = \Sigma_\alpha m_\alpha D_\alpha$, one has the character $\chi(g) = \Sigma_\alpha m_\alpha \chi_\alpha(g)$. The multiplicity m_α of an irreducible representation $D_\alpha(G)$ in the reduction of a representation $D(G)$ is found using the orthogonality relations. One finds

$$\frac{1}{N}\sum_{g\in G} \chi(g)\chi_\alpha^*(g) = \frac{1}{N}\sum_{g\in G}\sum_{\beta=1}^{r} m_\beta\chi_\beta(g)\chi_\alpha^*(g) = \sum_{\beta=1}^{r} m_\beta\delta_{\alpha\beta} = m_\alpha . \tag{1.18}$$

From this follows

PROPOSITION 1.6. The characters of two representations $D_1(G)$ and $D_2(G)$ are equal if and only if the representations are equivalent.

Proof. We have already seen that the characters of equivalent representations are equal. For irreducible representations it follows from eq. (1.16) that nonequivalent irreducible representations have different characters. Finally, the multiplicities and the irreducible components for reducible representations are uniquely determined by eq. (1.18).

By eq. (1.18), which can also be written as

$$m_\alpha = \frac{1}{N} \sum_{i=1}^{r} n_i \chi(C_i) \chi_\alpha^*(C_i) , \tag{1.19}$$

it is possible to determine the reduction of a representation $D(G)$ with character $\chi(G)$. One can determine whether the representation is irreducible or not by the irreducibility criterion in the following proposition.

PROPOSITION 1.7. $D(G)$ is irreducible if and only if

$$\sum_{i=1}^{r} n_i |\chi(C_i)|^2 = N . \tag{1.20}$$

Proof. When the representation $D(G)$ can be written as a sum of irreducible components $D(G) = \Sigma_\alpha m_\alpha D_\alpha(G)$, it follows from eq. (1.16) that

$$\sum_{i=1}^{r} n_i |\chi(C_i)|^2 = \sum_{i=1}^{r} \sum_{\alpha,\beta}^{r} n_i m_\alpha m_\beta \chi_\alpha(C_i) \chi_\beta^*(C_i) = N \sum_{\alpha,\beta}^{r} m_\alpha m_\beta \delta_{\alpha\beta} = N \sum_{\alpha=1}^{r} m_\alpha^2 .$$

If $D(G)$ is irreducible, there is only one α such that $m_\alpha = 1$ and $m_\beta = 0$ for any $\beta \neq \alpha$. If $D(G)$ is reducible $\Sigma_{\alpha=1}^{r} m_\alpha^2 > 1$. This proves the proposition.

To find another relation for representations we consider the regular representation, which is in general reducible. The character of an element $g \in G$, i.e. the number of elements g_i for which $gg_i = g_i$, is zero for $g \neq e$, and it is the order N of G for $g = e$. Hence for the regular representation

$$\chi(g) = 0 \quad \text{for } g \neq e ,$$

$$\chi(e) = N = d = \text{dimension of the representation} .$$

Consider the reduction of this representation into its irreducible components $D_1(G), D_2(G), \ldots$ of dimension $d_1, d_2, \ldots$ respectively. The multiplicities of the irreducible components are found from eq. (1.19).

$$m_\alpha = (1/N) \sum_{g \in G} \chi(g) \chi_\alpha^*(g) = (1/N)\, \chi(e) \chi_\alpha^*(e) = d_\alpha .$$

Hence in the decomposition of the regular representation an irreducible representation $D_\alpha(G)$ occurs as many times as its dimension d_α. Because

$$\chi(e) = \sum_{\alpha=1}^{r} m_\alpha \chi_\alpha(e)$$

one has

$$N = \chi(e) = \sum_{\alpha=1}^{r} m_\alpha \chi_\alpha(e) = \sum_{\alpha=1}^{r} |\chi_\alpha(e)|^2 = \sum_{\alpha=1}^{r} d_\alpha^2 \,.$$

This is called *Burnside's formula*. We now come to the following fundamental proposition.

PROPOSITION 1.8. The number of nonequivalent irreducible representations of a finite group G is equal to the number of its conjugacy classes.

Proof. The equivalence of representations is an equivalence relation (cf. § 1.2). Therefore, we can form the equivalence classes of irreducible representations. From each equivalence class we take one representative $D_\alpha(G)$. The functions $D_\alpha(g)_{ij}$, with $\alpha = 1, 2, \ldots$ and $i,j = 1, \ldots, d_\alpha$, are functions on the group G. There are $\Sigma_\alpha d_\alpha^2 = N$ of these functions. With respect to the positive definite Hermitian function defined on the vector space V of functions on G by

$$F(f_1, f_2) = \sum_{g\in G} f_1(g)^* f_2(g)$$

the N functions $D_\alpha(g)_{ij}$ are orthogonal, for from eq. (1.15) one has

$$F(D_{\alpha ij}, D_{\beta kl}) = \sum_{g\in G} D_\alpha^*(g)_{ij} D_\beta(g)_{kl} = \delta_{\alpha\beta}\, \delta_{ik}\, \delta_{jl} \frac{N}{d_\alpha}$$

where $D_{\alpha ij}$ is the function $D_\alpha(g)_{ij}$ on G. We recall that the dimension of the vector space V is N, as discussed in § 1.7. This means that the N functions form a basis in V. A subspace is formed by all class functions. An arbitrary class function $f(g)$ can be expanded in basis functions with complex coefficients $f_{\alpha ij}$:

$$f(g) = \sum_{\alpha ij} f_{\alpha ij} D_\alpha(g)_{ij} \,.$$

As $f(g)$ is a class function, one has

$$f(g) = \sum_{\alpha ij} (1/N) \sum_{h\in G} f_{\alpha ij} D_\alpha(hgh^{-1})_{ij}$$

$$= \sum_{\alpha ijkl} (1/N) \sum_{h\in G} f_{\alpha ij} D_\alpha(h)_{ik} D_\alpha(g)_{kl} D_\alpha(h^{-1})_{lj} \,.$$

Using eq. (1.14) one arrives at

$$f(g) = \sum_{\alpha ijkl} (1/N) f_{\alpha ij} D_\alpha(g)_{kl} \frac{N}{d_\alpha} \delta_{ij} \delta_{kl} = \sum_{\alpha=1}^{r} \sum_{i=1}^{d_\alpha} \frac{f_{\alpha ii}}{d_\alpha} \chi_\alpha(g) .$$

Therefore, each class function can be written as a linear combination of the character functions $\chi_\alpha(g)$ for the irreducible representations. This means that there are at least r (= number of conjugacy classes) of these functions, or in other words that there are at least r nonequivalent irreducible representations. On the other hand we have seen that the number of equivalence classes of irreducible representations is at most r.

COROLLARY. The functions $D_\alpha(g)_{ij}$ with $\alpha = 1, ..., r$ and $i,j = 1, ..., d_\alpha$ form an orthogonal basis in the space of all functions on G. The characters $\chi_\alpha(g)$ with $\alpha = 1, ..., r$ form an orthogonal basis in the subspace of all class functions.

As the character is a class function, the character of a representation of a group G with r classes is given by r complex numbers. On the other hand, according to the last proposition, there are r nonequivalent irreducible representations. Therefore, the characters of the nonequivalent irreducible representations may be given by a $r \times r$ matrix, called the *character table* of the group. The αj-element of this matrix is the character $\chi_\alpha(C_j)$. Consider the matrices M and M' defined by

$$M_{\alpha j} = \chi_\alpha(C_j) , \qquad M'_{i\beta} = \frac{n_i}{N} \chi^*_\beta(C_i) .$$

These $r \times r$ matrices are inverses one of the other as one sees from

$$(MM')_{\alpha\beta} = \sum_j \chi_\alpha(C_j) \chi^*_\beta(C_j) \frac{n_j}{N} = \delta_{\alpha\beta} .$$

Consequently, one has also

$$\delta_{ij} = (M'M)_{ij} = \sum_\alpha \frac{n_i}{N} \chi_\alpha(C_j) \chi^*_\alpha(C_i) . \qquad (1.21)$$

PROPOSITION 1.9. If c_{ijk} are the class multiplication constants of G, and n_i the order of the class C_i, one has

$$n_i \chi_\alpha(C_i) n_j \chi_\alpha(C_j) = d_\alpha \sum_{k=1}^{r} c_{ijk} n_k \chi_\alpha(C_k) .$$

Proof. For the irreducible representation $D_\alpha(G)$ one defines a matrix $D_\alpha(\mathcal{C}_i)$ by

$$D_\alpha(\mathcal{C}_i) = \sum_{g\in C_i} D_\alpha(g) .$$

As $g\mathcal{C}_i g^{-1} = \mathcal{C}_i$ for a class sum $\mathcal{C}_i$ and for any $g \in G$, one has $D_\alpha(g)\,D_\alpha(\mathcal{C}_i) = D_\alpha(\mathcal{C}_i)\,D_\alpha(g)$ for any $g \in G$. Now $D_\alpha(\mathcal{C}_i)$ commutes with all the elements of an irreducible representation. From Schur's lemma it follows that $D_\alpha(\mathcal{C}_i) = \lambda_i \mathbb{1}$.

The complex constant λ_i can be found comparing $\operatorname{tr} D_\alpha(\mathcal{C}_i) = \lambda_i d_\alpha$ and $\operatorname{tr} D_\alpha(\mathcal{C}_i) = \Sigma_{g\in C_i} \operatorname{tr} D_\alpha(g) = n_i\, \chi_\alpha(C_i)$. Therefore, $\lambda_i = n_i\, \chi_\alpha(C_i)/d_\alpha$. From the definition of class multiplication constants $D_\alpha(\mathcal{C}_i)\, D_\alpha(\mathcal{C}_j) = \Sigma_k\, c_{ijk} D_\alpha(\mathcal{C}_k)$, or $\lambda_i \lambda_j = \Sigma_k\, c_{ijk} \lambda_k$. Substituting the values for λ_i one obtains the desired formula.

We sum up the important relations found for the character table.

$$\frac{1}{N} \sum_{i=1}^{r} n_i\, \chi_\alpha(C_i)\, \chi_\beta^*(C_i) = \delta_{\alpha\beta} \tag{1.22}$$

$$\frac{1}{N} \sum_{\alpha=1}^{r} \chi_\alpha(C_i)\, \chi_\alpha^*(C_j) = \frac{1}{n_i}\, \delta_{ij} \tag{1.23}$$

$$n_i\, \chi_\alpha(C_i)\, n_j\, \chi_\alpha(C_j) = d_\alpha \sum_{k=1}^{r} c_{ijk} n_k\, \chi_\alpha(C_k) . \tag{1.24}$$

A special case of eq. (1.23) is

$$\sum_{\alpha=1}^{r} d_\alpha^2 = N . \tag{1.25}$$

Using eqs. (1.22)–(1.24) one can always construct the character table of a given group. It is a straightforward, but often tedious calculation. To do this with the help of an electronic computer, programs have been made as described in Flodmark and Blokker [1967], Brott [1966], and McKay [1968]. However, for small groups the character table may often be found using only some of the relations. Examples will be given later, in §3.5 and Ch. 3, §2.2. For small groups one can often obtain the number and the dimensions of the irreducible representations directly from eq. (1.25).

The character of the sum representation $(D_1 \oplus D_2)(G)$ is given by $\chi = \chi_1 + \chi_2$. It is evidently reducible. The character of the product representation is given by $\chi(g) = \chi_1(g)\chi_2(g)$. In general this representation is reducible. The multiplicities m_α in $D_1 \otimes D_2 = \Sigma_\alpha m_\alpha D_\alpha$ are given by eq. (1.18).

$$m_\alpha = (1/N) \sum_{i=1}^{r} n_i \chi_1(C_i)\chi_2(C_i)\chi_\alpha^*(C_i)\,.$$

The character of the outer Kronecker product $D_1(G) \times D_2(H)$, which is a representation of the direct product $G \times H$ is $\chi(g,h) = \chi_1(g)\chi_2(h)$. If $D_1(G)$ and $D_2(H)$ are irreducible representations the representation $(D_1 \times D_2)(G \times H)$ is also irreducible. We will come back to this point in Ch. 3, §2.3.

1.3.4. Example: the representations of the group of permutations of three elements

We will treat as an example for which the results of the foregoing sections are worked out in more detail, the group considered in §1.8. This group, denoted by S_3, is isomorphic to the group of three-dimensional rotations which transform an equilateral triangle into itself. As these rotations are non-singular linear transformations of the three-dimensional space, they form a representation of S_3. Choosing a basis $\boldsymbol{e}_1, \boldsymbol{e}_2, \boldsymbol{e}_3$ (fig. 1.2) one obtains the elements ($\boldsymbol{e}_3$ is perpendicular to $\boldsymbol{e}_1$ and $\boldsymbol{e}_2$)

$$D(e) = \epsilon = \begin{pmatrix} 1 & 0 & 0 \\ 0 & 1 & 0 \\ 0 & 0 & 1 \end{pmatrix} \qquad D(a) = \alpha = \begin{pmatrix} -1 & -1 & 0 \\ 1 & 0 & 0 \\ 0 & 0 & 1 \end{pmatrix}$$

$$D(a^2) = \alpha^2 = \begin{pmatrix} 0 & 1 & 0 \\ -1 & -1 & 0 \\ 0 & 0 & 1 \end{pmatrix} \qquad D(c) = \beta = \begin{pmatrix} 0 & 1 & 0 \\ 1 & 0 & 0 \\ 0 & 0 & -1 \end{pmatrix}$$

$$D(ac) = \alpha\beta = \begin{pmatrix} -1 & -1 & 0 \\ 0 & 1 & 0 \\ 0 & 0 & -1 \end{pmatrix} \qquad D(a^2c) = \alpha^2\beta = \begin{pmatrix} 1 & 0 & 0 \\ -1 & -1 & 0 \\ 0 & 0 & -1 \end{pmatrix}.$$

It is a faithful representation of S_3, because all matrices are different. The representation is reducible, as one can check also with the reducibility criterion eq. (1.20). One has $\Sigma_{i=1}^3 n_i |\chi(C_i)|^2 = 1 \times 3^2 + 2 \times 0^2 + 3 \times (-1)^2 = 12 = 2N$. The two-dimensional component

$$D_3(S_3) = \left\{ \begin{pmatrix} 1 & 0 \\ 0 & 1 \end{pmatrix}, \begin{pmatrix} -1 & -1 \\ 1 & 0 \end{pmatrix}, \begin{pmatrix} 0 & 1 \\ -1 & -1 \end{pmatrix}, \begin{pmatrix} 0 & 1 \\ 1 & 0 \end{pmatrix}, \begin{pmatrix} -1 & -1 \\ 0 & 1 \end{pmatrix}, \begin{pmatrix} 1 & 0 \\ -1 & -1 \end{pmatrix} \right\}$$

is seen to be irreducible from the same criterion. The regular representation is given by the matrices

$$D(e) = \begin{pmatrix} 1 & 0 & 0 & 0 & 0 & 0 \\ 0 & 1 & 0 & 0 & 0 & 0 \\ 0 & 0 & 1 & 0 & 0 & 0 \\ 0 & 0 & 0 & 1 & 0 & 0 \\ 0 & 0 & 0 & 0 & 1 & 0 \\ 0 & 0 & 0 & 0 & 0 & 1 \end{pmatrix} \quad D(a) = \begin{pmatrix} 0 & 0 & 1 & 0 & 0 & 0 \\ 1 & 0 & 0 & 0 & 0 & 0 \\ 0 & 1 & 0 & 0 & 0 & 0 \\ 0 & 0 & 0 & 0 & 0 & 1 \\ 0 & 0 & 0 & 1 & 0 & 0 \\ 0 & 0 & 0 & 0 & 1 & 0 \end{pmatrix} \quad D(c) = \begin{pmatrix} 0 & 0 & 0 & 1 & 0 & 0 \\ 0 & 0 & 0 & 0 & 0 & 1 \\ 0 & 0 & 0 & 0 & 1 & 0 \\ 1 & 0 & 0 & 0 & 0 & 0 \\ 0 & 0 & 1 & 0 & 0 & 0 \\ 0 & 1 & 0 & 0 & 0 & 0 \end{pmatrix}$$

and their products.

The character table can be found as follows. From eq. (1.25) one has $d_1^2 + d_2^2 + d_3^2 = 6$, which has the unique solution (up to a change of order) $d_1 = d_2 = 1$, and $d_3 = 2$. For the first row $\chi_1(C_1) = \chi_1(C_2) = \chi_1(C_3) = 1$. For the second row one has $\chi_2(a) = \chi_2(a^2)$, but as the representation is one-dimensional $\chi_2(a^2) = \chi_2(a)^2$. Hence $\chi_2(a) = 1 = \chi_2(C_2)$, because the character of a one-dimensional representation can not be 0. Using eq. (1.22) with $\beta = 1$ one obtains $1 + 2\chi_2(C_2) + 3\chi_2(C_3) = 0$. Therefore $\chi_2(C_3) = -1$. Using eq. (1.23) with $j = 1$ one has $1 + \chi_2(C_i) + 2\chi_3(C_i) = 0$, from which it follows that $\chi_3(C_2) = -1$, and $\chi_3(C_3) = 0$. The character table of S_3 is table 1.2. For its construction we needed only a few of the relations (1.22)–(1.24). The remaining relations are readily verified. It is to be noted that for a general group one has to proceed in a more systematic way. However, for almost all crystallographic groups the character table can be found in a similar ad hoc way.

With the aid of the character table we can reduce an arbitrary representation into its irreducible components. For the regular representation with $\chi = (6, 0, 0)$ one has

$$m_\alpha = \frac{1}{N}\sum_{i=1}^{3} n_i \chi_\alpha(C_i)\chi^*(C_i) = \frac{1}{6}\chi_\alpha(C_1)\,6 = d_\alpha, \quad \text{or}$$

$$D_{\text{reg}}(S_3) = D_1(S_3) \oplus D_2(S_3) \oplus 2D_3(S_3)\,.$$

For the product representation $(D_\alpha \otimes D_\beta)(S_3)$ one has

$$m_\gamma = \frac{1}{6}\sum_{i=1}^{3} n_i \chi_\gamma^*(C_i)\,\chi_\alpha(C_i)\,\chi_\beta(C_i)\,.$$

Table 1.2
Character table of S_3

	$C_1 = e$	$C_2 = [a, a^2]$	$C_3 = [c, ac, a^2c]$
D_1	1	1	1
D_2	1	1	-1
D_3	2	-1	0

This gives

$$D_1 \otimes D_1 = D_1 \qquad D_2 \otimes D_1 = D_2 \qquad D_3 \otimes D_1 = D_3$$

$$D_1 \otimes D_2 = D_2 \qquad D_2 \otimes D_2 = D_1 \qquad D_3 \otimes D_2 = D_3$$

$$D_1 \otimes D_3 = D_3 \qquad D_2 \otimes D_3 = D_3 \qquad D_3 \otimes D_3 = D_1 \oplus D_2 \oplus D_3 \, .$$

1.3.5. Projective representations

For physics a useful generalization of the concept of representation is that of projective representation. In a linear vector space the one-dimensional subspaces are the straight lines through the origin. They are the elements of the set called *projective space*. By a nonsingular linear transformation of the vector space V, one-dimensional subspaces are mapped onto one-dimensional subspaces. Moreover, linear transformations of V which differ only by a factor give the same mapping of the projective space. This leads us to consider transformations differing by a factor as "the same" in some sense.

Consider a mapping P of a group G into $GL(V)$. We will denote the corresponding matrices by $P(g)$. We call this mapping a *projective representation*, if for some complex number $\omega(g,h)$ one has

$$P(g)P(h) = \omega(g,h)P(gh) \qquad (\text{any } g, h \in G) \, . \tag{1.26}$$

The function ω which assigns to each pair of elements $g, h \in G$ a complex number is called a *factor system*. The theory of these projective representations for finite groups was given by Schur (Schur [1904], [1907]). They are also called ray or multiplier representations. A particular case is given by $\omega(g,h) = 1$ for all $g, h \in G$. Then P is an *ordinary representation*, also called *vector representation*, discussed in the foregoing sections. As $P(gh)$ and

$P(g)P(h)$ only differ by a factor, they determine the same transformation of the projective space. This transformation is unchanged if we replace $P(g)$ by $u(g)P(g)$ with $u(g) \neq 0$ a complex function on the group. We call the projective representations $P(G)$ and $P'(G) = u(G)P(G)$ *associated representations*. We will choose a projective representation always in such a way, that $P(e)$ is the unit matrix. For the factor system ω this means that $\omega(g,e) = \omega(e,g) = 1$ for any $g \in G$. Moreover, in this case one derives from eq. (1.26)

$$P(g^{-1}) = \omega(g,g^{-1})P(g)^{-1} = \omega(g^{-1},g)P(g)^{-1} . \tag{1.27}$$

As an example, consider the mapping of the direct product $C_2 \times C_2$ of two cyclic groups of order two into the group $GL(2, C)$, given by

$$P(e) = \begin{pmatrix} 1 & 0 \\ 0 & 1 \end{pmatrix} \quad P(a) = \begin{pmatrix} 0 & 1 \\ 1 & 0 \end{pmatrix} \quad P(b) = \begin{pmatrix} 0 & i \\ -i & 0 \end{pmatrix} \quad P(ab) = \begin{pmatrix} 1 & 0 \\ 0 & -1 \end{pmatrix} .$$

The group is generated by a and b with relations $a^2 = b^2 = e$, $ab = ba$, and is of order four. The mapping P is a projective representation of $C_2 \times C_2$ with factor system $\omega(g,g) = \omega(g,e) = \omega(e,g) = 1$ for all g, $\omega(a,ab) = \omega(b,a) = \omega(ab,a) = \omega(ab,b) = i$, and $\omega(a,b) = \omega(b,ab) = -i$. To the defining relations correspond the following relations between the matrices: $P(a)^2 = \mathbb{1}$, $P(b)^2 = \mathbb{1}$, $P(a)P(b) = -P(b)P(a)$.

Not every function ω can occur as a factor system. A necessary condition follows from the associativity of the matrix multiplication. Because $[P(g)P(h)]P(l) = P(g)[P(h)P(l)]$ one has

$$\omega(g,h)\,\omega(gh,l) = \omega(g,hl)\,\omega(h,l) \qquad \text{for all } g,h,l \in G . \tag{1.28}$$

On the other hand one can show that any ω satisfying eq. (1.28) can occur as a factor system for a projective representation. For this and for other results mentioned in this section we refer to Schur. Because the product of two functions satisfying eq. (1.28) also satisfies this relation, the factor systems form a multiplicative Abelian group, denoted by $Z^2(G)$. The inverse of ω is ω^{-1}, the unit element is $\omega(g,h) = 1$ for all $g,h \in G$. If P and P' are associated projective representations there is a function $u(G)$ such that $P'(g) = u(g)P(g)$. The corresponding factor systems are then related by

$$\omega'(g,h) = \frac{u(g)\,u(h)}{u(gh)}\,\omega(g,h) . \tag{1.29}$$

Two such factor systems are called *associated factor systems.* In particular a factor system associated with the (trivial) factor system of an ordinary representation has the form $\omega(g,h) = u(g)u(h)u(gh)^{-1}$. The factor systems of this form set up a subgroup of $Z^2(G)$ denoted by $B^2(G)$. The relation (1.29) gives an equivalence relation between factor systems. The equivalence classes are the elements of the factor group $Z^2(G)/B^2(G)$. This group is called the *Schur's multiplicator* of G, denoted by $M(G)$. In our example of the projective representation of the group $C_2 \times C_2$ it is easily verified that the factor system is not equivalent with the trivial one. Hence the multiplicator of $C_2 \times C_2$ has at least two elements.

For projective representations one can also introduce the notions of equivalence and irreducibility in a way similar to that for ordinary representations. Two projective representations P and P' are *equivalent* if they are equivalent as sets of transformations. This means, if there is a non-singular matrix S such that $P'(g) = S^{-1}P(g)S$ for any $g \in G$. It follows immediately that equivalent representations have the same factor systems. For a given factor system ω, this equivalence relation gives a division into equivalence classes of the projective representations with this factor system. Two projective representations P and P' are called *similar* if they satisfy the relation $P'(g) = u(g)S^{-1}P(g)S$ for some function $u(g)$, some nonsingular S, and for all $g \in G$. Equivalent representations are similar, as are associated representations. The set of all projective representations can be divided into similarity classes. Each similarity class can be divided into subclasses with the same factor systems, and each such class into further subclasses of equivalent representations.

PROPOSITION 1.10. For a finite group G of order N each factor system is associated with one consisting entirely of Nth roots of 1.

Proof. From eq. (1.28) it follows that for any $g, h \in G$ one can write

$$\omega(g,h)^N = \left(\frac{\omega(g,hl)\,\omega(h,l)}{\omega(gh,l)}\right)^N = \prod_{l\in G}\left(\frac{\omega(g,hl)\,\omega(h,l)}{\omega(gh,l)}\right).$$

Now define a function $f(g) = \Pi_l\, \omega(g,l)$. Then one has the relation

$$\omega(g,h)^N = f(g)\,f(h)/f(gh)\,.$$

As this factor system is associated with the trivial one, the factor system

$$\omega'(g,h) = \left(\frac{f(gh)}{f(g)f(h)}\right)^{1/N} \omega(g,h)$$

is associated with $\omega(g,h)$ and $(\omega'(g,h))^N = 1$. In the following we always will take a factor system which consists of Nth roots of 1. This means that the elements $\omega(g,h)$ are of modulus one.

PROPOSITION 1.11. Every projective representation of a finite group G with a factor system of modulus one has an equivalent unitary representation.

Proof. As in §3.1 we consider a positive definite Hermitian function f on the representation space V. We define another function f' by

$$f'(\boldsymbol{x},\boldsymbol{y}) = \sum_{g\in G} f(P_g\boldsymbol{x},P_g\boldsymbol{y}) \,.$$

This function f' is invariant because

$$f'(P_h\boldsymbol{x},P_h\boldsymbol{y}) = \sum_{g\in G} f(P_gP_h\boldsymbol{x},P_gP_h\boldsymbol{y})$$

$$= \sum_{g\in G} \omega^*(g,h)\,\omega(g,h)\,f(P_{gh}\boldsymbol{x},P_{gh}\boldsymbol{y}) = f'(\boldsymbol{x},\boldsymbol{y})$$

as $|\omega(g,h)| = 1$. With respect to the orthonormal basis for which the matrix of f' is the unit matrix the matrices $P(h)$ become unitary for all $h \in G$.

A projective representation on V is *reducible* if V contains an invariant subspace different from both V and $\boldsymbol{0}$. Because a projective representation of a finite group is equivalent to a unitary one, the orthogonal complement of the invariant subspace (the subspace of elements of V orthogonal to all elements of the invariant subspace) is also invariant. Hence a reducible representation of a finite group is *fully reducible*. Therefore, a projective representation of a finite group is decomposable in irreducible components, all with the same factor system. On the other hand, for two projective representations P and P' with the same factor system ω the direct sum $P(G) \oplus P'(G)$ again gives a projective representation $(P\oplus P')(G)$ with this factor system. The inner Kronecker product $P(G) \otimes P'(G)$ of two projective representations with factor systems ω and ω' respectively forms the *product representation*. As $[(P\otimes P')(g_1)]\,[(P\otimes P')(g_2)] = [P(g_1) \otimes P'(g_1)]\,[P(g_2) \otimes P'(g_2)] = [P(g_1)P(g_2)] \otimes [P'(g_1)P'(g_2)] = \omega(g_1,g_2)\,\omega'(g_1,g_2)[P(g_1g_2) \otimes P'(g_1g_2)] = \omega(g_1,g_2)\,\omega'(g_1,g_2)\,(P\otimes P')(g_1g_2)$ the factor system of the product representation is the product of the two factor systems. The same is true for the outer Kronecker $P(G) \times P'(H)$ which is a projective representation of the direct product $G \times H$ with a factor system which is the product of the two factor systems. Here this means that if the factor systems for $P(G)$, $P'(H)$, and $P(G) \times P'(H)$ are ω, ω', and ω'' respectively, and if $a = (g,h)$, $a' = (g',h')$, one has $\omega''(a,a') = \omega(g,g')\,\omega'(h,h')$.

Orthogonality relations can also be formulated for projective representations. Suppose that $P_1(G)$ and $P_2(G)$ are irreducible projective representations of G with the same factor system ω and of dimensions d_1 and d_2 respectively. For an arbitrary $d_1 \times d_2$ matrix R one defines

$$S = \sum_{g \in G} \frac{P_1(g) R P_2(g^{-1})}{\omega(g, g^{-1})} .$$

For an element $h \in G$ one has

$$SP_2(h) = \sum_{g \in G} P_1(g) R P_2(g^{-1}h)\, \omega(g^{-1}, h)/\omega(g, g^{-1})$$

$$= P_1(h) \sum_{g \in G} P_1(h^{-1}g) R P_2(g^{-1}h) \frac{\omega(g^{-1}, h)\, \omega(h^{-1}, g)}{\omega(g, g^{-1})\, \omega(h, h^{-1})} .$$

From eq. (1.28) one sees that

$$\frac{\omega(g^{-1}, h)\, \omega(h^{-1}, g)}{\omega(h, h^{-1})\, \omega(g, g^{-1})} = \frac{\omega(g^{-1}, h)\, \omega(g^{-1}h, h^{-1})}{\omega(g^{-1}, e)\, \omega(h, h^{-1})} \frac{\omega(h^{-1}, g)}{\omega(g, g^{-1})\, \omega(g^{-1}h, h^{-1})}$$

$$= 1/\omega(g^{-1}h, h^{-1}g) .$$

Therefore, one has

$$SP_2(h) = P_1(h) \sum_{g \in G} \frac{P_1(h^{-1}g) R P_2(g^{-1}h)}{\omega(g^{-1}h, h^{-1}g)} = P_1(h) S$$

because of eq. (1.27). As both P_1 and P_2 are irreducible, one concludes with Schur's lemma that either $S = 0$ (and $P_1 \not\sim P_2$) or S is nonsingular (and $P_1 \sim P_2$). In the first case one concludes like in §3.2 that $\Sigma_{g \in G} P_\alpha(g)_{ij} P_\beta(g)^{-1}_{kl} = 0$. In the other case we take $P_1 = P_2$. Then S is a scalar multiple of the unit matrix: $S = \lambda \mathbb{1}$. Moreover,

$$\mathrm{tr}\, S = \sum_{g \in G} \mathrm{tr}\, P_1(g) R P_2(g)^{-1} = \lambda d_1 .$$

Then one has proved: *for two irreducible, projective representations $P_\alpha(G)$ and $P_\beta(G)$ with the same factor system one has the orthogonality relations*

$$\sum_{g \in G} P_\alpha(g)_{ij} P_\beta(g)^{-1}_{kl} = \frac{N}{d_\alpha} \delta'_{\alpha\beta} \delta_{il} \delta_{jk} . \qquad (1.30)$$

For a unitary representation eq. (1.30) becomes

$$\sum_{g\in G} P_\alpha(g)_{ij} P_\beta(g)^*_{lk} = \frac{N}{d_\alpha} \delta'_{\alpha\beta} \delta_{il} \delta_{jk} .$$

The *character* of a projective representation is defined by $\chi(g) = \mathrm{tr}\, P(g)$. Again, the characters of equivalent representations are the same, but in this case the character is not a class function as $\chi(hgh^{-1}) = \mathrm{tr}\, P(hgh^{-1}) = \omega(h,h^{-1})\,\omega(h,g)^{-1}\omega(hg,h^{-1})^{-1}\,\mathrm{tr}\, P(g) = \omega(h,h^{-1})\,\omega(h,g)^{-1}\omega(hg,h^{-1})^{-1}\chi(g)$. For characters corresponding to irreducible representations with the same factor system one again has orthogonality relations

$$\sum_{g\in G} \chi_\alpha(g)\,\chi_\beta(g^{-1})/\omega(g,g^{-1}) = N\,\delta_{\alpha\beta} \tag{1.31}$$

and for characters of unitary irreducible representations with the same ω

$$\sum_{g\in G} \chi_\alpha(g)\,\chi^*_\beta(g) = N\,\delta_{\alpha\beta} \,. \tag{1.32}$$

A reducible projective representation can be decomposed into irreducible components with the same ω. The multiplicities of these components are

$$m_\alpha = \sum_{g\in G} \chi(g)\,\chi^*_\alpha(g) \tag{1.33}$$

as follows from the orthogonality relations.

To determine the projective representations of a group G one can use a procedure due to Schur. First we consider the following situation. Suppose that a group R has an Abelian subgroup A such that all elements of A commute with all elements of R. As A is an invariant subgroup, one can consider the factor group R/A. We now suppose that R/A is isomorphic to G. With the canonical epimorphism $\pi : R \to R/A \simeq G$ any representation T of G gives a representation D of R by $D(h) = T(\pi h)$ for any $h \in R$. It is a representation, since $D(h_1)\,D(h_2) = T(\pi h_1)\,T(\pi h_2) = T(\pi(h_1 h_2)) = D(h_1 h_2)$. Now consider on the other hand an *irreducible* representation $D(R)$. As π is an epimorphism from R onto G, for each $g \in G$ one can choose one element $r(g) \in R$ such that $\pi r(g) = g$. Now we define $P(g) = D(r(g))$. For the product of two matrices P one has $P(g_1)P(g_2) = D(r(g_1))\,D(r(g_2))$. Because $\pi(r(g_1)r(g_2)) = \pi r(g_1)\,\pi r(g_2) = g_1 g_2$, the product $r(g_1g_2)\,r(g_1)^{-1}r(g_2)^{-1}$ is mapped by π on $e \in G$. Hence there is an element $a(g_1,g_2) \in A$ such that $r(g_1)\,r(g_2) = a(g_1,g_2)\,r(g_1g_2)$. As $a(g_1,g_2)$ is an element of A and therefore commutes with all elements of R, the matrix $D(a)$ commutes with all matrices of $D(R)$.

According to Schur's lemma it is a multiple of the unit matrix: $D(a(g_1,g_2)) = \omega(g_1,g_2)\mathbb{1}$, for some complex number ω depending on g_1, and g_2. Then $P(g_1)P(g_2) = D(a(g_1,g_2))P(g_1g_2) = \omega(g_1,g_2)P(g_1g_2)$. This means that P is a projective representation of G. In this way every irreducible representation of R gives for a fixed choice of $r(G)$ a projective representation of G. Now Schur has shown the following proposition. For its proof we refer to Schur [1904, 1907].

PROPOSITION 1.12. For each finite group G, there is at least one group R, containing an Abelian subgroup A of elements commuting with all elements of R, such that

i) R/A is isomorphic to G,

ii) A is isomorphic to the multiplicator $M(G)$,

iii) each projective representation of G can be obtained from the irreducible representations of R in the way described above.

The group R in the proposition is called a representation group. In general, the representation group of G is not unique. Its order is the product of the orders of G and of $M(G)$. Point iii) in the proposition has to be understood in the following way. For a fixed choice of the mapping $r: G \to R$ the non-equivalent irreducible representations give, for each class of associated factor systems, one set of nonequivalent irreducible projective representations.

As an example consider again the direct product $C_2 \times C_2$. This group G is of order four. It is generated by a and b with relations $a^2 = b^2 = e$, and $ab = ba$. Its elements are e, a, b, and ab. Its multiplicator has two elements and a representation group is the group of order 8, generated by α and β with defining relations $\alpha^4 = \beta^2 = (\alpha\beta)^2 = \epsilon$. It has a subgroup A consisting of ϵ and α^2. The homomorphism π maps ϵ and α^2 on e, (α, α^3) on a, $(\beta, \alpha^2\beta)$ on b, and $(\alpha\beta, \alpha^3\beta)$ on ab. Now one can choose $r(e) = \epsilon$, $r(a) = \alpha$, $r(b) = \beta$, $r(ab) = \alpha\beta$. The group R has the character table given in table 1.3. The four

Table 1.3
Character table for the representation group R of $C_2 \times C_2$.

R	ϵ	$[\alpha, \alpha^3]$	$[\alpha^2]$	$[\beta, \alpha^2\beta]$	$[\alpha\beta, \alpha^3\beta]$
D_1	1	1	1	1	1
D_2	1	−1	1	1	−1
D_3	1	1	1	−1	−1
D_4	1	−1	1	−1	1
D_5	2	0	−2	0	0

representations $D_1, ..., D_4$ give projective representations with trivial factor system. These are the four irreducible, ordinary representations of the Abelian group $C_2 \times C_2$ of order four. For $D_5(R)$ one finds a projective representation, which can be chosen to be

$$P(e) = \begin{pmatrix} 1 & 0 \\ 0 & 1 \end{pmatrix} \quad P(a) = \begin{pmatrix} 0 & 1 \\ -1 & 0 \end{pmatrix} \quad P(b) = \begin{pmatrix} 0 & 1 \\ 1 & 0 \end{pmatrix} \quad P(ab) = \begin{pmatrix} 1 & 0 \\ 0 & -1 \end{pmatrix}.$$

It is an irreducible representation with nontrivial factor system. One has the relations $P(a)^2 = -\mathbb{1}$, $P(b)^2 = \mathbb{1}$, $P(a)P(b) = -P(b)P(a)$. The associated representation $P'(e) = P(e)$, $P'(a) = iP(a)$, $P'(b) = P(b)$, $P'(ab) = P(ab)$ has the same matrix relations as the projective representation of $C_2 \times C_2$ discussed in the beginning of this section. These two two-dimensional representations are similar. Notice that the four representations $D_1, ..., D_4$ give non-equivalent representations, but these representations are associated because one can be obtained from the other by elementswise multiplication with a complex number.

For more details concerning projective representations we refer to Schur [1904], [1907], Rudra [1965], Harter [1969], and Janssen [1972].

CHAPTER II

GROUP THEORY AND QUANTUM MECHANICS

In the preceding chapter we discussed the basic mathematical theory needed for group theoretical treatment of our physical problems. We will now turn to a subject which has already more connection with the mathematics one encounters in books on quantum mechanics. As is well known, the states of physical systems are usually described by vectors in a Hilbert space. Such a Hilbert space is also a complex linear vector space. It will turn out that this vector space can be considered as the carrier space of representations of certain groups. Therefore we start with a discussion of representations in a Hilbert space. Afterwards we define symmetry groups of operators on this space. The most important symmetry group will be that of the Hamiltonian operator. The properties of the representations of these groups will lead to physical consequences that we look at in the second section of this chapter. This will bring us to topics, like classification of eigenstates, selection rules, perturbation theory, and the construction of invariants. Like in the first chapter we will not give an exhaustive treatment of the theory. For more details we refer to other books. The classical text is the book by Wigner, Wigner [1959]. Other works on applications of group theory on quantum mechanics are Hamermesh [1962], Heine [1960], Lomont [1959], Ljubarski [1962], and Jansen and Boon [1967].

2.1. Representations in a Hilbert space

2.1.1. Hilbert space

First we give a definition of the notion of *Hilbert space*, which plays such an important role in quantum mechanics. A Hilbert space is a complex linear vector space with an inner product: for any two ψ_1, ψ_2, elements of $\mathcal{H}$, is given a complex number $\langle\psi_1|\psi_2\rangle$ such that

1) $\langle\psi_1|\psi_2+\psi_3\rangle=\langle\psi_1|\psi_2\rangle+\langle\psi_1|\psi_3\rangle$,

2) $\langle\psi_1|\alpha\psi_2\rangle=\alpha\langle\psi_1|\psi_2\rangle$ for any complex α,

3) $\langle\psi_1|\psi_2\rangle=\langle\psi_2|\psi_1\rangle^*$,

4) $\langle\psi|\psi\rangle\geqslant 0$, and $\langle\psi|\psi\rangle=0$ if and only if $\psi=0$.

Moreover, $\mathcal{H}$ must be complete, which means that for any sequence $\psi_1,\psi_2,\ldots$ with the property that $\lim\langle\psi_n-\psi_m|\psi_n-\psi_m\rangle=0$ when $n,m\to\infty$, there exists an element $\psi\in\mathcal{H}$ such that $\lim\langle\psi_n-\psi|\psi_n-\psi\rangle=0$ when $n\to\infty$.

The real number $\langle\psi|\psi\rangle$ is called the *norm* of ψ. A Hilbert space can be of finite, as well as of infinite dimension.

A mapping P of $\mathcal{H}$ into itself is called an *operator* on $\mathcal{H}$. The operator P is linear if $P(\alpha\psi_1+\beta\psi_2)=\alpha P(\psi_1)+\beta P(\psi_2)$ for any $\alpha,\beta\in\mathbb{C}$. It is *antilinear*, if $P(\alpha\psi_1+\beta\psi_2)=\alpha^*P(\psi_1)+\beta^*P(\psi_2)$ for any $\alpha,\beta\in\mathbb{C}$.[†]

An example of a Hilbert space is given by the set of square integrable functions of one variable. Here the inner product is defined by

$$\langle\psi_1|\psi_2\rangle=\int dx\,\psi_1^*(x)\,\psi_2(x)\,.$$

A linear operator on this space is the derivation d/dx, which maps $\psi(x)$ on $d\psi/dx$. An antilinear operator is the operator which maps $\psi(x)$ on $\psi^*(x)$, as $(\alpha\psi_1+\beta\psi_2)^*=\alpha^*\psi_1^*+\beta^*\psi_2^*$. The operator is the complex conjugation operator. We notice that the sum of two (anti-)linear operators is again an (anti-) linear operator. The product of two antilinear operators is linear, the product of a linear and an antilinear operator is antilinear. Here, by product of the operators P and Q we mean the operator (PQ) defined by $(PQ)\psi=P(Q\psi)$.

The *inverse* of an operator P does not always exist. If it exists, it is denoted by P^{-1}. The *adjoint* $P^\dagger$ of a linear operator P is defined by $\langle\psi_1|P\psi_2\rangle=\langle P^\dagger\psi_1|\psi_2\rangle$ for all $\psi_1,\psi_2\in\mathcal{H}$. If for a linear operator $P=P^\dagger$, the operator is *Hermitian* or *self-adjoint*. A *unitary* operator P is a linear operator for which the inverse exists and $P^\dagger P=PP^\dagger=\mathbb{1}$. An operator is *antiunitary* if it is antilinear, has an inverse and $\langle P\psi_1|P\psi_2\rangle=\langle\psi_1|\psi_2\rangle^*$.

We mentioned here only some of the most important notions and we did not go into detail. For a more complete treatment of Hilbert spaces and operators on it we refer to the mathematical literaure, or to books on quantum mechanics, like Messiah [1961]. The theory of Hilbert spaces is so im-

[†] We do not treat notions like bounded operators, unbounded operators, domain of an operator and so on. Properly speaking, in the example given one has to discuss the domain of the operator d/dx. However, we will not speak about this aspect.

portant for quantum mechanics, because in the usual formulation the states of a quantum mechanical system can be described by vectors in a Hilbert space. We restrict ourselves to normalized states associated with vectors of norm 1. Then there is still not a one-to-one correspondence between states and vectors, as vectors differing by a phase factor (a complex number of absolute value 1) describe the same state. The *observables* are associated with Hermitian operators. Then the expectation value of the observable A in the state ψ is given by $\langle \psi | A \psi \rangle / \langle \psi | \psi \rangle$. The transition probability from a state ψ_1 to a state ψ_2 is given by $|\langle \psi_1 | \psi_2 \rangle|$. If ψ is an eigenvector of the Hermitian operator A, i.e. if $A\psi = a\psi$ for some real number a, the *expectation value* of A in the state ψ is a. The vectors ψ which are eigenvectors of A with eigenvalue a form a Hilbert subspace of $\mathcal{H}$. If A has a discrete spectrum the space $\mathcal{H}$ can be decomposed in the direct sum of the eigenspaces of A with different eigenvalue. This means that there is a basis of $\mathcal{H}$ consisting of eigenvectors of A. If $A_1, A_2, ..., A_n$ is a complete set of commuting observables, there is a basis of $\mathcal{H}$ consisting of simultaneous eigenvectors of all the operators $A_1, ..., A_n$. That the set is complete means that each element of the basis is characterized by the n eigenvalues of the operators on this basis element.

2.1.2. Representations in a Hilbert space

Since a Hilbert space is a linear vector space, it can carry a linear representation of a group G if there is a homomorphism of G into the group of non-singular linear operators on $\mathcal{H}$. Suppose T_g is the linear operator corresponding to the element $g \in G$. We denote the group of all operators T_g with $g \in G$ by T_G. If in $\mathcal{H}$ a subspace $\mathcal{H}_d$ of dimension d exists which is invariant under T_G (we say also that it is G-invariant), this subspace carries a d-dimensional representation of G. Suppose that $\psi_1, ..., \psi_d$ is a basis of $\mathcal{H}_d$. Then for an $\psi \in \mathcal{H}_d$ the element $T_g\psi$ is a linear combination of the basis elements. In particular

$$T_g \psi_i = \sum_{j=1}^{d} D(g)_{ji} \psi_j \ . \tag{2.1}$$

For $g_1 g_2 = g$ one has on one side

$$T_{g_1} T_{g_2} \psi_i = T_g \psi_i = \sum_{j=1}^{d} D(g)_{ji} \psi_j$$

and on the other side

$$T_{g_1} T_{g_2} \psi_i = T_{g_1} \sum_{k=1}^{d} D(g_2)_{ki} \psi_k = \sum_{k=1}^{d} \sum_{j=1}^{d} D(g_2)_{ki} D(g_1)_{jk} \psi_j$$

because T_g is a linear operator. It follows that $D(g_1 g_2) = D(g_1) D(g_2)$, i.e. the matrices of $D(G)$ form a matrix representation of G of dimension d. The vectors $\psi_1, ..., \psi_d$ are called *basis functions* for this representation. One uses the name basis functions instead of basis vectors as the vectors of $\mathcal{H}$ correspond to wave functions. In the following we suppose again, as we did in Ch. 1, §3, that G is a finite group of order N. Then one can easily indicate a finite-dimensional, G-invariant subspace of $\mathcal{H}$. If ψ is an element of $\mathcal{H}$ the elements $T_{g_1} \psi, ..., T_{g_N} \psi$ span a subspace which is clearly invariant under G. Moreover, its dimension is at most N, the order of G.

If the representation space $\mathcal{H}_d$, spanned by basis functions $\psi_1, ..., \psi_d$, has a G-invariant subspace ($\neq \mathcal{H}_d, \neq 0$) it is reducible; if there is no such subspace, $\mathcal{H}_d$ and the representation are irreducible.

Suppose that $\mathcal{H}_d$ is irreducible. When the operators T_g are unitary, the corresponding matrices $D(g)$ are unitary if the basis is orthonormal. One has

$$\begin{aligned} \langle \psi_i | \psi_j \rangle &= \langle T_g \psi_i | T_g \psi_j \rangle \qquad \text{(for any } g \in G) \\ &= (1/N) \sum_{g \in G} \langle T_g \psi_i | T_g \psi_j \rangle \\ &= (1/N) \sum_{g \in G} \sum_{k,l=1}^{d} D(g)^*_{ki} D(g)_{lj} \langle \psi_k | \psi_l \rangle \\ &= (1/d) \sum_{k,l=1}^{d} \delta_{kl} \delta_{ij} \langle \psi_k | \psi_l \rangle = (1/d) \delta_{ij} \sum_{k=1}^{d} \langle \psi_k | \psi_k \rangle \,. \end{aligned}$$

Therefore, if $\psi_i' = \psi_i / (d^{-1} \Sigma_k \langle \psi_k | \psi_k \rangle)^{1/2}$ one has

$$\langle \psi_i' | \psi_j' \rangle = \delta_{ij} \,. \tag{2.2}$$

In the following we will assume basis functions of irreducible, unitary representations to be orthogonal and normalized.

2.1.3. Expansion in basis functions of irreducible representations

Now we will show that one can decompose $\mathcal{H}$ into the direct sum of subspaces carrying irreducible representations of the group G. In order to see this, consider an arbitrary function ψ, i.e. a vector in $\mathcal{H}$. By speaking about functions instead of vectors we have a better correspondence with the terminology of quantum mechanics. The set $T_G\psi$ spans a subspace of $\mathcal{H}$, which is finite-dimensional and G-invariant. It carries a representation of G which is, in general, reducible. Suppose the reduction of the matrix representation $D(G)$ is performed by a matrix U. Then one has $UD(G)U^{-1} = D'(G) = \Sigma_{\alpha=1}^{r} m_\alpha D_\alpha(G)$, where $D_\alpha(G)$ are irreducible representations. If $\psi_1, ..., \psi_d$ form a basis of $D(G)$, the elements $\psi_i' = U\psi_i$ $(i = 1, ..., d)$ form a basis for $D'(G)$. As $\mathcal{H}_d$ is the direct sum of $\Sigma_\alpha m_\alpha$ irreducible subspaces, the basis elements for $D'(G)$ can be characterized by three indices: one denoting the irreducible component, one labeling the subspaces which carry the same irreducible component, and one labeling the basis functions of the subspace specified by the two preceding labels. If d_α is the dimension of the component $D_\alpha(G)$, one can write

$$\{\psi_i', i = 1, ..., d\} = \{\psi_{\alpha lk}; \alpha = 1, ..., r; l = 1, ..., m_\alpha; k = 1, ..., d_\alpha\} .$$

The functions $\psi_{\alpha l1}, ..., \psi_{\alpha ld}$ form a basis for $D_\alpha(G)$. As the multiplicity of $D_\alpha(G)$ is m_α, the different spaces carrying this same representation are distinguished by the index l. Under the operator T_g the basis functions with given α and l transform according to

$$T_g\psi_{\alpha li} = \sum_{j=1}^{d} D_\alpha(g)_{ji}\psi_{\alpha lj} . \tag{2.3}$$

One says that the function $\psi_{\alpha lk}$ *belongs to the k-th row* of the irreducible representation $D_\alpha(G)$. To come to the result announced in the beginning, one can write, taking the arbitrary ψ as the first element ψ_1 of the basis,

$$\psi = \psi_1 = \sum_{j=1}^{d} U_{j1}\psi_j' = \sum_{\alpha=1}^{r}\sum_{l=1}^{m_\alpha}\sum_{k=1}^{d_\alpha} U_{\alpha lk,1}\psi_{\alpha lk} ,$$

where the triple (αlk) labels basis functions of the space spanned by $\psi_1, ..., \psi_d$. As the functions $\psi_{\alpha lk}$ are basis functions for irreducible representations, we have proved that any function can be expanded in such basis functions:

$$\psi = \sum_{\alpha lk} f_{\alpha lk}\psi_{\alpha lk} , \tag{2.4}$$

for arbitrary $\psi \in \mathcal{H}$, the coefficients $f_{\alpha lk}$ are complex numbers.

2.1.4. Construction of basis functions

According to eq. (2.4) each function ψ can be written as a sum of terms transforming under the operators T_g as the different rows of the various irreducible representations of G. To find these components we can make use of projection operator techniques. As a second result we will find functions which are basis functions of the irreducible representations. In order to do this we choose one representation from each equivalence class of irreducible representations of G. If the matrices of these representations are $D_\alpha(G)$, with $\alpha = 1, ..., r$, we define an operator

$$\rho^\alpha_{ij} = (d_\alpha/N) \sum_{g \in G} D^*_\alpha(g)_{ij} T_g \, , \qquad (i, j = 1, ..., d) \, , \tag{2.5}$$

where d_α is the dimension of $D_\alpha(G)$, and N the order of G. We now suppose that both T_g and $D_\alpha(g)$ are unitary.
PROPOSITION 2.1. The operators ρ^α_{ij} satisfy the relations

$$1) \ (\rho^\alpha_{ij})^\dagger = \rho^\alpha_{ji} \, ,$$

$$2) \ \rho^\alpha_{ij}\rho^\beta_{kl} = \delta_{\alpha\beta}\delta_{jk}\rho^\alpha_{il} \, . \tag{2.6}$$

Proof.

$$1) \ (\rho^\alpha_{ij})^\dagger = (d_\alpha/N) \sum_{g \in G} D_\alpha(g)_{ij} T^\dagger_g = (d_\alpha/N) \sum_{g \in G} D^*_\alpha(g^{-1})_{ji} T_{g^{-1}} = \rho^\alpha_{ji}$$

$$2) \ \rho^\alpha_{ij}\rho^\beta_{kl} = (d_\alpha d_\beta/N^2) \sum_{g,h \in G} D^*_\alpha(g)_{ij} D^*_\beta(h)_{kl} T_g T_h$$

$$= (d_\alpha d_\beta/N^2) \sum_{g,a \in G} D^*_\alpha(g)_{ij} D^*_\beta(g^{-1}a)_{kl} T_a \qquad \text{with } a = gh$$

$$= (d_\alpha d_\beta/N^2) \sum_{g,a \in G} D^*_\alpha(g)_{ij} [D_\beta(g^{-1}) D_\beta(a)]^*_{kl} T_a$$

$$= (d_\alpha d_\beta/N^2) \sum_{m=1}^{d_\beta} \sum_{g,a \in G} D^*_\alpha(g)_{ij} D_\beta(g)_{mk} D^*_\beta(a)_{ml} T_a$$

$$= (d_\alpha/N) \sum_{m=1}^{d_\alpha} \sum_{a \in G} \delta_{\alpha\beta}\delta_{im}\delta_{jk} D^*_\beta(a)_{ml} T_a$$

$$= (d_\alpha/N) \sum_{a \in G} D^*_\beta(a)_{il} T_a \delta_{\alpha\beta}\delta_{jk} = \delta_{\alpha\beta}\delta_{jk}\rho^\alpha_{il} \, .$$

Consequently for $i = j$ the operator ρ^α_{ii} is *Hermitian*. Moreover, it is *idempotent*: $\rho^\alpha_{ii}\rho^\alpha_{ii} = \rho^\alpha_{ii}$. Hence it is a projection operator. We recall that P is a *projection operator* if it is Hermitian and if $P^2 = P$. For $i \neq j$ the operator ρ^α_{ij} is *nilpotent*: $\rho^\alpha_{ij}\rho^\alpha_{ij} = 0$.

To see what the operator ρ^α_{ii} projects out, we consider the action of ρ^α_{ij} on an arbitrary function ϕ. Using eq. (2.4) we have

$$\phi = \sum_{\beta lk} f_{\beta lk}\psi_{\beta lk} ,$$

and

$$\begin{aligned}\rho^\alpha_{ij}\phi &= \sum_{\beta lk} f_{\beta lk}(d_\alpha/N) \sum_{g\in G} D^*_\alpha(g)_{ij} T_g \psi_{\beta lk} \\ &= \sum_{\beta lk} f_{\beta lk}(d_\alpha/N) \sum_{g\in G} D^*_\alpha(g)_{ij} \sum_{m=1}^{d_\beta} D_\beta(g)_{mk}\psi_{\beta lm} \\ &= \sum_{\substack{\beta lk\\ m}} f_{\beta lk}\,\delta_{\alpha\beta}\,\delta_{im}\,\delta_{jk}\,\psi_{\beta lm} = \sum_l f_{\alpha lj}\,\psi_{\alpha li} .\end{aligned}$$

From this we derive the following conclusions. The operator ρ^α_{ii} projects from ϕ the component which belongs to the i-th row of the representation $D_\alpha(G)$. The summation over α and i of these operators gives $\mathbb{1}$, because $\Sigma_{\alpha i}\,\rho^\alpha_{ii}\phi = \Sigma_{\alpha li} f_{\alpha li}\,\psi_{\alpha li} = \phi$ (eq. (2.4)) and consequently $\Sigma_{\alpha i}\,\rho^\alpha_{ii} = \mathbb{1}$. That ρ^α_{ii} really gives the αi-component is seen from the transformation property of $\rho^\alpha_{ij}\phi$ under the action of T_g:

$$\begin{aligned}T_g\rho^\alpha_{ij}\phi &= (d_\alpha/N) \sum_{h\in G} D^*_\alpha(h)_{ij} T_{gh}\,\phi = (d_\alpha/N) \sum_{h\in G} D^*_\alpha(g^{-1}h)_{ij} T_h\,\phi \\ &= (d_\alpha/N) \sum_{h\in G} \sum_{m=1}^{d_\alpha} D_\alpha(g)_{mi} D^*_\alpha(h)_{mj} T_h\,\phi \\ &= \sum_{m=1}^{d_\alpha} D_\alpha(g)_{mi}\rho^\alpha_{mj}\,\phi .\end{aligned} \tag{2.7}$$

Eq. (2.7) gives for $i = j$

$$T_g\rho^\alpha_{ii}\phi = \sum_m D_\alpha(g)_{mi}\rho^\alpha_{mi}\,\phi .$$

Therefore, for fixed j the functions $\rho^\alpha_{ij}\phi$ $(i = 1, ..., d_\alpha)$ form a basis for the irreducible representation $D_\alpha(G)$, and the function $\rho^\alpha_{ii}\phi$ belongs to the i-th row of this irreducible representation.

We recapitulate the results of this section. 1) The projection operator ρ_{ii}^{α} projects from an arbitrary function ϕ its component which transforms as the i-th row of the irreducible representation $D_{\alpha}(G)$. 2) To find a basis of the irreducible representation $D_{\alpha}(G)$ one can take an arbitrary function ϕ. Then the d_{α} functions ρ_{ij}^{α} with fixed j and $i = 1, ..., d_{\alpha}$ are either zero (if ϕ has no αi-component), or they form a basis for the representation. Notice that it is important to keep j fixed, as the functions $\rho_{ii}^{\alpha}\phi$ do not form a basis.

2.1.5. Example

Consider the Hilbert space of three-particle states in one dimension. The elements are functions $\phi(x_1,x_2,x_3)$ and the scalar product is given by $\langle\phi_1|\phi_2\rangle = \int dx_1dx_2dx_3\phi_1^*(x_1,x_2,x_3)\phi_2(x_1,x_2,x_3)$. For the group G we take the group of Ch. 1, §1.8, the permutation group of three elements: e, a, a^2, c, ac, a^2c. The operators T_g for the permutation $g = \binom{1\ 2\ 3}{i_1\ i_2\ i_3}$ are defined by $T_g\phi(x_1,x_2,x_3) = \phi(x_{i_1},x_{i_2},x_{i_3})$, which is denoted by $g\phi$.

We give two three-particle functions:

$$\phi_1(x_1,x_2,x_3)=\{x_1^2(x_1+x_2-x_3)+x_2^2(x_2+x_3-x_1)$$

$$+x_3^2(x_3+x_1-x_2)\}\exp(-x_1^2-x_2^2-x_3^2)\,,$$

$$\phi_2(x_1,x_2,x_3)=\phi_1(x_2,x_1,x_3)\,.$$

One easily checks the relations

$$a\phi_1 = a^2\phi_1 = \phi_1,\quad a\phi_2 = a^2\phi_2 = \phi_2,\quad c\phi_1 = (ac)\phi_1 = (a^2c)\phi_1 = \phi_2,$$

$$c\phi_2 = (ac)\phi_2 = (a^2c)\phi_2 = \phi_1\,.$$

Hence the space spanned by the functions ϕ_1 and ϕ_2 is invariant under G. It carries a two-dimensional representation of G with matrices (with ϕ_1 and ϕ_2 as basis) $D(e) = D(a) = D(a^2) = \begin{pmatrix}1&0\\0&1\end{pmatrix}$, $D(c) = D(ac) = D(a^2c) = \begin{pmatrix}0&1\\1&0\end{pmatrix}$. The representation is reducible as using eq. (1.20) one sees from the character $\chi = (2, 2, 2, 0, 0, 0)$. A basis which gives the matrices the reduced form is $\psi_1 = \phi_1 + \phi_2$, and $\psi_2 = \phi_1 - \phi_2$. The representation with respect to this basis is determined by $a\psi_1 = a^2\psi_1 = c\psi_1 = (ac)\psi_1 = (a^2c)\psi_1 = \psi_1, a\psi_2 = a^2\psi_2 = \psi_2$, and $c\psi_2 = (ac)\psi_2 = (a^2c)\psi_2 = -\psi_2$. This means that ψ_1 spans a one-dimensional representation denoted by $D_1(G)$ in Ch. 1, §3.4 and ψ_2 spans the one-dimensional representation $D_2(G)$.

In the same space of three-particle states we give an example of the projection operator technique. From an arbitrary function $\phi = \phi(x_1,x_2,x_3)$ one can construct basis functions for each of the irreducible representations $D_\alpha(G)$, ($\alpha = 1, 2, 3$). They are

$$\rho^1\phi = \{\phi(x_1,x_2,x_3) + \phi(x_2,x_3,x_1) + \phi(x_3,x_1,x_2) + \phi(x_2,x_1,x_3) + \phi(x_3,x_2,x_1) + \phi(x_1,x_3,x_2)\}/6$$

$$\rho^2\phi = \{\phi(x_1,x_2,x_3) + \phi(x_2,x_3,x_1) + \phi(x_3,x_1,x_2) - \phi(x_2,x_1,x_3) - \phi(x_3,x_2,x_1) - \phi(x_1,x_3,x_2)\}/6$$

$$\rho^3_{11}\phi = \{\phi(x_1,x_2,x_3) - \phi(x_2,x_3,x_1) - \phi(x_3,x_2,x_1) + \phi(x_1,x_3,x_2)\}/3$$

$$\rho^3_{21}\phi = \{\phi(x_2,x_3,x_1) - \phi(x_3,x_1,x_2) + \phi(x_2,x_1,x_3) - \phi(x_1,x_3,x_2)\}/3\,.$$

It is possible that some of these functions are zero. In that case we have to take another function which has a component in each representation. Notice that the function $\rho^1\phi$ is *fully symmetric* in its three indices, whereas the second one, $\rho^2\phi$, is *fully antisymmetric*. The first function can describe bosons, because the wave functions of bosons have to be fully symmetric. The second function can describe fermions. However, according to the symmetrization postulate, which states that wave functions of identical particles are either symmetric, or antisymmetric, the functions $\rho^3_{11}\phi$ and $\rho^3_{21}\phi$ do not describe systems of identical particles. The operator ρ^1 is the projection operator on the space of symmetric wave functions, the operator ρ^2 that on the space of antisymmetric wave functions.

2.1.6. *Projective representations on a Hilbert space*

In quantum mechanics to each vector of the Hilbert space $\mathcal{H}$ corresponds a state of the system. This correspondence is not one-to-one because all vectors differing only by a complex factor give the same state. Therefore, for quantum mechanics the important objects are the one-dimensional subspaces of $\mathcal{H}$, called the *vector rays* of the Hilbert space. These rays are the elements of the projective Hilbert space $\boldsymbol{\mathcal{H}}$, to be compared with the projective space discussed in Ch. 1, §3.5. The ray $\boldsymbol{\psi}$ is the one-dimensional subspace to which ψ belongs. Although $\boldsymbol{\mathcal{H}}$ is not a linear vector space, we can define a kind of inner product, more precisely a mapping which assigns to each pair Ψ_1, Ψ_2 a real number

$$(\Psi_1, \Psi_2) = \frac{|\langle \psi_1 | \psi_2 \rangle|}{(\langle \psi_1 | \psi_1 \rangle \langle \psi_2 | \psi_2 \rangle)^{1/2}}. \tag{2.8}$$

The reason for this definition will become clear in the next section. We notice that it corresponds to the transition probability between ψ_1 and ψ_2. As the ray of ψ is the same as the ray of $\alpha\psi$, it is readily verified that the definition eq. (2.8) does not depend on the choice of the vectors ψ_1 and ψ_2 from their respective rays. If one chooses $\alpha\psi_1$ and $\beta\psi_2$ instead of ψ_1 and ψ_2, one has $|\langle \alpha\psi_1 | \beta\psi_2 \rangle| / (\langle \alpha\psi_1 | \alpha\psi_1 \rangle \langle \beta\psi_2 | \beta\psi_2 \rangle)^{1/2} = (\Psi_1, \Psi_2)$. Now we can consider mappings of $\boldsymbol{\mathcal{H}}$ onto itself which leave invariant this "inner product". We call these mappings *automorphisms* of $\boldsymbol{\mathcal{H}}$.

Let A be a unitary or antiunitary operator on the Hilbert space $\mathcal{H}$. The operator $\boldsymbol{A}$ defined on $\boldsymbol{\mathcal{H}}$ by $\boldsymbol{A}\Psi = (A\psi)$ is an automorphism of $\boldsymbol{\mathcal{H}}$, as one easily verifies. On the other hand Wigner has shown that every automorphism of $\boldsymbol{\mathcal{H}}$ can be obtained in this way from either a unitary, or an antiunitary operator on $\mathcal{H}$. Moreover, any two unitary or antiunitary operators giving the same automorphism differ only by a scalar factor. The set of all operators αA differing only by the scalar factor α is called the *operator ray* $\boldsymbol{A}$. The operator rays of unitary operators form a subgroup (of index two) of the group of automorphisms of $\boldsymbol{\mathcal{H}}$. A homomorphism of a group G into this subgroup is a projective representation of G in $\boldsymbol{\mathcal{H}}$. Now consider a subspace $\mathcal{H}_d$ of dimension d, which is invariant under all the (linear) operators corresponding to the rays of the projective representation on $\boldsymbol{\mathcal{H}}$. Then the projective space $\boldsymbol{\mathcal{H}}_d$ carries a projective representation in the sense of Ch. 1, §3.5. If we choose in each operator ray of the representation a unitary operator, the set of operators satisfies eq. (1.26).

2.2. Symmetry and quantum mechanics

2.2.1. Symmetry transformations

A state of a physical system corresponds to a ray in a Hilbert space. A representative chosen from this ray is described by its coordinates with respect to a basis of eigenfunctions of a complete set of commuting observables. This set of commuting observables describes a series of measurements in a certain reference system. In another reference system the description would be different. Let ψ be the state vector (determined up to a phase factor) for one observer, and ψ' the state vector for another observer. Suppose that the reference systems are related by a transformation g, which is a transformation in space and time. The transformation g induces a transformation of states $\psi \to \psi' = T_g \psi$. Of course the physical content of the description must remain the same. This means that transition probabilities $|\langle \psi | \varphi \rangle|$ have the same value in both reference systems. (Here we only consider transformations of the spatial coordinates, time reversal, and time translations.) The condition $|\langle \psi' | \varphi' \rangle| = |\langle \psi | \varphi \rangle|$ means that g induces an automorphism of the projective Hilbert space $\mathcal{H}$. According to Wigner's theorem (§ 1.6) there always exists a unitary or antiunitary operator T_g which induces this automorphism. It turns out that only when time-reversal is involved we have an antiunitary operator. In all other cases the operators are unitary.

As an example we consider a particle without spin. Its wave function is given by $\psi(\boldsymbol{x})$ for one observer, by $\psi'(\boldsymbol{x}')$ for the other one. In this case we can choose for the operator T_g the so-called substitution operator P_g, defined by its action

$$P_g \psi(\boldsymbol{x}) = \psi(g^{-1}\boldsymbol{x}) . \tag{2.9}$$

In that case $\psi'(\boldsymbol{x}') = (P_g \psi)(g\boldsymbol{x}) = \psi(\boldsymbol{x})$, which means that the value of ψ' in the point described by $\boldsymbol{x}'$ for one observer is the same as the value of ψ in the same point described by $\boldsymbol{x}$ for the other. Of course the same transformation of the projective space, i.e. the same mapping of the rays, is given by the operator $\exp(i\varphi) P_g$, which differs only by a phase factor from P_g. The operator P_g is indeed a unitary operator as

$$\langle \varphi' | \psi' \rangle = \int d\boldsymbol{x} \varphi'^*(g^{-1}\boldsymbol{x})\, \psi'(g^{-1}\boldsymbol{x}) = \int d\boldsymbol{x} |g| \varphi^*(\boldsymbol{x})\, \psi(\boldsymbol{x}) = \langle \varphi | \psi \rangle ,$$

where the Jacobian $|g|$ is assumed to be unity. This is the case for all transformations we will consider. Moreover, we notice that for the elements g of a group of transformations G, the operators P_g form a representation as

$$P_g P_h \psi(\boldsymbol{x}) = P_h \psi(g^{-1}\boldsymbol{x}) = \psi(h^{-1}g^{-1}\boldsymbol{x}) = \psi((gh)^{-1}\boldsymbol{x}) = P_{gh} \psi(\boldsymbol{x}) .$$

If an observable is associated with the Hermitian operator A for observer one, the same quantity for observer two is associated with the operator $A' = T_g A T_g^{-1}$, because $\langle \varphi | A | \psi \rangle = \langle T_g \varphi | T_g A T_g^{-1} | T_g \varphi \rangle = \langle \varphi' | A' | \psi' \rangle$. If the two operators A and A' are the same, i.e. if $A = T_g A T_g^{-1}$ or $[T_g, A] = 0$, the operator T_g is called a *symmetry operator* for A, and g is called a symmetry of the system. More in general, one defines a symmetry operator for A as a unitary or antiunitary operator P commuting with A. A special kind of symmetry operators are those arising from space (and time) transformations. These are called *geometric symmetry operators*. The other symmetry operators can not be interpreted as arising from transformations connecting different reference systems. The product of two symmetry operators is again a symmetry operator. So we can consider a group of symmetry operators for A. A symmetry group of the operator A is a group of unitary or antiunitary operators commuting with A. If the symmetry operator is induced by a space-time transformation, we call it an *invariance operator* for A. It is the product of a substitution operator and an operator such that the combination commutes with A. A group of invariance operators (geometric symmetry operators) is called an *invariance group* or invariance operator group for A. The group of space-time transformations which induces the invariance operator group is also called an invariance group for A. The importance of the symmetry operators follows from the following property. If P is a symmetry operator for A, it transforms eigenfunctions of A into eigenfunctions with the same eigenvalue: if $A|\psi\rangle = a|\psi\rangle$, then $AP|\psi\rangle = PA|\psi\rangle = aP|\psi\rangle$. For an operator which is not a symmetry operator one has $PAP^\dagger|\psi\rangle = a|\psi'\rangle$. This simply means, that for geometric operators P the description of one observer can be translated into the description of the other. For a symmetry operator it is not only a transformation: both observers see the same system.

Suppose the geometric transformation g gives the relation between the reference systems of observers 1 and 2. A state is described by the first one with the ray Ψ, by the other one with Ψ'. As transition probabilities must be the same for both observers, one has $(\Psi_1, \Psi_2) = (\Psi'_1, \Psi'_2)$ for any pair Ψ_1, Ψ_2 (cf. eq. (2.8)). This means that to g corresponds an automorphism of the projective Hilbert space $\mathcal{H}$. According to Wigner's theorem there is a (anti-) unitary operator T_g such that $T_g\Psi = \Psi'$. As $T_gT_h = (T_{gh})$ for two transformations g and h, one has $T_gT_h = \omega(g,h)T_{gh}$. This means that for a symmetry group G of space transformations the operators T_G form a projective representation of G.

In the general definition of symmetry operator for an operator A we considered all unitary and antiunitary operators commuting with A. If $[P,A] = 0$ for an arbitrary phase factor $\exp(i\varphi)$ also $[\exp(i\varphi)P,A] = 0$, and $\exp(i\varphi)P$ is also a symmetry operator. Therefore, the symmetry group, which is the group of all symmetry operators, is never a finite group. As both P and $\exp(i\varphi)P$ determine the same automorphism of the projective space, the symmetry group determines a subgroup G of the automorphism group of $\mathcal{H}$. Choosing one operator from each operator ray in this group G, one obtains a projective representation of G. However, it is nearly always possible to choose the phases of the operators in such a way that choosing one or two operators from each ray of G, one obtains a group of operators on $\mathcal{H}$. This group G' is a subgroup of the symmetry group, and is homomorphic to G: to each element of G correspond a number of elements of G'. In the case of spinless particles to each symmetry transformation g corresponds an operator ray $\boldsymbol{T}_g$, and in each ray one can choose one substitution operator P_g such that these form a group. For particles with spin, as we will see later in Ch. 5, it will be possible to choose two operators from each ray such that the resultant set of operators forms a group. In general, we can choose a number of operators from each ray to form a group. We take this number as small as possible. As we will see later on, the consequences of the existence of this smaller group are the same as those for the complete symmetry group of all operators with all phases commuting with A.

2.2.2. Symmetry operators for the Hamiltonian operator

Although the concept of symmetry group of an operator is useful for an arbitrary operator, that for the Hamilton operator is especially important. All unitary and antiunitary operators commuting with the Hamilton operator H form the *symmetry group* of H. Any subgroup of this group is a group of symmetry operators for H. Among these subgroups is the group of all invariance operators. Consider the most simple case: a particle without spin in a potential $V(\boldsymbol{x})$. Then $H = \boldsymbol{p}^2/2m + V(\boldsymbol{x})$. Let us look for the substitution operators commuting with H. First we remark that the commutation of a substitution operator P_g with a function $f(\boldsymbol{x})$ depending only on the position coordinates is given by

$$P_g f(\boldsymbol{x}) P_g^{-1} = f(g^{-1}\boldsymbol{x}) ,$$

because acting with this operator on a wave function $\psi(\boldsymbol{x})$ gives $P_g f(\boldsymbol{x}) P_g^{-1} \psi(\boldsymbol{x}) = P_g f(\boldsymbol{x}) \psi(g\boldsymbol{x}) = f(g^{-1}\boldsymbol{x}) \psi(\boldsymbol{x})$. For the commutation of P_g with the momentum operator one has

$$P_g p_i P_g^{-1} = \sum_{j=1}^{3} R_{ji} p_j ,$$

where R is the homogeneous part of g given by its action on $\boldsymbol{x}$ by $(g\boldsymbol{x})_i = \Sigma_j R_{ij} x_j + t_i$. Then $P_g p_i P_g^{-1} \psi(\boldsymbol{x}) = -\hbar i P_g(\partial/\partial x_i) P_g^{-1} \psi(\boldsymbol{x}) = -\hbar i P_g(\partial/\partial x_i) \psi(g\boldsymbol{x}) = -\hbar i (\partial/\partial (g^{-1}\boldsymbol{x})_i) \psi(\boldsymbol{x}) = -\hbar i \Sigma_j (\partial/\partial x_j) \psi(\boldsymbol{x}) (\partial x_j/\partial (g^{-1}\boldsymbol{x})_i) = -\hbar i \Sigma_j R_{ji} (\partial/\partial x_j) \psi(\boldsymbol{x}) = \Sigma_j R_{ji} p_j \psi(\boldsymbol{x})$.

If the matrix R is orthogonal one has $P_g p_i P_g^{-1} = \Sigma_j R_{ij}^{-1} p_j$ or $P_g \boldsymbol{p} P_g^{-1} = R^{-1}\boldsymbol{p}$. This means that for the commutation of P_g with the Hamilton operator H one has

$$P_g(\boldsymbol{p}^2/2m + V(\boldsymbol{x}))P_g^{-1} = (R^{-1}\boldsymbol{p})^2/2m + V(g^{-1}\boldsymbol{x}) .$$

If R is orthogonal $(R^{-1}\boldsymbol{p})^2 = \boldsymbol{p}^2$. Hence H commutes with P_g if the homogeneous part of g is orthogonal and if the potential $V(\boldsymbol{x})$ is invariant under g. The invariance group of H is then determined as the group of transformations g with R orthogonal and leaving V invariant. For particles with internal degrees of freedom, like spin, the operator T_g is, in general, not plainly a substitution operator. We will come to this in Ch. 5.

Until further notice we will consider only the symmetry operators of H which are substitution operators. These form the group G. When $\mathcal{H}_d$ is a subspace of the Hilbert space belonging to an eigenvalue E of H, one has for an eigenvector $\psi \in \mathcal{H}_d : HP_g\psi = P_gH\psi = EP_g\psi$ for any $g \in G$. This means that $\mathcal{H}_d$ is an invariant subspace of $\mathcal{H}$. If the operators P_g are unitary (and not antiunitary) this space carries a representation of G. The dimension of this representation is the dimension of $\mathcal{H}_d$, i.e. the degeneracy of the energy level E. In general, such a representation is reducible.

Assume that $\mathcal{H}_d$ is an irreducible representation space of G. Then $\mathcal{H}_d$ is an irreducible representation space of G. This $\mathcal{H}_d$ is also a representation space for any subgroup K of G. When the representation of G corresponds to the matrices $D(g_1), ..., D(g_N)$, the matrices for K are $D(K) \subseteq D(G)$. In general, this representation is reducible. The smaller K is, the more representations, found in this way, are reducible. The eigenvalue belonging to an eigenspace which yields an irreducible representation of G is said to be *naturally degenerate* with respect to G. Is G a symmetry group of H, a level is *accidentally degenerate* with respect to G if its eigenvalue corresponds to an eigenspace which carries a reducible representation of G. When this occurs the degeneracy can be removed by a change in the Hamiltonian H without changing its symmetry. With respect to a subgroup of the full symmetry group, e.g. for the symmetry group of substitution operators, one can easily have accidental degeneracy. A well known example is the Hamiltonian of an electron in the Coulomb potential. If one considers the substitution symmetry, the symmetry group is the orthogonal group $O(3)$. With respect to this symmetry the levels are accidentally degenerate. However, Fock has shown that there is in fact $O(4)$ symmetry and with respect to this group the levels are naturally degenerate. In the following we suppose that there is no accidental degeneracy.

The Hamiltonian H has eigenvalues E_i, where the index i runs through a discrete set of parameters. The corresponding eigenspaces $\mathcal{H}_{\alpha i}$ are representation spaces for irreducible representations of the symmetry group G, consisting of unitary operators T_g $(g \in G)$ commuting with H. Choosing a set of basis functions which are orthogonal and normalized one has also a basis for the representation. Hence the basis functions are labelled by three indices: $\psi_{\alpha lk}$ $(\alpha = 1, ..., r;\ l = 1, 2, ...;\ k = 1, ..., d_\alpha)$, where l is a label distinguishing between eigenspaces with the same α. The eigenspaces and corresponding eigenvalues are denoted by $\mathcal{H}_{\alpha l}$ and $E_{\alpha l}$ respectively. In this way the symmetry group of the Hamiltonian may be used for the classification of eigenfunctions and eigenvalues. The degeneracy of $E_{\alpha l}$ is given by the dimension d_α of $\mathcal{H}_{\alpha l}$. According to § 1.2 the eigenfunctions transform under the group G as $T_g \psi_{\alpha lk} = \Sigma_m D_\alpha(g)_{mk} \psi_{\alpha lm}$. A well known example is the classification of the eigenfunctions in a rotationally symmetric potential. There the eigenfunctions are denoted by ψ_{nlm}, where n is the principal quantum number, l the eigenvalue of the operator L^2 (which denotes the representation D_l of the symmetry group, which is the three-dimensional rotation group), and m the eigenvalue of L_3, which numbers the $(2l+1)$ eigenfunctions which are the basis functions of the representation.

A Hermitian operator commuting with H has an expectation value which does not depend on the time, because $(d/dt)\langle\psi|P|\psi\rangle = i\langle\psi|[H,P]|\psi\rangle/\hbar$. Hence any Hermitian operator commuting with H gives a constant of motion. The symmetry group consists of unitary operators. A unitary operator is only Hermitian if it is of order two: $U^{-1} = U^\dagger = U$. On the other hand if the operator P is Hermitian and commutes with H, also the unitary operator $\exp(iPt)$ commutes with H for every value of the real parameter t. This is an element of an infinite group (t can vary continuously). Therefore, there is a correspondence between continuous symmetry groups and constants of motion, as one knows very well e.g. for the rotation group. As we will consider here only finite groups, the symmetry operators do not give rise, in general, to constants of motion.

As the Hilbert space $\mathcal{H}$ is the direct sum of eigenspaces of the Hamiltonian H, and for each eigenspace one can give a basis consisting of basis functions of irreducible representations, a basis for $\mathcal{H}$ is given by the functions $\psi_{\alpha lk}$. These functions are simultaneous eigenfunctions of the operators H (with eigenvalue $E_{\alpha l}$) and ρ^α_{ii}, because the projection operator ρ^α_{ii} acts on $\psi_{\beta lk}$ as $\rho^\alpha_{ii}\psi_{\beta lk} = \delta_{\alpha\beta}\delta_{ik}\psi_{\beta lk}$. Moreover, the operators ρ^α_{ii} commute with H because they are linear combinations of the operators T_g commuting with H. This resembles the description of Dirac. In Dirac's formulation a basis of $\mathcal{H}$ is formed by simultaneous eigenfunctions of a complete set of commuting operators. However, if accidental degeneracy occurs, an eigenfunction is not completely determined by its eigenvalues for H and ρ^{α}_{ii}. An eigenvalue E fixes the space $\mathcal{H}$, but if this space carries a reducible representation of G with two identical irreducible components, say $D_{\alpha}(G)$, the i-th basisfunctions in both irreducible subspaces are eigenfunctions of ρ^{α}_{ii} with eigenvalue 1. In the case where no accidental degeneracy occurs the basisfunctions are completely determined by their eigenvalues under H and ρ^{α}_{ii}, of course up to a phase factor.

2.2.3. *Symmetry and perturbation theory*

A second region, where group theory is useful in quantum mechanics is perturbation theory, where group theory can make predictions about the splitting of degenerate energy levels under a perturbation. One considers a Hamiltonian H which can be decomposed as $H = H_{\text{o}} + \lambda H_i$ in the Hamiltonian of the unperturbed system and an interaction part λH_i which is supposed to be small compared with H_{o}. We will denote the symmetry group of H_{o} by G_{o}, that of H by G. Usually one takes H_{o} in such a way that it has a larger symmetry. In general G is a subgroup of G_{o}, but this is not necessarily the case, as we shall see in Ch. 6, §3, where we will discuss the symmetry of a system in a magnetic field.

For a level E^{o} of H_{o} the eigenfunctions of H_{o} transform according to a d-dimensional representation $D(G_{\text{o}})$ which is, barring accidental degeneracy, irreducible. As one knows from the theory of perturbation of degenerate levels, an orthonormal basis of the eigenspace of E^{o} is in general not a good zeroth order approximation for the eigenfunctions of H. One has to perform first a unitary transformation on this basis. To see what this has to do with the symmetry we consider the splitting of the level by the perturbation. Under λH_i the eigenvalue E^{o} splits up into a number of levels $E_1, ..., E_n$. To each of these levels E_{α} belongs a representation $D_{\alpha}(G)$ which we will assume to be irreducible. In each subspace $\mathcal{H}_{\alpha}$ one may choose an orthonormal basis $\psi_{\alpha k}$ $(k = 1, ..., d_{\alpha})$. For vanishing interaction, i.e. in the limit $\lambda \to 0$ one has $\lim E_{\alpha} = E^{\text{o}}$ and $\lim \psi_{\alpha k} = \psi^{\text{o}}_{\alpha k}$. Then $\psi^{\text{o}}_{\alpha k}$ $(\alpha = 1, ..., n; k = 1, ..., d_{\alpha})$ form an orthonormal basis for the eigenspace of the level E^{o}. The functions of this basis are apparently good zeroth order approximations by definition.

In the space $\mathcal{H}_{\alpha}$ belonging to the eigenvalue E_{α} of H an irreducible representation of G is given by

$$T_g \psi_{\alpha k} = \sum_{l=1}^{d_{\alpha}} D_{\alpha}(g)_{lk} \psi_{\alpha l} \qquad (g \in G) .$$

In the limit $\lambda \to 0$ one obtains

$$T_g\psi^{\mathrm{o}}_{\alpha k} = \sum_{l=1}^{d_\alpha} D_\alpha(g)_{lk}\psi^{\mathrm{o}}_{\alpha l} \qquad (g\in G)\,.$$

As the function $\psi^{\mathrm{o}}_{\alpha k}$ ($\alpha = 1, ..., n$ and $k = 1, ..., d_\alpha$) form an orthonormal basis in the space belonging to the eigenvalue E^{o} of H_{o}, this space carries a representation of G. However, if the level splits into more than one level, the representation is reducible. Here we used the fact that the functions $\psi_{\alpha k}$ which are linearly independent, remain so in the limit. Then the functions $\psi^{\mathrm{o}}_{\alpha k}$ span for fixed α a G-invariant subspace of the eigenspace with eigenvalue E^{o}. On the other hand, this eigenspace carries an irreducible representation of G_{o} and consequently also a representation of the subgroup G. In general, this representation is reducible: if a basis $\phi_1, ..., \phi_d$ gives a representation

$$T_g\phi_i = \sum_j D^{\mathrm{o}}(g)_{ji}\phi_j \qquad (\text{any } g\in G)\,,$$

the basis $\psi^{\mathrm{o}}_{11}, ..., \psi^{\mathrm{o}}_{\alpha k}, ..., \psi^{\mathrm{o}}_{nd_n}$ gives a representation $D(G) = \Sigma_\alpha m_\alpha D^{\mathrm{o}}_\alpha(G)$, which is equivalent to $D^{\mathrm{o}}(G)$. The basis $\psi^{\mathrm{o}}_{\alpha k}$ in the eigenspace belonging to the eigenvalue E^{o} is the basis which brings the representation $D^{\mathrm{o}}(G)$ in reduced form, and which consists of functions which are good zeroth order approximations to the eigenfunctions of H. As the representations $D_\alpha(G)$ are the irreducible components of the reducible representation $D^{\mathrm{o}}(G)$, or of its equivalent representation $D(G)$, one can determine to which irreducible representations of G correspond the levels of H into which splits the level E^{o} of H_{o}. Among other things this means that the degeneracies of the sublevels are determined from the dimensions of the irreducible components of $D(G)$.

Now we drop the assumption that the levels correspond to irreducible representations of the symmetry groups. Then by accident the energy levels denoted by $E_{\alpha l}$ and $E_{\beta k}$ can coincide. This means that for this level we have accidental degeneracy, and that the eigenspace of the Hamiltonian carries a reducible representation of the symmetry group. One can have the following four situations.

1) The groups G and G_{o} are the same and E^{o} is naturally degenerate. This means that the symmetry is not lowered by the perturbation. As both D and D^{o} are irreducible representations of $G_{\mathrm{o}} = G$, there is no level splitting.

2) G is a proper subgroup of G_{o} and E^{o} is naturally degenerate. The representation $D(G_{\mathrm{o}})$ is irreducible, but $D(G)$ may be reducible: $D(G) \sim \Sigma^n_{\alpha=1} D^{\mathrm{o}}_\alpha(G)$. In this case the level splits up into n sublevels. By accident some of these levels may coincide and then we have accidental degeneracy. The degeneracies of the sublevels are given by the dimensions d_α of the irreducible components $D^{\mathrm{o}}_\alpha(G)$, which are the irreducible components of the representation $D(G_{\mathrm{o}})$ restricted to the subgroup G.

3) $G_o = G$, and E^o is an accidentally degenerate level. Then the eigenspace is a reducible space with respect to G_o and the corresponding representation $D(G_o)$ is reducible: $D(G_o) = \Sigma_{\alpha=1}^{p} D_\alpha(G_o)$. As $G = G_o$ only the representations $D_1(G), ..., D_p(G)$ occur as irreducible components of $D(G)$. The level E^o may split up into p sublevels. Again by accident some of these sublevels may coincide. This means that in general the accidental degeneracy is partially or completely lifted, such that the natural degeneracy remains.
4) If $G \subset G_o$ and E^o is accidentally degenerate, one can write

$$D(G) = \sum_{\alpha=1}^{p} D_\alpha(G) = \sum_{\alpha=1}^{p} \sum_{i=1}^{n_\alpha} D_{\alpha i}(G) \, ,$$

where $D_\alpha(G_o)$ are the irreducible components of $D(G_o)$ and $D_{\alpha i}(G)$ are the irreducible components of $D_\alpha(G)$. The level E^o now splits up into as many sublevels as there are irreducible components $D_{\alpha i}(G)$. Some of the levels may again coincide.

Concluding one can find the *splitting of an energy level* E^o under a perturbation H_i, if the symmetry group of H is a subgroup of that of H_o. To find the splitting one has to reduce the restriction of the representation $D(G_o)$, carried by the eigenspace of E^o, to the subgroup G (i.e. $D(G)$) into its irreducible components. Each irreducible component corresponds to an energy level of H. The degeneracy of this level is the dimension of the representation. The reduction gives the maximal number of energy sublevels, because some levels may coincide.

2.2.4. *Selection rules*

In the last section we considered the splitting of an energy level under a perturbation. We found a method of making qualitative predictions about this splitting. However, the method said nothing about the magnitude of splitting. In quantum mechanics the transition from one eigenstate ψ_1 of the Hamiltonian H_o to another one ψ_2, and the energy shift under influence of a perturbation H_i is determined by the matrix element $\langle\psi_1|H_i|\psi_2\rangle$. Group theory alone is not sufficient to calculate this matrix element, but it is very useful to facilitate these calculations by general statements. A central role here is taken by the following proposition.
PROPOSITION 2.2. Basis functions belonging to different irreducible representations, or to different rows of identically the same irreducible representations are orthogonal.
Proof. Suppose that $\psi_{\alpha i}$ is a basis function of the i-th row of the irreducible

representation $D_\alpha(G)$ in $\mathcal{H}$ and $\psi'_{\beta j}$ a basis function of $D_\beta(G)$ in $\mathcal{H}'$. Then the inner product of two such functions is

$$\langle\psi_{\alpha i}|\psi'_{\beta j}\rangle = \langle T_g\psi_{\alpha i}|T_g\psi'_{\beta j}\rangle = \frac{1}{N}\sum_{g\in G}\langle T_g\psi_{\alpha i}|T_g\psi'_{\beta j}\rangle$$

$$= \frac{1}{N}\sum_{g\in G}\sum_{k,l} D^*_\alpha(g)_{ki} D_\beta(g)_{lj}\langle\psi_{\alpha k}|\psi'_{\beta l}\rangle$$

$$= \frac{1}{d_\alpha}\sum_{k,l}\delta'_{\alpha\beta}\delta_{kl}\delta_{ij}\langle\psi_{\alpha k}|\psi'_{\beta l}\rangle = \delta'_{\alpha\beta}\delta_{ij}\frac{1}{d_\alpha}\sum_{k=1}^{d_\alpha}\langle\psi_{\alpha k}|\psi'_{\alpha k}\rangle .$$

This proves the proposition. Notice that the inner product does not depend on the row, but only on the representation α and on $\mathcal{H}$ and $\mathcal{H}'$.

The proposition enables us to determine whether matrix elements between two functions vanish on symmetry grounds or not. To see this, consider a basis $\psi_1, ..., \psi_{d_\alpha}$ for an irreducible representation $D_\alpha(G)$ of a group G. Further consider an operator A which commutes with T_g for any $g \in G$. Then $A\psi_1, ..., A\psi_{d_\alpha}$ is a basis for the same representation because

$$T_g A\psi_{\alpha i} = A\,T_g\psi_i = \sum_{j=1}^{d_\alpha} D_\alpha(g)_{ji} A\,\psi_j .$$

This means that a matrix element of A between two basis functions of irreducible representations of the symmetry group of A is

$$\langle\psi_{\alpha i}|A|\psi'_{\beta j}\rangle = \langle\psi_{\alpha i}|A\psi'_{\beta j}\rangle = \delta'_{\alpha\beta}\delta_{ij}\frac{1}{d_\alpha}\sum_k\langle\psi_{\alpha k}|A|\psi'_{\alpha k}\rangle . \qquad (2.10)$$

Therefore, the matrix element vanishes between functions which belong to nonequivalent representations or to different rows of identically the same representation.

For an arbitrary matrix element $\langle\psi_1|A|\psi_2\rangle$ one can use the decomposition by the projection operators ρ^α_{ij} (cf. § 1.4). One has

$$\psi_1 = \sum_{\alpha i}\rho^\alpha_{ii}\psi_1 ,$$

$$\psi_2 = \sum_{\beta j}\rho^\beta_{jj}\psi_2 .$$

Suppose that A is an operator such that $[T_g, A] = 0$ for any $g \in G$. Then the matrix element is given by

$$\langle\psi_1|A|\psi_2\rangle = \sum_{\alpha\beta ij} \langle\rho_{ii}^\alpha\psi_1|A|\rho_{jj}^\beta\psi_2\rangle$$

$$= \sum_{\alpha\beta ij} \delta'_{\alpha\beta}\,\delta_{ij}\langle\psi_{1\alpha}\|A\|\psi_{2\alpha}\rangle\,,$$

where $\langle\psi_{1\alpha}\|A\|\psi_{2\alpha}\rangle = (1/d_\alpha)\,\Sigma_k\,\langle\rho_{kk}^\alpha\psi_1|A|\rho_{kk}^\alpha\psi_2\rangle$. This means that the matrix element vanishes if the decompositions of ψ_1 and ψ_2 have no common terms.

As an example consider an interaction $V(\boldsymbol{r})$ which is invariant under the parity transformation P defined by $PV(\boldsymbol{r})P^\dagger = V(-\boldsymbol{r})$. Invariance means that $PV(\boldsymbol{r})P^\dagger = V(\boldsymbol{r})$. The group G is here the group of order two consisting of the unit matrix and its negative. Its irreducible representations are given by $D_\pm(P) = \pm 1$. Hence the projection operators are ρ^+ defined by $\rho^+\psi(\boldsymbol{r}) = \frac{1}{2}[\psi(\boldsymbol{r})+\psi(-\boldsymbol{r})]$, and ρ^- defined by $\rho^-\psi(\boldsymbol{r}) = \frac{1}{2}[\psi(\boldsymbol{r})-\psi(-\boldsymbol{r})]$. The proposition about matrix elements says that the matrix element of $V(\boldsymbol{r})$ between symmetric and antisymmetric functions vanishes.

2.2.5. Tensor operators

In the preceding sections we considered operators which were invariant under certain transformations. However, this is a special case. In quantum mechanics also operators play a role which are not invariant, but which transform in a specific way under Hilbert space operators. As an example we mention the momentum operator $\boldsymbol{p}$. Under the substitution operators P_R corresponding to rotations R this operator is not invariant, but transforms as a vector. This means the following. Consider a rotation R. In the three-dimensional space this corresponds, with respect to an orthonormal basis, to a matrix with elements R_{ij}. This matrix is orthogonal ($\tilde{R} = R^{-1}$), and all three-dimensional orthogonal matrices form a (faithful) irreducible representation of the group of orthogonal transformations. If one determines the commutation of P_R with the components of $\boldsymbol{p}$, one obtains

$$P_R\,p_i\,P_R^{-1} = \sum_{j=1}^{3} R_{ji}\,p_j \qquad (i = 1, 2, 3)\,.$$

We call an operator which transforms in this way a *vector operator*.

We generalize this definition in the following way. An *irreducible tensor operator* for the irreducible representation $D_\alpha(G)$ of a group G is a set of d_α

(the dimension of $D_\alpha(G)$) operators such that the commutation relations with the operators T_g are given by

$$T_g O_i T_g^{-1} = \sum_{j=1}^{d_\alpha} D_\alpha(g)_{ji} O_j \qquad (i = 1, ..., d_\alpha) , \tag{2.11}$$

where $D_\alpha(g)$ is the matrix representing T_g. When one does not specify the group G one usually has in mind the orthogonal group in 3 dimensions. As one knows (see e.g. Boerner [1967]) the irreducible representation of this group are denoted by $D^{(l)}$, which has dimension $2l+1$.

A special case of a tensor operator is one which transforms according to the trivial representation. In this case $T_g O T_g^{-1} = O$, or $[T_g, O] = O$. Such an operator is called a *scalar operator*. It is nothing but the invariant operator from the preceding sections. The operator T_g is a symmetry operator for O.

A *vector operator* T for the rotation group in 3 dimensions is an operator which transforms according to $D^{(1)}$ of this group. It has $2l+1 = 3$ components T_1, T_2, T_3. An example is the operator $\boldsymbol{p}$.

When the operators $O_1^\alpha, ..., O_{d_\alpha}^\alpha$ are the components of an irreducible tensor operator for the representation $D_\alpha(G)$, and $O_1^\beta, ..., O_{d_\beta}^\beta$ those of one for $D_\beta(G)$, the operators $O_i^\alpha O_j^\beta$ transform according to

$$T_g O_i^\alpha O_j^\beta T_g^{-1} = T_g O_i^\alpha T_g^{-1} T_g O_j^\beta T_g^{-1} = \sum_{k,l} D_\alpha(g)_{ki} D_\beta(g)_{lj} O_k^\alpha O_l^\beta \, .$$

Hence the operators $O_i^\alpha O_j^\beta$ transform according to the tensor product $(D_\alpha \otimes D_\beta)(G)$. This is in general not an irreducible representation. If the representation is reducible, one can make linear combinations of the operators $O_i^\alpha O_j^\beta$ which belong to irreducible representations. Although the three-dimensional rotation group is not finite, we choose this group again as an example, because this gives tensor operators which are well known from quantum mechanics. If T_1, T_2, T_3 are the components of a vector operator, the operators $T_i T_j$ belong to the product representation $D^{(1)} \otimes D^{(1)}$, which is reducible into three components, of dimension 1, 3 and 5. The operator $T_1T_1 + T_2T_2 + T_3T_3$ is a scalar operator, belonging to the representation $D^{(0)}$. The 3 operators $V_1 = \frac{1}{2}(T_2T_3 - T_3T_2)$, $V_2 = \frac{1}{2}(T_1T_3 - T_3T_1)$, and $V_3 = \frac{1}{2}(T_1T_2 - T_2T_1)$ are the three components of a vector operator. Finally there are five other operators which form the components of a five-dimensional tensor operator.

In § 2.4 we found that matrix elements of an operator which is invariant under a group G, vanish between functions belonging to nonequivalent irreducible representations or to different rows of identically the same representation. To generalize this result we consider the transformation properties of the function $O_i^\alpha \psi_{\beta k}$, where O_i^α is an irreducible tensor operator belonging to $D_\alpha(G)$, and $\psi_{\beta k}$ is a basis function belonging to the k-th row of $D_\beta(G)$. Then one has

$$T_g O_i^\alpha \psi_{\beta k} = T_g O_i^\alpha T_g^{-1} T_g \psi_{\beta k} = \sum_{j,l} D_\alpha(g)_{ji} D_\beta(g)_{lk} O_j^\alpha \psi_{\beta l}$$

$$= \sum_{j,l} (D_\alpha \otimes D_\beta)(g)_{jl,ik} O_j^\alpha \psi_{\beta l} \,. \tag{2.12}$$

Therefore, the $d_\alpha d_\beta$ functions $O_i^\alpha \psi_{\beta k}$ form a basis for the product representation $(D_\alpha \otimes D_\beta)(G)$. This is, in general, a reducible representation, i.e. there is a matrix S such that $S\,[D_\alpha \otimes D_\beta](g) = D(g)\,S$, where $D(g)$ is the direct sum $\Sigma_\gamma m_\gamma D_\gamma(g)$. The matrix S determines a basis transformation (eq. (1.4)). When $\psi_{\gamma lj}$ ($\gamma = 1, ..., r; l = 1, ..., m_\gamma; j = 1, ..., d_\gamma$) are basis functions which bring the matrices into the reduced form $D(g)$, one has

$$O_i^\alpha \psi_{\beta k} = \sum_{\gamma lj} S_{\gamma lj,\alpha\beta ik} \psi_{\gamma lj} \,. \tag{2.13}$$

To what extent are the functions $\psi_{\gamma lj}$ determined by the requirement that they bring $[D_\alpha \otimes D_\beta](G)$ in reduced form? Suppose S' also gives such a basis, i.e. $S'\,[D_\alpha \otimes D_\beta](G) = D(G)S'$ as well. Then $S'S^{-1}D(G)S = D(G)S'$, or $U = S'S^{-1}$ commutes with every matrix $D(g)$. If $D(G)$, and thus $[D_\alpha \otimes D_\beta](G)$, were irreducible, U would be a multiple of the unit matrix according to Schur's lemma. In that case the basis functions are determined up to a common factor. When $D(G)$ is not irreducible, one can write its matrices in block form

$$D(g) = \begin{pmatrix} D_{11}(g) & 0 \,\ldots\ldots\ldots\ldots\, 0 \\ 0 \ldots 0 & D_{22}(g) \quad 0 \ldots 0 \\ \ldots\ldots\ldots\ldots\ldots\ldots & \\ 0 \,\ldots\ldots\ldots\ldots\, 0 & D_{rr}(g) \end{pmatrix}$$

where $D_{ii}(G)$ is the i-th irreducible component. Write U in the same block form

$$U = \begin{pmatrix} U_{11} & U_{12} \,\ldots\ldots\ldots\, U_{1n} \\ \ldots\ldots\ldots\ldots\ldots\ldots & \\ U_{n1} \,\ldots\ldots\ldots\ldots & U_{nn} \end{pmatrix}$$

such that U_{ij} is a $d_i \times d_j$ matrix. The condition $UD(g) = D(g)\,U$ then reads

$U_{ij}D_{jj}(G) = D_{ii}(G)\,U_{ij}$. As a consequence of Schur's lemma

$$U_{ij} = \begin{cases} 0 & \text{if } D_{ii} \nsim D_{jj} \\ \lambda_{ij}\mathbb{1} & \text{if } D_{ii} = D_{jj}\,. \end{cases}$$

We recall that from each equivalence class of irreducible representations we have taken one representation, such that, if $D_{ii} \sim D_{jj}$, one means $D_{ii} = D_{jj}$. Hence, if all components of $D(G)$ are different, U is a diagonal matrix and the basis functions are determined up to a factor which is common to all functions belonging to the same component. It does not depend on the row of the function. If an irreducible component occurs with a multiplicity greater than 1, one can make linear combinations of the basis elements belonging to the same row of the various equal components. However, the linear combinations must be the same for all rows.

An example of the last case is the following. Suppose that $D(G) = D_\alpha(G) \oplus D_\alpha(G)$, and that a basis which gives this reduced matrix representation is given by $\psi_{\alpha 1}, \ldots, \psi_{\alpha d_\alpha}, \psi'_{\alpha 1}, \ldots, \psi'_{\alpha d_\alpha}$. Then another basis which gives the same matrix representation is given by $\lambda\psi_{\alpha i} + \mu\psi'_{\alpha i}$, $\rho\psi_{\alpha i} + \sigma\psi'_{\alpha i}$ $(i = 1, \ldots, d_\alpha)$, where λ, μ, ρ, and σ do not depend on i, and where $\lambda\sigma - \mu\rho \neq 0$ in order to have a nonsingular transformation. It gives rise to the same matrices because

$$T_g(\lambda\psi_{\alpha i} + \mu\psi'_{\alpha i}) = \sum_j \lambda D_\alpha(g)_{ji}\psi_{\alpha j} + \mu D_\alpha(g)_{ji}\psi'_{\alpha j}$$
$$= \sum_j D_\alpha(g)_{ji}[\lambda\psi_{\alpha j} + \mu\psi'_{\alpha j}]\,,$$
$$T_g(\rho\psi_{\alpha i} + \sigma\psi'_{\alpha i}) = \sum_j D_\alpha(g)_{ji}[\rho\psi_{\alpha j} + \sigma\psi'_{\alpha j}]\,.$$

The corresponding matrix U is

$$U = \begin{pmatrix} \lambda\mathbb{1} & \mu\mathbb{1} \\ \rho\mathbb{1} & \sigma\mathbb{1} \end{pmatrix}.$$

Now we choose for a basis $\psi_{\alpha\beta,ik}$ $(i = 1, \ldots, d_\alpha;\ k = 1, \ldots, d_\beta)$ of $[D_\alpha \otimes D_\beta](G)$ a basis transformation which reduces this representation.

$$\psi_{\alpha\beta,ik} = \sum_{\gamma l j} \left(\begin{smallmatrix}\alpha & \beta \\ i & k\end{smallmatrix}\middle|\begin{smallmatrix}\gamma & \\ j & l\end{smallmatrix}\right)\psi_{\gamma lj}\,. \tag{2.14}$$

The coefficients $\left(\begin{smallmatrix}\alpha & \beta \\ i & k\end{smallmatrix}\middle|\begin{smallmatrix}\gamma & \\ j & l\end{smallmatrix}\right)$ are the *(generalized) Clebsch–Gordan coefficients*.

They are determined by the representations $D_\alpha(G)$ up to factors $A_{\gamma l}$. For the matrix S in eq. (2.13) one has

$$S_{\gamma lj,\alpha\beta ik} = A_{\gamma l}\begin{pmatrix}\alpha & \beta & | & \gamma \\ i & k & | & j\ l\end{pmatrix}.$$

Using this relation one can determine the matrix element

$$\langle\psi_{\alpha i}|O_j^\beta|\psi_{\gamma k}\rangle = \sum_{\delta lm}\begin{pmatrix}\beta & \gamma & | & \delta \\ j & k & | & m\ l\end{pmatrix} A_{\delta l}\langle\psi_{\alpha i}|\psi_{\delta lm}\rangle.$$

In the proof of proposition 2.2 we have seen that $\langle\psi_{\alpha i}|\psi'_{\delta lm}\rangle = \delta_{\alpha\gamma}\,\delta_{im}\,B_{\alpha l}$, where B does not depend on the row index m. Hence

$$\langle\psi_{\alpha i}|O_j^\beta|\psi'_{\gamma k}\rangle = \sum \begin{pmatrix}\beta & \gamma & | & \alpha \\ j & k & | & i\ l\end{pmatrix} C_{\alpha l}, \tag{2.15}$$

where $C_{\alpha l}$ is a constant which does not depend on the row index. Eq. (2.15) is a generalization of the *Wigner–Eckart theorem*, which is often used in quantum mechanics. A special case is eq. (2.10), where the invariant operator transforms with the trivial representation, which means $\beta = j = 1$. Then $\begin{pmatrix}1 & \gamma & | & \alpha \\ 1 & k & | & i\ 1\end{pmatrix} = \delta_{\alpha\gamma}\delta_{ik}$, and therefore $\langle\psi_{\alpha i}|A|\psi'_{\gamma k}\rangle = \delta_{\alpha\gamma}\delta_{ik}C_{\alpha 1}$, where now $C_{\alpha 1} = (1/d_\alpha)\Sigma_k\langle\psi_{\alpha k}|A|\psi'_{\alpha k}\rangle$ depends only on α, as seen in eq. (2.10).

The Clebsch–Gordan coefficients have the following properties.

1) $\begin{pmatrix}\alpha & \beta & | & \gamma \\ i & k & | & j\ l\end{pmatrix} = \begin{pmatrix}\beta & \alpha & | & \gamma \\ k & i & | & j\ l\end{pmatrix}$,

2) $\begin{pmatrix}\alpha & \beta & | & \gamma \\ i & k & | & j\ l\end{pmatrix}$ vanishes if in the product $[D_\alpha \otimes D_\beta](G)$ the irreducible component D_γ does not occur, or if l is larger than the multiplicity of D_γ.

The Wigner–Eckart theorem eq. (2.15) in combination with the properties of the Clebsch–Gordan coefficients makes it possible to predict the vanishing of matrix elements of tensor operators between basis functions of irreducible representations.

2.2.6. *Invariants*

Let T_G be an irreducible representation of G in a space $\mathcal{H}$. Unless this is the trivial representation no element $\neq 0$ of $\mathcal{H}$ remains invariant under G. However, if G is a finite group, one can construct a second order invariant. This is a Hermitian form on $\mathcal{H}$ which is invariant under G. To obtain these invariants we consider the space of complex linear functions on $\mathcal{H}$. These form the dual space $\mathcal{H}^d$. Any element ψ of $\mathcal{H}$ gives a linear function by $\phi \to \langle\psi|\phi\rangle (\forall\phi\in\mathcal{H})$. It is readily verified that the mapping $\mathcal{H} \to \mathcal{H}^d$ which assigns the function $\langle\psi|$ to the element ψ is an isomorphism. We define in

$\mathcal{H}^d$ a representation T^d_G of G by

$$\langle T^d_g \psi | \phi \rangle = \langle \psi | T_{g^{-1}} \phi \rangle \qquad (\text{all } \phi \in \mathcal{H}) .$$

It is a representation because $\langle T^d_g T^d_h \psi | \phi \rangle = \langle T^d_h \psi | T_{g^{-1}} \phi \rangle = \langle \psi | T_{(gh)^{-1}} \phi \rangle = \langle T^d_{gh} \psi | \phi \rangle$. It is called the *adjoint or contragredient representation*. When $\mathcal{H}$ carries the representation $D_\alpha(G)$ the matrices of the adjoint representation are $D^d_\alpha(g) = \tilde{D}_\alpha(g^{-1})$. For a unitary representation this is $D_\alpha(g)^*$. The product space $\mathcal{H}^d \otimes \mathcal{H}$ carries the representation $D^d_\alpha \otimes D_\alpha$ with character $\chi(g) = \chi^*_\alpha(g)\, \chi_\alpha(g)$. It is, in general, a reducible representation which contains the trivial representation exactly once, because the multiplicity of this representation is

$$n_1 = \frac{1}{N} \sum_{g \in G} \chi^*(g)\, \chi(g) = 1 .$$

The unique invariant Hermitian form is constructed as in Ch. 1, §3.1:

$$(\psi | \phi) = \frac{1}{N} \sum_{g \in G} \langle T_g \psi | T_g \phi \rangle .$$

For a unitary representation this is exactly $\langle \psi | \phi \rangle$.

Let A be an irreducible scalar operator. This operator is invariant if and only if $T_g A T_g^{-1} = A$ for any $g \in G$. However, again one can construct invariant operators from arbitrary irreducible tensor operators. Suppose that A_i and B_j are irreducible tensor operators transforming according to the same irreducible representation $D(G)$ of G. This means that $T_g A_i T_g^{-1} = \Sigma_j D(g)_{ji} A_j$ and $T_g A^\dagger_i T_g^{-1} = \Sigma_j D(g)^*_{ji} A^\dagger_j$ (we assume T_g to be unitary). The tensor operator $A^\dagger_i B_j$ transforms according to the representation $D^d(G) \otimes D(G)$ which contains the trivial representation exactly once. The component transforming according to this trivial representation is $\Sigma_i A^\dagger_i B_i$. It is invariant because $\Sigma_i T_g A^\dagger_i B_i T_g^{-1} = \Sigma_{ijk} D(g)^*_{ji} D(g)_{ki} A^\dagger_j B_k = \Sigma_i A^\dagger_i B_i$ when $D(G)$ is unitary. If $D(g)$ is not unitary, it is equivalent to a unitary representation, because G is finite. Then $D(G) = S^{-1} D'(G) S$ with $D'(G)$ unitary. An example is the tensor operator $\boldsymbol{p}$. It is an irreducible tensor operator of the cubic group. The only invariant operator one can construct is $p_1^2 + p_2^2 + p_3^2 = \boldsymbol{p}^2$.

CHAPTER III

CRYSTALLOGRAPHIC POINT GROUPS

Following the general discussion of the role of symmetry in quantum mechanics the present chapter and the following one will deal with examples of symmetry groups. We start with groups of transformations which leave a point fixed. Groups of this kind, called point groups occur as symmetry groups of atoms and ions in molecules and in crystals. The symmetry group of an atom in a crystal has particular properties. Such a group is called crystallographic point group. The definition and properties of these groups are given in section one. As we have seen, the representations of symmetry groups play an important role for the physical consequences of symmetry. The representations are treated in section two. In the third section we study atoms or ions in a crystal from a viewpoint of point group symmetry. In this crystal field theory, the use and the limitations of symmetry considerations as discussed in the preceding chapter can be seen.

3.1. Crystallographic point groups

3.1.1. The orthogonal group

A large class of transformations occurring as symmetries of physical systems is formed by transformations in the three-dimensional space which leave a point fixed. To be more precise, we consider a three-dimensional *Euclidean vector space*, which is a real linear vector space with an inner product (cf. Ch. 1, §2). The group of nonsingular linear transformations is denoted by $GL(3, \mathbb{R})$. The inner product of two vectors $\boldsymbol{x}$ and $\boldsymbol{y}$ is denoted by $(\boldsymbol{x}, \boldsymbol{y})$ or sometimes by $\boldsymbol{x} \cdot \boldsymbol{y}$. A subgroup of $GL(3, \mathbb{R})$ is formed by the elements which leave the norm $(\boldsymbol{x}, \boldsymbol{x})$ invariant for any $\boldsymbol{x}$ in the space. This subgroup is the orthogonal group denoted by $O(3)$. After a choice of a basis the elements of $GL(3, \mathbb{R})$ and of $O(3)$ are three-by-three matrices. With respect to an orthonormal basis the elements of $O(3)$ are orthogonal matrices: $A\widetilde{A} = \mathbb{1}$.

When we take determinants of this relation we get $\det(A)\det(\widetilde{A}) = 1$, or $\det(A) = \pm 1$. The orthogonal transformations with determinant +1 (this does not depend on the basis) are called *rotations*, and form a subgroup of index two denoted by $SO(3)$. Any element of this rotation group $SO(3)$ can by a suitable choice of basis be brought into the form

$$\begin{pmatrix} \cos\phi & \sin\phi & 0 \\ -\sin\phi & \cos\phi & 0 \\ 0 & 0 & 1 \end{pmatrix}. \tag{3.1}$$

It is evident that both $O(3)$ and $SO(3)$ are infinite groups.

An element of $O(3)$ with determinant -1 is $-\mathbb{1}$. We denote this element by I. All matrices of $O(3)$ which are not in $SO(3)$ are the product of a rotation with I. As I is of the order two and commutes with all elements of $O(3)$, the group $O(3)$ is the direct product $SO(3) \times C_2$, where C_2 is the cyclic group of order two. The elements of $O(3)$ of order two which are not in $SO(3)$ are called *reflections*. An example of such a reflection is the element I, also called *central inversion*. The other elements in $O(3)$ with determinant -1 are products of I with a rotation through an angle $\phi \neq \pi$. These are called *roto-reflections*.

Any rotation has an axis. This means that, if we consider a sphere of unit radius around the origin, any rotation transforms this sphere into itself and there are exactly two points on the sphere which are left invariant. On the other hand, a rotation is determined by its axis and the angle of rotation ϕ. The axis of a rotation is an eigenvector with eigenvalue one. The two other eigenvalues are $\exp(\pm i\phi)$. Two rotations through the same angle are conjugated by a rotation which transforms the axes into each other. If the axes are v_1 and v_2, and the rotations R_1 and R_2, and R a rotation such that $\boldsymbol{v}_1 = R\boldsymbol{v}_2$, then $RR_2R^{-1}R\boldsymbol{v}_2 = RR_2\boldsymbol{v}_2 = R\boldsymbol{v}_2$, or $RR_2R^{-1}\boldsymbol{v}_1 = \boldsymbol{v}_1$. Moreover, for R_1 and RR_2R^{-1} the eigenvalues $(1, e^{+i\phi})$, hence the rotation angle ϕ, are the same.

3.1.2. Lattices

Another group of transformations of the three-dimensional space is the group of *translations*. When a point $\boldsymbol{x}$ has components ξ_1, ξ_2, ξ_3, the action of a translation $\boldsymbol{t} = (\tau_1, \tau_2, \tau_3)$ is defined by $\boldsymbol{x} \to \boldsymbol{x} + \boldsymbol{t} = (\xi_1 + \tau_1, \xi_2 + \tau_2, \xi_3 + \tau_3)$. The translations form an Abelian group $T(3)$. One can define a scalar multiplication by $\alpha\boldsymbol{t} = (\alpha\tau_1, \alpha\tau_2, \alpha\tau_3)$ for any $\alpha \in \mathbb{R}$. Then $T(3)$ has the structure of a linear vector space isomorphic to $\mathbb{R}^3$, the space of triplets of real numbers.

Suppose $\boldsymbol{a}_1, \boldsymbol{a}_2, \boldsymbol{a}_3$ are three linearly independent translations. We can speak about linear independence since $T(3)$ has the structure of a vector space. The elements $\boldsymbol{a}_1, \boldsymbol{a}_2, \boldsymbol{a}_3$ generate a subgroup of $T(3)$ with elements $n_1\boldsymbol{a}_1 + n_2\boldsymbol{a}_2 + n_3\boldsymbol{a}_3$, where n_1, n_2, n_3 are integers. Such a subgroup of $T(3)$ is called a *lattice group* U. It is a group of translations acting on the space V. Starting from one point in V and operating with U on this point one obtains a subset of V which is called a *lattice* Λ. Although Λ is a set of points of V (the points of V are also called vectors, since V is a vector space), and U is a group of translations, it is clear that they are closely related. For many purposes we can identify Λ and U by choosing the null vector of V as the point on which U acts. In the same way we can identify the space V and the group of translations $T(3)$. However, sometimes it is convenient to distinguish between points and operations. It is for that reason that we have introduced the distinct notions of lattice and of lattice group.

3.1.3. Point groups

Each subgroup of $O(3)$ is called a *point group*, because it leaves the origin invariant. Point groups occur as symmetry groups of molecules and of atoms in a crystal. In the latter case the point group has an additional property: it transforms a lattice into itself. Any point group which does so for some lattice Λ is called a *crystallographic point group*. When K is a crystallographic point group, there is a lattice Λ such that for every $\boldsymbol{x} \in \Lambda$ and any $\alpha \in K$ the point $\alpha\boldsymbol{x}$ belongs to Λ. We write this as $K\Lambda = \Lambda$. In this chapter we will be concerned mainly with crystallographic point groups.

Two point groups which are isomorphic have the same multiplication table. One can say that they are both isomorphic to an abstract group which is given by its multiplication table. This group is called the *abstract point group*. It is an arbitrary element from the isomorphism class and not necessarily a subgroup of $O(3)$. The different abstract point groups are representatives of the isomorphism classes of point groups.

With respect to an orthonormal basis $\boldsymbol{e}_1, \boldsymbol{e}_2, \boldsymbol{e}_3$ the element α of a point group K is an orthogonal matrix $A(\alpha)$. Moreover, for a crystallographic point group there is a lattice which is left invariant. If this lattice Λ is obtained by the action of U on a point $\boldsymbol{x}_o$, also U is left invariant. Take $\alpha \in K$, $\boldsymbol{a} \in U$. Then $\alpha\boldsymbol{a} = \alpha(\boldsymbol{x}_o + \boldsymbol{a} - \boldsymbol{x}_o) = \alpha(\boldsymbol{x}_o + \boldsymbol{a}) - \alpha\boldsymbol{x}_o = \boldsymbol{x}_o + \boldsymbol{a}' - \boldsymbol{x}_o - \boldsymbol{a}''(\boldsymbol{a}', \boldsymbol{a}'' \in U)$ $\in U$. With respect to a basis of U the point group elements are nonsingular matrices $\phi(\alpha)$. When a basis $\boldsymbol{a}_1, \boldsymbol{a}_2, \boldsymbol{a}_3$ is obtained from $\boldsymbol{e}_1, \boldsymbol{e}_2, \boldsymbol{e}_3$ by a basis transformation S, then one has $\phi(\alpha) = SA(\alpha)S^{-1}$ for any $\alpha \in K$. Furthermore, the entries of the matrix $\phi(\alpha)$ are integers because $\alpha a_i = \Sigma_j \phi(\alpha)_{ji} a_j$ belongs to the lattice. Hence one can say that a crystallographic point group is a three-dimensional representation of an abstract point group K and this representation is equivalent to one with orthogonal matrices and also to one with integral matrices.

Two point groups which are equivalent as representations of an abstract point group are called *geometrically equivalent*. This means that they are related by a nonsingular transformation $S \in GL(3, \mathbb{R})$, or in other words they are conjugate subgroups of $GL(3, \mathbb{R})$. Both point groups are subgroups of $O(3)$. There is a theorem (see Burckhardt [1967], p. 47) which states that two geometrically equivalent point groups are orthogonally equivalent. This means that there is an orthogonal transformation which links both groups. Hence geometrically equivalent point groups are conjugate subgroups of $O(3)$. The equivalence classes of the geometric equivalence relation are called the *geometrical crystal classes*. Since two geometrically equivalent point groups are isomorphic each isomorphism class contains complete geometrical crystal classes. In three dimensions – the only case we consider here – the 18 isomorphism classes can be subdivided into 32 geometrical crystal classes. A derivation of these classes will be given in the next section. Finally we notice the following. If two point groups are conjugated by an orthogonal transformation they are also conjugated by a rotation: either the conjugating matrix is a rotation, or it is a rotation multiplied with I, in which case the rotation part also conjugates the point groups. We can interpret this as follows. If one point group describes a group of orthogonal transformations in one basis, the other group describes the same transformations in a basis obtained from the first by the conjugating rotation.

Although we have given here the definitions for point groups for the three-dimensional space only, they can be generalized in a straightforward way to arbitrary dimension. For spaces up to dimension four the isomorphism classes and the geometrical crystal classes have been determined. The number of abstract crystallographic point groups (isomorphism classes) for dimension $n = 1, 2, 3$, and 4 is equal to 2, 9, 18, and 118 respectively. The number of crystal classes is 2, 10, 32, and 227 respectively. For the four-dimensional classes see Hurley [1968].

3.1.4. Derivation of abstract point groups and geometrical crystal classes

To determine the isomorphism classes of the crystallographic point groups we consider first those groups which are subgroups of $SO(3)$. We call these groups, consisting entirely of rotations, *point groups of the first kind*. Each element can by a suitable basis choice be written as in eq. (3.1). The character of the element in this representation is $1 + 2\cos\phi$. As this character is invariant under similarity transformations, it is the same in a representation with integral matrices, for which the character is of course integral. Hence $1 + 2\cos\phi$ must be an integer. This means that there are only five possibilities for ϕ: $0°, 60°, 90°, 120°$, and $180°$ (the character being $3, 2, 1, 0, -1$ respectively). It follows that any axis of rotation in a crystallographic point group is 1-, 2-, 3-, 4-, or 6-fold.

Lemma. A crystallographic point group is of finite order.

Proof. Consider a crystallographic point group K. There is a lattice left invariant by K. Let a_1, a_2, a_3 be a basis of this lattice. Now construct a sphere around the origin which encloses the basis vectors. Inside the sphere there are only a finite number of vectors. Each element $\alpha \in K$ gives a permutation of these vectors. If α leaves fixed all vectors inside the sphere, it leaves the lattice pointwise fixed, in which case α must be the identity. Therefore, K has only a finite number of elements, because there is only a finite number of permutations of a finite number of vectors.

So the crystallographic point groups are among the finite subgroups of $O(3)$. We will determine here all finite subgroups of $O(3)$ and start with the finite subgroups of $SO(3)$. Suppose that K is such a group, not only consisting of the unit element. Any element $g \neq \mathbb{1}$ of the group has an axis of rotation which intersects the unit sphere around the origin in two points, the *poles* of the rotation. We introduce an equivalence relation between the poles of the elements of K. We call p_1 and p_2 equivalent if there is an element $g \in K$ such that $p_1 = gp_2$. The equivalence classes of poles are denoted by $C_1, ..., C_m$. When the pole p is in class C_i, and when K_p is the subgroup of K leaving the pole p invariant, one can decompose K into cosets of K_p by $K = K_p + g_2K_p + ... + g_rK_p$. The different poles in class C_i are $p, g_2p, ..., g_rp$. Therefore, the number of cosets, which is equal to the number of poles in class C_i depends only on the class, not on the choice of the pole p. For two poles p_1 and p_2 in C_i the invariance groups K_{p_1} and K_{p_2} are related by $K_{p_2} = gK_{p_1}g^{-1}$ if $p_2 = gp_1$. This means that the order of K_p also depends only on the class. It is denoted by n_i. Then the order of K is $N = n_ir_i$ for any i. In K there are $N-1$ elements different from $\mathbb{1}$. We now count the number of elements corresponding to the various poles. For a pole p in the class C_i there are n_i-1 elements different from $\mathbb{1}$ in the group K_p, for the r_i poles in C_i there are $r_i(n_i-1)$ of these elements. In all there are $\Sigma_i r_i(n_i-1)$ elements $\neq \mathbb{1}$. However, each element has two poles. Hence $N - 1 = \frac{1}{2}\Sigma_i r_i(n_i-1)$ or

$$2\left(1-\frac{1}{N}\right)=\sum_{i=1}^{m}\left(1-\frac{1}{n_i}\right). \tag{3.2}$$

The eq. (3.2) will give us all possible finite subgroups of $SO(3)$. From $N \geqslant n_i \geqslant 2$ we have the inequalities $2 > 2(1-1/N) \geqslant 1$ and $1 > (1-1/n_i) \geqslant \frac{1}{2}$. Hence m can take only the values 2 and 3. We consider now the different cases.

1) $m=2$. In this case eq. (3.2) gives

$$\frac{2}{N}=\frac{1}{n_1}+\frac{1}{n_2}.$$

Moreover, $1/n_1 \geqslant 1/N$, $1/n_2 \geqslant 1/N$. Hence $n_1 = n_2 = N \geqslant 2$.

2) $m=3$. This case leads to

$$\frac{1}{n_1}+\frac{1}{n_2}+\frac{1}{n_3}=1+\frac{2}{N}>1.$$

We now suppose $n_1 \geqslant n_2 \geqslant n_3$. Then necessarily $n_3 = 2$ in order to satisfy the inequality. Because $2/n_2 \geqslant 1/n_1 + 1/n_2 = 1/2 + 2/N > 1/2$, n_2 can take the values 2 and 3. For $n_2 = n_3 = 2$ one has $1/n_1 = 2/N$ or $N = 2n_1 \geqslant 4$. For $n_2 = 3, n_3 = 2$ there are three possibilities. Either the triple $(n_1, n_2, n_3) = (3, 3, 2)$, or $(4, 3, 2)$, or $(5, 3, 2)$. The order N follows from eq. (3.2): it is 12, 24, or 60 respectively. Taking into account the restrictions on n_i for crystallographic point groups one has the following possibilties.

$m=2$, $n_1 = 2, 3, 4$, or 6, $n_2 = n_1 = N$,

$m=3$, $n_1 = 2, 3, 4$, or 6, $n_2 = n_3 = 2$, $N = 2n_1$,

$m=3$, $n_1 = 3$, $n_2 = 3$, $n_3 = 2$, $N = 12$,

$m=3$, $n_1 = 4$, $n_2 = 3$, $n_3 = 2$, $N = 24$.

The corresponding groups are described in the next section. They belong together with the group consisting only of the identity to 11 isomorphism classes. The finite subgroups which cannot occur as crystallographic point groups can appear as symmetry groups *e.g.* of molecules. Although not important for crystals, we shall enumerate them. They are

$m=2,\ n_1=5,7,8,\ldots,\ n_2=n_1=N,$

$m=3,\ n_1=5,7,8,\ldots,\ n_2=n_3=2,\ N=2n_1,$

$m=3,\ n_1=5,\ n_2=3,\ n_3=2,\ N=60.$

They are respectively the cyclic groups of order 5, 7, 8, ..., the dihedral groups of order 10, 14, 16, 18, ..., and the icosahedral group which is isomorphic to the so-called alternating group A_5.

A *point group of the second kind* is a point group which contains elements with determinant -1. Such a point group K contains a subgroup of elements with determinant $+1$. As the product of two elements with determinant -1 is an element with determinant $+1$, this subgroup, denoted by K_+, is of index two. The coset which contains all the elements with determinant -1 is denoted by K_-. One can write $K = K_+ + K_-$. We will construct now another subgroup of $O(3)$, related to K, but which is in addition a subgroup of $SO(3)$. We multiply all elements of K_- with the central inversion I. If $I \in K$ and $\beta \in K_-$, one has $I\beta \in K$ and thus $I\beta \in K_+$. In that case $IK_- = K_+$. On the other hand, if $\beta \in K_-$ and $I\beta \in K_+$, it follows that $I = I\beta\beta^{-1} \in K$. Therefore, if K does not contain I, IK_- and K_+ have no elements in common. In that case a new subgroup of $O(3)$ defined by $K' = K_+ + IK_-$ is a point group of the first kind isomorphic to K. The isomorphism is given by

$$\phi(\alpha) = \alpha \in K_+ \subset K' \quad \text{for} \quad \alpha \in K_+ \subset K,$$

$$\phi(\alpha) = I\alpha \in K' \quad \text{for} \quad \alpha \in K_- \subset K.$$

Since the groups K and K' are isomorphic, point groups of the second kind not isomorphic to one of the first kind must contain I. If a group contains I, it is the direct product of its rotation subgroup with the cyclic group of order two: $K = K_+ \times C_2$, because K contains K_+ and the group generated by I as subgroups which have only the identity in common, and I commutes with all elements of K_+. In this way one finds another 7 abstract crystallographic point groups. A more complete derivation can be found in Burckhardt [1967]. There it is proved that the 11 classes found previously correspond to 11 isomorphism classes and 11 geometrical crystal classes.

Because the 11 different classes of finite subgroups correspond to 11 different geometrical crystal classes, there are 11 geometrical crystal classes of the first kind. The classes of the second kind which contain I consist of direct products of a group of the first kind and C_2. So there are also 11 of them. Finally, the groups of the second kind which do not contain I can be found in the following way. Consider one group from each geometrical crystal class of the first kind, and determine all possible subgroups of index two. Multiply the coset of such a subgroup by I. The union of the subgroup and its coset multiplied by I gives a group of the second kind. Among the groups found in this way several are still geometrically equivalent. Elimination of equivalent groups gives the remaining crystal classes of the second kind. A complete discussion can be found in Burckhardt [1967]. The results are discussed in the next section.

3.1.5. The abstract crystallographic point groups and the geometrical crystal classes

Since any crystallographic point group is either isomorphic to one of the first kind, or to the direct product of a group of the first kind with C_2, it is convenient to treat first the isomorphism classes of the groups of the first kind. They were discussed in the preceding section. Including the group of order one we have the following possibilities.

1) The *cyclic groups* C_n for $n = 1, 2, 3, 4$, or 6. The group C_n is an Abelian group of order n generated by an n-fold rotation. This group has two classes of poles: both with one pole of an n-fold rotation ($n_1 = n_2 = n$). As an abstract group it is generated by α with defining relation $\alpha^n = \epsilon$. The elements of the group are $\alpha, \alpha^2, ..., \alpha^n = \epsilon$.

2) The *dihedral groups* D_n for $n = 2, 3, 4$, or 6. It is the group of three-dimensional rotations transforming a regular n-gon into itself. It is a group of order $2n$ with a cyclic subgroup of order n. This group has 3 classes of poles: one with 2 poles of an n-fold rotation and two with n poles of 2-fold rotations. It is generated by an n-fold rotation and a 2-fold rotation with axes perpendicular to the axis of the first rotation (fig. 3.1a). As an abstract group it is generated by α and β with defining relations $\alpha^n = \beta^2 = (\alpha\beta)^2 = \epsilon$. As from these relations it follows that $\beta\alpha = \alpha^{-1}\beta$ the elements of the group are $\alpha, \alpha^2, ..., \alpha^n = \epsilon, \beta, \alpha\beta, \alpha^2\beta, ..., \alpha^{n-1}\beta$. The conjugacy classes of the different groups are given in table 3.1. We recall that rotations in the same conjugacy class have the same order.

3) The *tetrahedral group* T is the group of rotations transforming a regular tetrahedron into itself (fig. 3.1b). It has order 12. The group has 3 classes of poles: two with 4 poles of 3-fold rotations and one with 6 poles of 2-fold rotations. Its elements are 8 3-fold rotations, 3 2-fold rotations and the identity. As an abstract group it is generated by α and β with defining relations $\alpha^3 = \beta^2 = (\alpha\beta)^3 = \epsilon$. In terms of the generators, the elements can be written as $\alpha, \beta\alpha\beta, \beta\alpha, \alpha\beta$ (forming one conjugacy class of order 3 elements), $\alpha^2, \alpha\beta\alpha$,

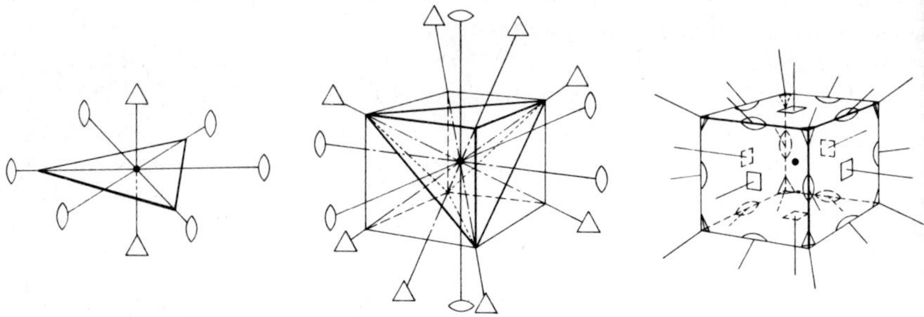

Fig. 3.1. Elements of the groups D_3, T and O.

Table 3.1
The dihedral groups D_n: conjugacy classes.

Defining relations: $\alpha^n = \beta^2 = (\alpha\beta)^2 = \epsilon$
The order of the elements in the i-th class is denoted by p_i.

D_2	ϵ $(p_1=1)$	α $(p_2=2)$	β $(p_3=2)$	$\alpha\beta$ $(p_4=2)$		
D_3	ϵ $(p_1=1)$	α,α^2 $(p_2=3)$	$\beta,\alpha\beta,\alpha^2\beta$ $(p_3=2)$			
D_4	ϵ $(p_1=1)$	α^2 $(p_2=2)$	α,α^3 $(p_3=4)$	$\beta,\alpha^2\beta$ $(p_4=2)$	$\alpha\beta,\alpha^3\beta$ $(p_5=2)$	
D_6	ϵ $(p_1=1)$	α^2,α^4 $(p_2=3)$	$\alpha^2\beta,\alpha^4\beta$ $(p_3=2)$	α^3 $(p_4=2)$	α,α^5 $(p_5=6)$	$\alpha\beta,\alpha^5\beta$ $(p_6$

$\alpha^2\beta$, $\beta\alpha^2$ (forming another class of order 3 elements), β, $\alpha^2\beta\alpha$, $\alpha\beta\alpha^2$ (forming a class of order 2 elements) and ϵ. So there are 4 classes. The multiplication table is given in table 3.2. Matrices for the rotations with respect to an orthogonal basis are given in table 4.4.

4) The *octahedral group O* is the group of rotations which transform a cube into itself (fig. 3.1c). The order of the group is 24. According to the preceding section the group has 3 classes of poles: one with 6 poles of 4-fold rotations, one with 8 poles of 3-fold rotations and one with 12 poles of 2-fold rotations. Its elements are 6 four-fold rotations, 8 three-fold rotations, 3 two-fold rotations which are squares of four-fold rotations, 6 more two-fold rotations and the identity. The multiplication table is given in table 3.2. Matrices for these rotations are given in table 4.4. The abstract group is generated by α and β with defining relations $\alpha^4 = \beta^3 = (\alpha\beta)^2 = \epsilon$. The group has five classes which are given in table 3.2. All four-fold rotations belong to one class, as

le and classes of the tetrahedral and octahedral groups

	p_i	1	p_2=3								p_3=2			p_4=4						p_5=2					
classes	p_i	1	p_2=3				p_3=3				p_4=2														
T	g_i	1	2	3	4	5	6	7	8	9	10	11	12	13	14	15	16	17	18	19	20	21	22	23	24
	1	1	2	3	4	5	6	7	8	9	10	11	12	13	14	15	16	17	18	19	20	21	22	23	24
	2	2	6	9	7	8	1	12	10	11	5	3	4	20	16	24	21	13	15	23	17	14	19	22	18
β	3	3	8	7	9	6	12	1	11	10	4	2	5	22	18	13	17	24	19	14	21	23	15	20	16
	4	4	9	6	8	7	10	11	1	12	3	5	2	18	22	23	15	14	21	24	19	13	17	16	20
	5	5	7	8	6	9	11	10	12	1	2	4	3	16	20	14	19	23	17	13	15	24	21	18	22
	6	6	1	11	12	10	2	4	5	3	8	9	7	17	21	18	14	20	24	22	13	16	23	19	15
α	7	7	11	1	10	12	5	3	2	4	9	8	6	15	19	22	24	16	14	18	23	20	13	21	17
β	8	8	12	10	1	11	3	5	4	2	6	7	9	21	17	16	23	22	13	20	24	18	14	15	19
2	9	9	10	12	11	1	4	2	3	5	7	6	8	19	15	20	13	18	23	16	14	22	24	17	21
	10	10	4	5	2	3	9	8	7	6	1	12	11	14	13	21	22	19	20	17	18	15	16	24	23
α^2	11	11	5	4	3	2	7	6	9	8	12	1	10	23	24	17	20	15	22	21	16	19	18	13	14
$\beta\alpha$	12	12	3	2	5	4	8	9	6	7	11	10	1	24	23	19	18	21	16	15	22	17	20	14	13
	13	13	21	19	15	17	18	16	22	20	14	24	23	10	1	2	8	3	9	5	6	4	7	11	12
	14	14	15	17	21	19	20	22	16	18	13	23	24	1	10	4	7	5	6	3	9	2	8	12	11
	15	15	20	18	22	16	14	24	13	23	19	17	21	9	7	11	2	1	4	12	5	10	3	8	6
	16	16	24	13	14	23	17	19	21	15	20	22	18	2	5	7	12	8	1	9	11	6	10	4	3
	17	17	16	22	18	20	24	14	23	13	21	15	19	8	6	1	5	11	3	10	2	12	4	9	7
	18	18	13	24	23	14	21	15	17	19	22	20	16	3	4	9	1	6	12	7	10	8	11	5	2
	19	19	22	16	20	18	23	13	24	14	15	21	17	7	9	10	3	12	5	1	4	11	2	6	8
	20	20	14	23	24	13	15	21	19	17	16	18	22	5	2	6	10	9	11	8	1	7	12	3	4
	21	21	18	20	16	22	13	23	14	24	17	19	15	6	8	12	4	10	2	11	3	1	5	7	9
	22	22	23	14	13	24	19	17	15	21	18	16	20	4	3	8	11	7	10	6	12	9	1	2	5
	23	23	19	21	17	15	22	20	18	16	24	14	13	12	11	5	9	4	8	2	7	3	6	1	10
	24	24	17	15	19	21	16	18	20	22	23	13	14	11	12	3	6	2	7	4	8	5	9	10	1

olumn expresses the elements of O in generators $\alpha = g_{13}$ and $\beta = g_2$, the second column expresses the f T in generators $\alpha = g_2$ and $\beta = g_{10}$, the defining relations are $\alpha^4 = \beta^3 = (\alpha\beta)^2 = \epsilon$ for O and $(\alpha\beta)^3 = \epsilon$ for T.

nts $g_1, ..., g_{12}$ form the group T the Cayley table of which is given in the outlined part of the table.

of the elements in the i-th class is given by p_i.

for any pair of axes of these rotations there is a three-fold rotation in the group which transforms one into the other. Moreover, any two-fold rotation perpendicular to a four-fold axis transforms a 90° rotation into a rotation with inverse sense, which is the inverse element of the first rotation.

The *point groups of the second kind* containing I are direct products of a point group of the first kind with C_2. Because one has the isomorphisms $C_1 \times C_2 \cong C_2$, $C_2 \times C_2 \cong D_2$, $C_3 \times C_2 \cong C_6$ and $D_3 \times C_2 \cong D_6$ the groups of the second kind give only 7 new isomorphism classes: $C_4 \times C_2$, $C_6 \times C_2$, $D_2 \times C_2$, $D_4 \times C_2$, $D_6 \times C_2$, $T \times C_2$, and $O \times C_2$. So there are 18 isomorphism classes of crystallographic point groups, or 18 abstract crystallographic point groups. The rotation groups C_n, D_n $(n = 1, 2, 4)$, T and O leave invariant a lattice with orthonormal basis $\boldsymbol{e}_1, \boldsymbol{e}_2, \boldsymbol{e}_3$, C_n, D_n $(n = 3, 6)$ one with $\boldsymbol{e}_1$, $\frac{1}{2}\boldsymbol{e}_1 + \frac{1}{2}\sqrt{3}\boldsymbol{e}_2, \boldsymbol{e}_3$. As I leaves invariant any lattice, one sees that indeed all these groups are crystallographic.

The 32 geometric crystal classes can be given by one representative point group from each class. The elements of such a point group are orthogonal transformations. If one takes a sphere around the origin, and a point $\boldsymbol{x}$ on the sphere in such a way that for any element α of the group K the point $\alpha\boldsymbol{x}$ is different from $\boldsymbol{x}$, the group K can be visualized by projection of the sphere on a tangent plane. When the sphere is given by $x^2 + y^2 + z^2 = 1$, the projection of the point x_0, y_0, z_0 on the plane $z = -1$ is the point (x_0, y_0). If $z_0 > 0$, one denotes the point of the projection with an open circle, if $z_0 < 0$, by a cross. In fig. 3.2a such a projection is given for a group from the iso-

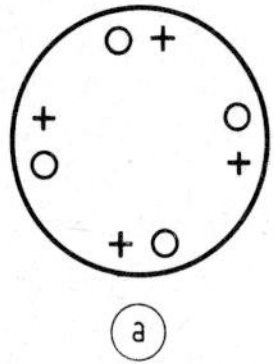

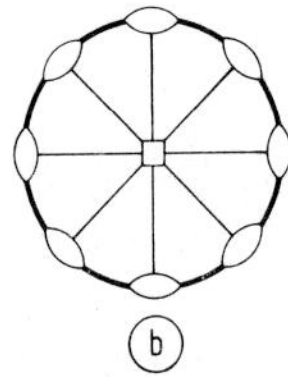

Fig. 3.2. Diagrams for the rotation group 422.

morphism class D_4. The point group of the first kind is generated by a 90° rotation around the z-axis and a 180° rotation around the x-axis. The group is of order 8. As all points on the sphere obtained from the starting point by the action of the point group are different, there are 8 points in the projection. The points with circles lie on the upper hemisphere, the points with crosses on the lower hemisphere.

To denote geometrical crystal classes one uses either the *international symbol* (Hermann–Maugin notation), or the *Schoenflies symbol*. For the international symbol an n-fold rotation is denoted by the number n, the product of an n-fold rotation with I by $\bar{n}$ with a bar on the number. A mirror plane, i.e. a reflection which is not the central inversion, through a rotation axis, i.e. a reflection which leaves this axis invariant, is denoted by the letter m next to the number indicating the order of the rotation. A mirror plane perpendicular to an n-fold axis by n/m. The central inversion is denoted by $\bar{1}$.

A second way to visualize a point group is to indicate in the projection of the unit sphere the projection of the poles of rotations, the poles of rotations occurring combined with I to roto-reflections, and the intersection of mirror planes with the sphere. An example is given in fig. 3.2b for the same group as that used in fig. 3.2a. Notice that a group is not uniquely determined by the first kind of diagram, but that it is so by the second kind. For example, the diagram of the point group m shows two points which could also be obtained from each other by a rotation. This ambiguity is not present in the second kind of diagram where the mirror plane is indicated. Finally, we remark that by the second kind of diagram a crystal class is completely determined, and that each group of a crystal class can be represented by the same diagram as two groups from the same class are conjugated by a rotation which only rotates the unit sphere. The 32 crystal classes are given in table 3.3. The groups are visualized in the diagrams of table 3.4.

3.2. Representations of crystallographic point groups

3.2.1. Representations of cyclic groups

As the cyclic group C_n is an Abelian group, each element forms a class. Thus there are n classes and consequently n nonequivalent irreducible representations. For the crystallographic point groups we denote the i-th irreducible representation by $\Gamma_i(G)$. The n irreducible representations of C_n are $\Gamma_1, ..., \Gamma_n$. We have already seen that the irreducible representations of an Abelian group are one-dimensional, as follows also from $\Sigma_i d_i^2 = n$. When

Table 3.3
The 32 geometric crystal classes.

Abstract group	Order	Point groups		
		1st kind	2nd kind without I	2nd kind with I
C_1	1	$1 \;= C_1$	–	–
C_2	2	$2 \;= C_2$	$m \;= C_s$	$\bar{1} \;= C_i$
C_3	3	$3 \;= C_3$	–	–
C_4	4	$4 \;= C_4$	$\bar{4} \;= S_4$	–
$D_2 \cong C_2 \times C_2$	4	$222 = D_2$	$2mm = C_{2v}$	$2/m \;= C_{2h}$
$C_6 \cong C_3 \times C_2$	6	$6 \;= C_6$	$\bar{6} \;= C_{3h}$	$\bar{3} \;= S_6$
D_3	6	$32 \;= D_3$	$3m \;= C_{3v}$	–
$C_4 \times C_2$	8	–	–	$4/m \;= C_{4h}$
D_4	8	$422 = D_4$	$4mm = C_{4v}$	–
			$\bar{4}2m \;= D_{2d}$	
$D_2 \times C_2$	8	–	–	$mmm \;= D_{2h}$
$D_6 \cong D_3 \times C_2$	12	$622 = D_6$	$6mm = C_{6v}$	$\bar{3}m \;= D_{3v}$
			$\bar{6}m2 \;= D_{3h}$	
T	12	$23 \;= T$	–	–
$C_6 \times C_2$	12	–	–	$6/m \;= C_{6h}$
$D_4 \times C_2$	16	–	–	$4/mmm = D_{4h}$
O	24	$432 = O$	$\bar{4}3m \;= T_d$	–
$D_6 \times C_2$	24	–	–	$6/mmm = D_{6h}$
$T \times C_2$	24	–	–	$m3 \;= T_h$
$O \times C_2$	48	–	–	$m3m \;= O_h$

The symbols for the crystal classes are given both in the notation of the international tables of X-ray crystallography (at the left in each column) and in the Schoenflies notation (at the right).

Table 3.4
The geometric crystal classes in 3 dimensions (diagrams).
(Left Schoenflies notation, right international symbol)

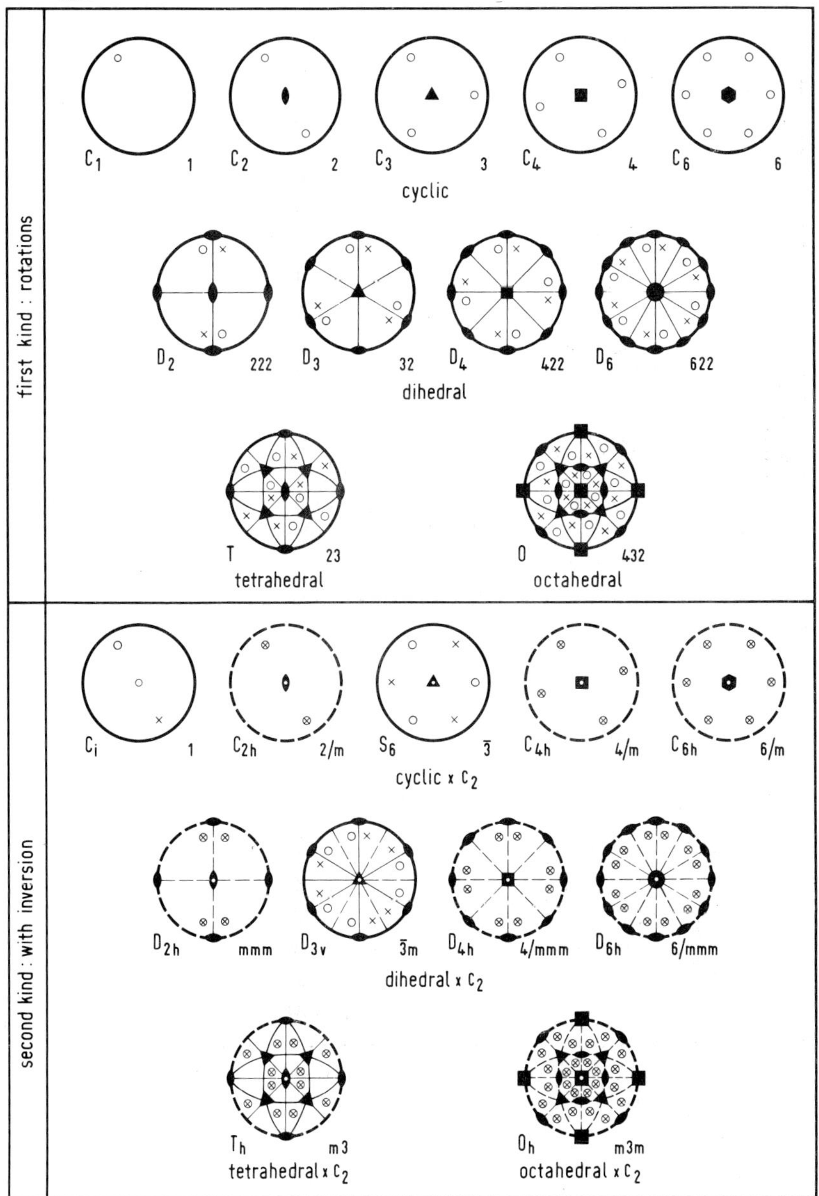

Table 3.4. (continued)

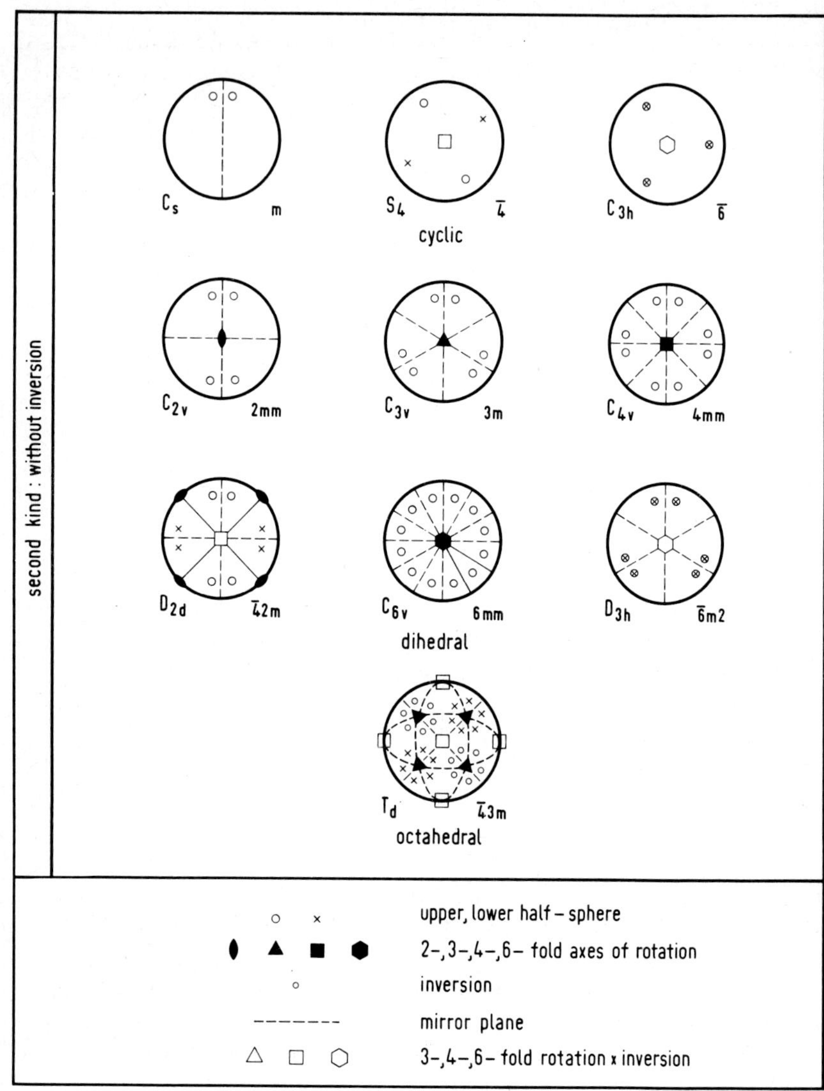

$\Gamma_i(\alpha)$ is the complex number representing the generator α of C_n in the i-th irreducible representation, one has $\Gamma_i(\alpha)^n = \Gamma_i(\alpha^n) = \Gamma_i(\epsilon) = 1$. So $\Gamma_i(\alpha)$ is a n-th root of unity. There are exactly n of them. Therefore, the representations are given in the following table. As the representations are one-dimensional, characters and representing matrices are the same.

	ϵ	α	α^2		α^{n-1}
Γ_1	1	1	1	...	1
Γ_2	1	$\exp(2\pi i/n)$	$\exp(4\pi i/n)$	...	$\exp(-2\pi i/n)$
---	---	---	---	---	---
Γ_{n-1}	1	$\exp(-4\pi i/n)$	$\exp(-8\pi i/n)$	...	$\exp(4\pi i/n)$
Γ_n	1	$\exp(-2\pi i/n)$	$\exp(-4\pi i/n)$	...	$\exp(2\pi i/n)$

The character tables for the cyclic crystallographic groups are given in the appendix. As the point groups are real three-dimensional representations of these abstract groups, they can be decomposed into irreducible components. *E.g.* for C_2 the point group 2 is equivalent to $\Gamma_1 \oplus \Gamma_2 \oplus \Gamma_2$, the point group m is equivalent to $\Gamma_1 \oplus \Gamma_1 \oplus \Gamma_2$, and the point group $\bar{1}$ is equivalent to $\Gamma_2 \oplus \Gamma_2 \oplus \Gamma_2$. The representation $\Gamma_1 \oplus \Gamma_1 \oplus \Gamma_1$ is not faithful.

As the multiplicator of a cyclic group is trivial (see Schur [1904]), each projective representation of a cyclic group is similar to an ordinary one with trivial factor system.

3.2.2. Representations of the dihedral and polyhedral groups

As these groups are of low order and have a simple structure, one can obtain the character table in a nonsystematic way using only some of the properties of the characters. An example is given for D_3, which is isomorphic to the group of permutations of 3 elements, in Ch. 1, §3.4. In an analogous way one can find the character table for D_4. For the other dihedral groups one has $D_2 \cong C_2 \times C_2$ and $D_6 \cong D_3 \times C_2$. The representations of such direct products will be treated in the next section. The character tables of D_3 and D_4 can be found in the appendix.

The multiplicator of D_3 is trivial, those of D_2, D_4, and D_6 are isomorphic to C_2 (see e.g. Döring [1959], where the projective representations of the crystallographic point groups are discussed). For the latter 3 groups there are nontrivial factor systems. Explicit matrix representations for the projective representations of these and the other crystallographic point groups can be found in Hurley [1966].

The dimensions of the irreducible representations of the tetrahedral group T follow from eq. (1.25): $12 = 1 + d_2^2 + d_3^2 + d_4^2$, or $d_2 = d_3 = 1, d_4 = 3$. For the 3 one-dimensional representations one has $\chi(\beta) = \chi(\alpha^2\beta\alpha) = \chi(\alpha\beta\alpha^2)$. As the elements of the classes 1 and 4 form an Abelian subgroup with $\beta(\alpha^2\beta\alpha) = \alpha\beta\alpha^2$, one has $\chi(\beta) = 1$. Eq. (1.23) then gives for the 4-th column $1 + 1 + 1 + |\chi_4(C_4)|^2 = 4$ or $\chi_4(C_4) = \pm 1$. Also for the 3 one-dimensional representations one has $\chi(\alpha^2) = \chi(\alpha)^2$ and $\chi(\alpha)^3 = 1$. Hence $\chi(\alpha)$ is a third root of one. In order to have nonequivalent representations the characters $\chi_1(\alpha)$, $\chi_2(\alpha)$ and $\chi_3(\alpha)$ must be different. Therefore, $\chi_1(\alpha) = 1$, $\chi_2(\alpha) = \omega^2 = \exp(2\pi i/3)$ and $\chi_3(\alpha) = \omega^4$. The sum of the squares of the absolute values of the characters in the second and third column must be 3. Hence $\chi_4(\alpha) = \chi_4(\alpha^2) = 0$. Finally the orthogonality of the first and fourth row gives $\chi_4(\beta) = -1$. This leads to the character table which can be found in the appendix.

In an analogous way one finds the character table for the octahedral group O. This group of order 24 has five classes which implies that there are 2 one-dimensional, 1 two-dimensional, and 2 three-dimensional irreducible representations. The character table is given in the appendix. The multiplicators of both T and O are isomorphic to C_2. Hence both groups have two classes of nonassociated factor systems. Explicit projective matrix representations are given in Hurley [1966].

3.2.3. Representations of direct products

When G is a group with r conjugacy classes $C_1, ..., C_r$ and H a group with s classes $C_1', ..., C_s'$, the group $G \times H$ has rs classes. Suppose a and b belong to C_i, and g and h belong to C_j'. Then there are elements $c \in G$ and $f \in H$ such that $b = cac^{-1}$ and $h = fgf^{-1}$. Then the elements (a,g) and (b,h) belong to the same class of $G \times H$, as $(c,f)(a,g)(c,f)^{-1} = (b,h)$. On the other hand, if this relation holds, a and b belong to the same class of G and g and h to the same class of H. Therefore, the class C_{ij} of $G \times H$ consists of all pairs (a, b) with $a \in C_i$ and $b \in C_j'$. This means that there are indeed rs classes.

When $D_\alpha(G)$ is an irreducible representation of G and $D_\beta(H)$ an irreducible representation of H, the outer Kronecker product $D_\alpha(G) \times D_\beta(H)$ is a representation of $G \times H$ with character $\chi(a,b) = \chi_\alpha(a)\chi_\beta(b)$. Using eq. (1.20) we have

$$\sum_{(a,b)\in G\times H} |\chi(a,b)|^2 = \sum_{a\in G} \sum_{b\in H} |\chi_\alpha(a)|^2 |\chi_\beta(b)|^2 = N_1 N_2 = N\,,$$

where N_1 is the order of G, N_2 that of H and N that of $G \times H$, and one sees that $D_\alpha(G) \times D_\beta(H)$ is an irreducible representation $D_{\alpha\beta}(G\times H)$ of the direct product. There are rs irreducible representations of this kind, which is the total number of nonequivalent irreducible representations of $G \times H$. Moreover, the representations $D_{\alpha\beta}(G\times H)$ and $D_{\alpha'\beta'}(G\times H)$ are nonequivalent if $\alpha \neq \alpha'$, or $\beta \neq \beta'$, because from $\chi_\alpha(a)\chi_\beta(b) = \chi_{\alpha'}(a)\chi_{\beta'}(b)$ for any $(a,b) \in G \times H$ it follows that

$$1 = \frac{1}{N} \sum_{(a,b)\in G\times H} |\chi(a,b)|^2 = \frac{1}{N_1} \sum_{a\in G} \chi_\alpha^*(a)\chi_{\alpha'}(a) \frac{1}{N_2} \sum_{b\in H} \chi_\beta^*(b)\chi_{\beta'}(b)$$

$$= \delta_{\alpha\alpha'}\delta_{\beta\beta'} .$$

Consequently the rs nonequivalent irreducible representations of $G \times H$ are exactly the rs representations $D_\alpha(G) \times D_\beta(H)$.

Now one can obtain the character tables for the other crystallographic point groups which are the direct product of a group from the preceding sections and the cyclic group C_2. As an example consider the group $D_6 \cong D_3 \times C_2$. The character tables of D_3 and C_2 are

D_3	C_1	C_2	C_3
Γ_1	1	1	1
Γ_2	1	1	−1
Γ_3	2	−1	0

C_2	ϵ	α
Γ_1	1	1
Γ_2	1	−1

Then the character table of D_6 looks like

D_6	$C_1=C_{11}$	$C_2=C_{21}$	$C_3=C_{31}$	$C_4=C_{12}$	$C_5=C_{22}$	$C_6=C_{32}$
Γ_1	1	1	1	1	1	1
Γ_2	1	1	−1	1	1	−1
Γ_3	2	−1	0	2	−1	0
Γ_4	1	1	1	−1	−1	−1
Γ_5	1	1	−1	−1	−1	1
Γ_6	2	−1	0	−2	1	0

3.2.4. The irreducible representations of $O(3)$

The orthogonal group $O(3)$ is an infinite group. Therefore, the representation theory of Ch. 1, §3 cannot be applied. Especially expressions involving summation over group elements have to be reconsidered. It turns out that with some changes the theory for finite groups can be extended to a certain class of infinite groups, the compact Lie groups. In the following we will need only representations of the group $O(3)$ which is an example of a compact Lie group. In the present section we will mention some properties of these representations, but we will not go into details. For a more profound treatment we refer e.g. to Hamermesh [1962].

The subgroup $SO(3)$ has an infinite number of nonequivalent irreducible representations denoted by $D^{(l)}$, where $l = 1, 2, \ldots$. The dimension of $D^{(l)}$ is $2l + 1$. Apart from $l = 0$ these representations are faithful. The representation for $l = 1$ can be realized by the 3 by 3 real orthogonal matrices with determinant +1 (sometimes called the identical representation). The character of a representation is a class function. Two rotations are conjugate elements in $SO(3)$ if they describe rotations through the same angle. The character of a rotation ϕ is given by

$$\chi_l(\phi) \frac{\sin\,(l + \frac{1}{2})\phi}{\sin \frac{1}{2}\phi}\,. \tag{3.3}$$

Basis functions for the representation $D^{(l)}$ are the spherical harmonics Y_{lm} ($m = -l, \ldots, +l$).

The group $O(3)$ is the direct product $SO(3) \times C_2$ of the group of rotations with the group generated by the central inversion I. Therefore, exactly as in §2.3, for each representation $D^{(l)}$ of $SO(3)$ there are two representations of $O(3)$. One has $D_{\pm}^{(l)}(g) = D^{(l)}(g)$ and $D_{\pm}^{(l)}(Ig) = \pm D^{(l)}(g)$ for any $g \in SO(3)$.

3.3. Crystal field theory

3.3.1. Splitting of atomic energy levels by a crystal field

Consider an atom (or an ion – we will always speak here about atoms, although we always mean atoms or ions) with m electrons around a nucleus. We assume the nucleus to be fixed in space. The Hamiltonian of the electron system is

$$H=\sum_i \frac{p_i^2}{2m}+\sum_i V(r_i)+\sum_{i<j} V_{ij}+V_s\,,$$

where p_i is the momentum operator for the i-th particle, $V(r)$ the potential created by the charge of the nucleus, V_{ij} the interaction potential between the i-th and j-th electrons, and V_s the term describing spin effects like spin-orbit coupling. In this chapter we will neglect the last term, to keep the discussion simple. We will come back to this in Ch. 5, where we will treat spin effects. When the atom has a number of closed shells with n electrons outside, one can approximate the Hamiltonian in a natural way by

$$H=H_{\text{core}}+\sum_{i=1}^{n} \frac{p_i^2}{2m}+\sum_i V(r_i)+\sum_{i<j} V_{ij}\,,$$

where now $V(r)$ denotes the spherically symmetric potential created by the core electrons and the nucleus. The summation runs only over the outer electrons. Their wave functions are linear combinations of products of one-particle wave functions, or in other words they are elements of the n-fold tensor product of one-particle Hilbert spaces $\mathcal{H}\otimes\mathcal{H}\ \ldots\otimes\mathcal{H}$. The Hamiltonian in this space is

$$H=\sum_i \frac{p_i^2}{2m}+\sum_i V(r_i)+\sum_{i<j} V_{ij}\,. \tag{3.4}$$

We denote the first two terms by H_0. To determine the invariance group of H we first consider the case $n = 1$. Then we get the Hamiltonian of Ch. 2, §2.2, which has $O(3)$ as invariance group, because the potential $V(r)$ is spherically symmetric. Then the Hamiltonian H_0 has the n-fold direct product $O(3)^n = O(3)\times O(3)\times\ldots\times O(3)$ as invariance group, because on each coordinate one can apply an orthogonal transformation and these transformations can be different for the various coordinates. The group of operators constituting the invariance group consists of substitution operators $P_{g_1,\ldots,g_n}$ given by

$$P_{g_1,\ldots,g_n}\psi(r_1,\ldots,r_n)=\psi(g_1^{-1}r_1,\ldots,g_n^{-1}r_n)\,. \tag{3.5}$$

The group $O(3)^n$ is not the full symmetry group. We did not consider, *e.g.* the permutation symmetry which exists because the electrons are identical particles and their coordinates may be interchanged. When P_{ij} is a permutation operator defined by

$$P_{ij}\psi(r_1, ..., r_i, ..., r_j, ..., r_n) = \psi(r_1, ..., r_j, ..., r_i, ..., r_n)\,, \tag{3.6}$$

it commutes also with H_o. For the moment we only consider the invariance group $O(3)^n$.

The substitution operator $P_{g_1,...,g_n}$ commutes with the complete Hamiltonian H only if $g_1 = ... = g_n = g \in O(3)$. The reason is that V_{ij} is transformed into itself by a coordinate transformation only if the distance between the various electrons remains the same. The substitution operators with $g_1 = ...$ $... = g_n$ form a group homomorphic to $O(3)$. Therefore, the invariance group of H is $O(3)$. A symmetry operator is now P_g defined by $P_g\psi(r_1, ..., r_n) = \psi(g^{-1}r_1, ..., g^{-1}r_n)$. If we identify $g \in O(3)$ with the element $(g, g, ..., g) \in O(3)^n$, the symmetry group of H is a subgroup of that of H_o and one can apply the general procedure from Ch. 2, § 2.3. The eigenfunctions of H_o can be labeled by indices n_i, l_i, m_i, the principal quantum number, the orbital angular momentum and the z-component of this angular momentum, respectively, for the i-th particle. These indices also indicate the irreducible representation and the row of the representation of $O(3)^n$ to which the eigenfunction belongs. The irreducible representations of $O(3)^n$ are $D^{(l_1)} \otimes D^{(l_2)}$ ... $... \otimes D^{(l_n)}$ and the rows of this representation are labeled by $m_1, ..., m_n$ $(-l_i \leqslant m_i \leqslant +l_i)$. By the interaction the symmetry is lowered. The eigenfunctions of H belong to irreducible representations of $O(3)$. They are labeled by L (denoting the representation $D^{(L)}$) and M (denoting the row of the representation). This corresponds to the fact that for H the $l_1, ..., l_n$ are no longer good quantum numbers, but have to be replaced by the total orbital angular momentum L. Also, $m_1, ..., m_n$ are no longer good quantum numbers, but the z-component of L (i.e. M) remains good.

When one considers an atom in a crystal, an additional term occurs in the Hamiltonian describing the interaction of the electrons with the crystal. This interaction is very complicated. In crystal field theory one assumes that it can be described by a potential $W(\boldsymbol{r})$ created by the electric charges of the surrounding particles of the atom. We do not discuss the validity of this approximation. A discussion is given, *e.g.*, in Herzfeld and Meijer [1961]. Since the potential $W(\boldsymbol{r})$ is created by the surrounding charges in the crystal, the symmetry of $W(\boldsymbol{r})$ is no longer $O(3)$, but is the symmetry of the site of the atom. This is a crystallographic point group. Then the symmetry of the total Hamiltonian

$$H = H_o + \sum_j W(\boldsymbol{r}_j) + \sum_{i<j} V_{ij} \tag{3.7}$$

is this crystallographic point group K. As the symmetry of the Hamiltonian is lowered by the crystal potential, the energy levels split up further in general. This level splitting produced by the crystal potential was studied in detail in Bethe [1929]. The splitting can occur in several ways depending on the mutual strength of the terms in eq. (3.7). The term H_0 is always the most important one; the splitting by $W(\boldsymbol{r})$ can vary over a large range. One can distinguish three cases.

1) *Strong field*: the influence of $W(\boldsymbol{r})$ is stronger than the interaction between the electrons. This is usually found in covalent compounds of the transition-metal ions.

2) *Intermediate field*: the influence of $W(\boldsymbol{r})$ is less than that of the interaction, but greater than that of the spin term neglected in eq. (3.7). This case is found in ionic compounds of transition-metal ions.

3) *Weak field*: the splitting by $W(\boldsymbol{r})$ is even smaller than the splitting by the spin term V_S. In this chapter we only consider systems where the spin effects may be neglected. In Ch. 5 we will discuss these effects and the weak field case. Here we are only concerned with the first two cases.

3.3.2. *The simplest case: one electron*

We start with the case of an atom with only one electron outside the closed shells. In this case the difference between strong and intermediate field vanishes. The Hamiltonian is now

$$H = \frac{p^2}{2m} + V(r) + W(\boldsymbol{r}) = H_0 + W(\boldsymbol{r}) . \tag{3.8}$$

As we have already seen, the invariance group of H_0 is $O(3)$, that of H is a crystallographic point group K. An energy level of H_0 is denoted by the principal quantum number n and the orbital angular momentum l. The eigenfunctions of this level transform according to the representation $D^{(l)}$. The eigenfunctions of H belong to irreducible representations of K. As K is a subgroup of $O(3)$, the representation $D^{(l)}$ subduces a representation $D^{(l)}(K)$ of K which is, in general, reducible. When $\Gamma_1, ..., \Gamma_r$ are the irreducible representations of K, one has

$$D^{(l)}(K) = \sum_{i=1}^{r} m_i^{(l)} \Gamma_i ,$$

where $m_i^{(l)}$ is the multiplicity of Γ_i in $D^{(l)}(K)$. According to Ch. 2, § 2.3 the level E_{nl} splits up into $m_1^{(l)} + ... + m_r^{(l)}$ sublevels, each of which is d_i-fold degenerate.

As an example, consider an ion in a cubic lattice, where the crystal produces a field at the position of the ion which has the symmetry of the cube, i.e. $m3m$ symmetry. We restrict us to the subgroup 432 of rotations. As we will see below, this is sufficient. For 432 the character of the representation subduced from $D^{(l)}$ is given by

	C_1	C_2	C_3	C_4	C_5
ϕ	0°	120°	180°	90°	180°
$D^{(0)}$ S-state	1	1	1	1	1
$D^{(1)}$ P-state	3	0	−1	1	−1
$D^{(2)}$ D-state	5	−1	1	−1	1
$D^{(3)}$ F-state	7	1	−1	−1	−1
$D^{(4)}$ G-state	9	0	1	1	1
....	....	....	....	....	....

The character table of $432 \cong O$ is given in the appendix. From this the reduction of $D^{(l)}(K)$ is easily determined using eq. (1.19).

$$D^{(0)} = \Gamma_1$$

$$D^{(1)} = \Gamma_4$$

$$D^{(2)} = \Gamma_3 \oplus \Gamma_5$$

$$D^{(3)} = \Gamma_2 \oplus \Gamma_4 \oplus \Gamma_5$$

$$D^{(4)} = \Gamma_1 \oplus \Gamma_3 \oplus \Gamma_4 \oplus \Gamma_5$$

....

Hence an S-level or P-level is not split up in a crystal field with cubic symmetry. A D-level is split up into one two-fold and one three-fold degenerate level and so on. Moreover, the eigenfunctions of the level coming from a P-level transform according to Γ_4 under transformations of 432, etc.

We considered only rotations in this case, whereas the full invariance group also contains the central inversion I. That this procedure is justified, can be seen from the following reasoning. The point group K is a direct product $H \times C_2$. When Γ_α are the irreducible representations of H, those of K are $\Gamma_\alpha^{\pm}$. The multiplicity of the irreducible representation $\Gamma_\alpha^{\epsilon_1}$ in $D_{\epsilon_2}^{(l)}(K)$ (with

$\epsilon_1, \epsilon_2 = \pm 1$) is given by ($N$ is the order of K)

$$m_{\alpha\epsilon_1}^{\epsilon_2} = \frac{1}{N} \sum_{g \in H} \{\chi_l^*(g)\,\chi_\alpha(g) + \epsilon_1\epsilon_2\,\chi_l^*(g)\,\chi_\alpha(g)\}$$

$$= \begin{cases} \dfrac{2}{N} \sum_{g \in H} \chi_l^*(g)\,\chi_\alpha(g) & \text{if } \epsilon_1\epsilon_2 = +1 \\ 0 & \text{if } \epsilon_1\epsilon_2 = -1. \end{cases}$$

Hence even representations reduce into even representations and odd representations reduce into odd representations. The multiplicity is equal to the multiplicity of $\Gamma_\alpha(H)$ in $D^{(l)}(H)$.

3.3.3. Levels of pairs of nonequivalent electrons

Consider now an atom with two electrons outside closed shells. In order to avoid difficulties with the Pauli exclusion principle we assume that the two electrons are not equivalent. This means that they do not belong to the same representation space of an irreducible representation of the invariance group of H_0 (3.7) (either they have different principal quantum number, or they have different orbital angular momentum). When the i-th electron is in the state $\psi_{nlm}(r_i)$ belonging to the irreducible representation $D^{(l)}$ of $O(3)$, the two-particle state $\psi_{n_1l_1m_1}(r_1)\,\psi_{n_2l_2m_2}(r_2)$ belongs to the representation $D^{(l_1)} \times D^{(l_2)}$ of the symmetry group $O(3) \times O(3)$ of H_0.

When the atom is placed in a strong crystal field, the level of the i-th electron is split up according to

$$D^{(l_i)}(K) = \sum_{\alpha}^{r} m_\alpha^{(i)} \Gamma_\alpha(K)\,,$$

where $\Gamma_1, ..., \Gamma_r$ are the irreducible representations of K. The symmetry group of H_0 is $O(3) \times O(3)$. The symmetry is lowered to $K \times K$ by the crystal field, and further to K by the interaction between the electrons. The representation $D^{(l_1)} \times D^{(l_2)}$ of $O(3) \times O(3)$ subduces a representation of $K \times K$ which can be decomposed as

$$D^{(l_1)} \times D^{(l_2)}(K \times K) = \sum_{\alpha\beta} m_\alpha^{(1)} m_\beta^{(2)} \Gamma_{\alpha\beta}(K \times K)\,,$$

whereas the subduced representation $\Gamma_{\alpha\beta}(K)$ can be reduced into

$$\Gamma_{\alpha\beta}(K) = \sum_\gamma m_\gamma^{\alpha,\beta}\Gamma_\gamma(K)\ .$$

Therefore, the level E characterized by n_1, l_1, n_2, l_2 is split up into $\Sigma_{\alpha\beta}\, m_\alpha^{(1)} m_\beta^{(2)}$ sublevels characterized by α, β and each of these sublevels splits up into $\Sigma_\gamma\, m_\gamma^{\alpha,\beta}$ sublevels.

As an example, consider 2 d-electrons ($l = 2$) in different shells. The five-fold degenerate level of each electron is split up into two sublevels when placed in a crystal field with cubic symmetry, because $D^{(2)}(K) = \Gamma_3 \oplus \Gamma_5$. Then the 25-fold degenerate two-electron level is split up into four sublevels corresponding, respectively, to $\Gamma_{33}, \Gamma_{35}, \Gamma_{53}$ and Γ_{55}. Restricted to the group $K \cong O$, the splitting of these four levels is determined by the reductions

$$\Gamma_3 \otimes \Gamma_3 = \Gamma_1 \oplus \Gamma_2 \oplus \Gamma_3$$

$$\Gamma_3 \otimes \Gamma_5 = \Gamma_5 \otimes \Gamma_3 = \Gamma_4 \oplus \Gamma_5$$

$$\Gamma_5 \otimes \Gamma_5 = \Gamma_1 \oplus \Gamma_3 \oplus \Gamma_4 \oplus \Gamma_5\ ,$$

as is easily seen from the character table of O. The splitting of the energy levels by the subsequent switching on of the interactions W and V can be visualized in a level scheme (fig. 3.3).

When the atom is placed in a crystal field of intermediate strength, the symmetry $O(3) \times O(3)$ is lowered to $O(3)$ by the interaction between the electrons, and further to K by the crystal field potential. The representation of $O(3)$ subduced from $D^{(l_1)} \times D^{(l_2)}$ of $O(3) \times O(3)$ can be decomposed into irreducible representations $D^{(L)}$.

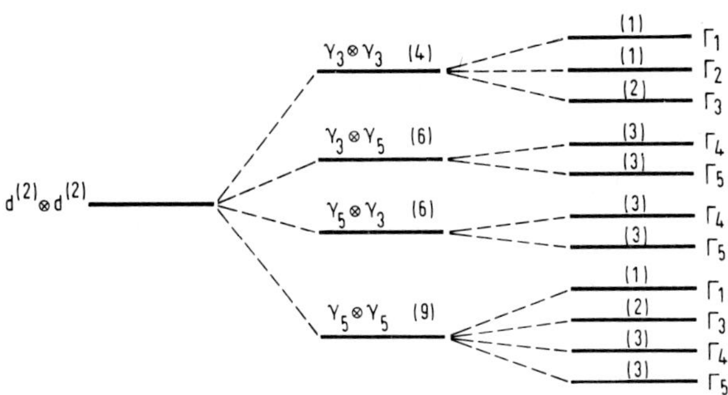

Fig. 3.3. Splitting of a level of 2 nonequivalent electrons in a strong crystal field of cubic symmetry.

$$D^{(l_1)} \otimes D^{(l_2)} = \sum_L m_L^{(l_1,l_2)} D^{(L)} .$$

The representation of K subduced from the representation $D^{(L)}$ of $O(3)$ decomposes into

$$D^{(L)}(K) = \sum_\alpha m_\alpha^{(L)} \Gamma_\alpha(K) .$$

Consequently, the $(2l_1+1)(2l_2+1)$-fold degenerate level characterized by n_1, l_1, n_2, l_2 is split up into $\Sigma_L\, m_L^{(l_1,l_2)}$ sublevels which are $(2L+1)$-fold degenerate and each of these levels is split up into $\Sigma_\alpha\, m_\alpha^{(L)}$ sublevels.

As an example we consider again an atom with two d-electrons in a field of cubic symmetry. From the interaction between the electrons, the 25-fold degenerate level splits up. From the theory of representations of $O(3)$ (cf. e.g. Hamermesh [1962]) one has the following decomposition.

$$D^{(2)} \otimes D^{(2)} = D^{(0)} \oplus D^{(1)} \oplus D^{(2)} \oplus D^{(3)} \oplus D^{(4)} ,$$

according to the general formula

$$D^{(l_1)} \otimes D^{(l_2)} = \sum_{|l_1-l_2|}^{l_1+l_2} D^{(L)} .$$

For the subsequent splitting by the crystal field one can use the decomposition of $D^{(L)}(K \cong O)$ given in the preceding section. The level splitting by the subsequent switching-on of the various terms in the Hamiltonian is visualized

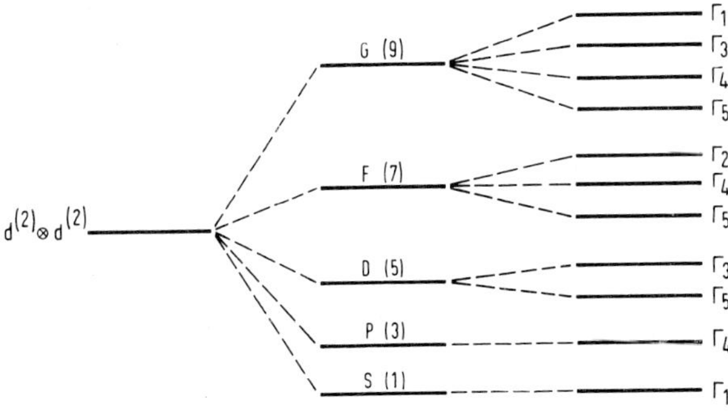

Fig. 3.4. Splitting of a level of 2 nonequivalent d-electrons in an intermediate crystal field of cubic symmetry.

by the level scheme in fig. 3.4. Of course, the number and the kind of the different sublevels is the same for a strong field as for an intermediate field, since they are found from a decomposition of the subduced representation $D^{(l_1)} \times D^{(l_2)}(K)$ into irreducible components; and this decomposition is unique, so can not depend on the steps of the different reductions. However, the distance of the levels and their ordering on the energy scale will be different for different fields. The magnitude of the splittings cannot be found in this way. One has to perform explicit calculations. Using perturbation theory one can also apply other group-theoretical methods as treated in Ch 2, §2.4. We will not treat this aspect here.

3.3.4. Levels of equivalent electrons and the influence of the Pauli principle

In the foregoing sections we did not take into account the fact that electrons are fermions. This was possible, as we will show, because we treated the case of electrons belonging to different levels. In general, however, we have to take into account the Pauli exclusion principle. The wave function of a system of electrons does not belong merely to the tensor product of one-particle Hilbert spaces, but it must be a totally antisymmetric wave function too. Consider first a system of two electrons in a centrally symmetric potential. We denote the Hamiltonian of the system by $H(1,2) = H_0(1,2) + V(1,2)$, where $H_0(1,2) = H(1) + H(2)$ and $V(1,2)$ is the interaction between the electrons. Apart from the group $O(3) \times O(3)$, the Hamiltonian $H_0(1,2)$ also has permutation symmetry, because one can interchange the particles without changing the Hamiltonian. The permutation operator P acts on $H(1,2)$ as $PH(1,2)P^{-1} = H(2,1)$. Hence the symmetry group S contains both $O(3) \times O(3)$ and the group of order two generated by P, but it is not the direct product of the two groups, as P does not commute with an element $(g,h) \in O(3) \times O(3)$ unless $g = h$. Let $\phi_{\alpha i}$ $(i = 1, ..., d_\alpha)$ denote the one-particle eigenfunctions of $H(1)$ or $H(2)$. A basis for the two-particle space is given by $\phi_{\alpha i}(1)\,\phi_{\beta j}(2)$.

First we consider two different levels denoted by α and β. Here we assume that the space generated by $\phi_{\alpha i}$ $(i = 1, ..., d_\alpha)$ is different from the space generated by $\phi_{\beta j}$. A basis for the eigenspace of the eigenvalue $E_\alpha + E_\beta$ of $H_0(1,2)$ is given by the functions $\phi_{\alpha i}(1)\phi_{\beta j}(2)$ and $\phi_{\beta j}(1)\,\phi_{\alpha i}(2)$. The dimension of this space is $2d_\alpha d_\beta$. This space is invariant under the group S generated by P and $O(3) \times O(3)$. It carries a representation of this group which, with respect to the basis given before, has the matrix form

$$D(g_1, g_2) = \begin{pmatrix} D_\alpha(g_1) \otimes D_\beta(g_2) & 0 \\ 0 & D_\alpha(g_2) \otimes D_\beta(g_1) \end{pmatrix}$$

$$D(P) = \begin{pmatrix} 0 & \mathbb{1} \\ \mathbb{1} & 0 \end{pmatrix} .$$

The given basis functions cannot be wave functions for the two-electron system as they are not antisymmetric. One can choose another basis, with functions $(\phi_{\alpha i}(1)\,\phi_{\beta j}(2) \pm \phi_{\alpha i}(2)\,\phi_{\beta j}(1))/\sqrt{2}$, which are either symmetric or antisymmetric. With respect to this basis, the representation is given by the matrices

$$D(g_1, g_2) = \tfrac{1}{2} \begin{pmatrix} D_\alpha(g_1)\otimes D_\beta(g_2) + D_\alpha(g_2)\otimes D_\beta(g_1) & D_\alpha(g_1)\otimes D_\beta(g_2) - D_\alpha(g_2)\otimes D_\beta(g_1) \\ D_\alpha(g_1)\otimes D_\beta(g_2) - D_\alpha(g_2)\otimes D_\beta(g_1) & D_\alpha(g_1)\otimes D_\beta(g_2) + D_\alpha(g_2)\otimes D_\beta(g_1) \end{pmatrix}$$

$$D(P) = \begin{pmatrix} \mathbb{1}_n & 0 \\ 0 & -\mathbb{1}_{n'} \end{pmatrix} \qquad (n = d_\alpha d_\beta = n') .$$

We notice that the symmetric and antisymmetric functions form a basis of the $2d_\alpha d_\beta$-dimensional space, because $\phi_{\alpha i}(1)\,\phi_{\beta j}(2)$ can never be equal to $\phi_{\beta j}(1)\,\phi_{\alpha i}(2)$ because the spaces of antisymmetric and of symmetric functions are not invariant under $O(3) \times O(3)$. This means that in the space of antisymmetric functions the invariance group can not be $O(3) \times O(3)$. The Pauli principle effectively acts as a repulsive force. Therefore, as soon as we consider fermions the symmetry is the same as when we introduce the interaction $V(1, 2)$. Now, the symmetry of $H(1, 2)$ is $O(3) \times C_2$, where C_2 is the group generated by P. The representation of this group subduced from the group S is given by the matrices

$$D(g) = D(g, g) = \begin{pmatrix} D_\alpha(g) \otimes D_\beta(g) & 0 \\ 0 & D_\alpha(g) \otimes D_\beta(g) \end{pmatrix}$$

$$D(P) = \begin{pmatrix} \mathbb{1} & 0 \\ 0 & -\mathbb{1} \end{pmatrix} .$$

The product representation $D_\alpha \otimes D_\beta$ can be decomposed into irreducible representations as before. For electrons, one has only to consider the subspace of antisymmetric functions which is now an invariant subspace, and so carries a representation of $O(3) \times C_2$. In this subspace one has

$$D(g) = D_\alpha(g) \otimes D_\beta(g) = \sum_\gamma m_{\gamma-} D_\gamma^-(g)$$

$$D(Pg) = -D(g)\,,$$

where the multiplicity $m_{\gamma-}$ (the multiplicities $m_{\gamma+}$ vanish all) is determined from the reduction of $(D_\alpha \otimes D_\beta)(G)$, i.e. entirely from the subgroup $O(3)$. Hence in this case one can forget about the Pauli principle for the determination of the level splitting. For a strong field the situation is analogous.

The situation is entirely different if the spaces spanned by $\phi_{\alpha i}$ and $\phi_{\beta j}$ are the same ($\alpha = \beta$). In this case a basis of two-particle functions at the level $2E_\alpha$ is given by $\phi_{\alpha i}(1)\,\phi_{\alpha j}(2)$. This d_α^2-dimensional space is invariant under the group S generated by P and $O(3) \times O(3)$. However, the $\frac{1}{2}d_\alpha(d_\alpha - 1)$-dimensional subspace of antisymmetric eigenfunctions is not an invariant subspace. A basis for this space is $(\phi_{\alpha i}(1)\,\phi_{\alpha j}(2) - \phi_{\alpha i}(2)\,\phi_{\alpha j}(1))/\sqrt{2}$ with $1 \leqslant i < j \leqslant d_\alpha$. A basis for the $\frac{1}{2}d_\alpha(d_\alpha + 1)$-dimensional subspace of symmetric functions is given by $(\phi_{\alpha i}(1)\,\phi_{\alpha j}(2) + \phi_{\alpha i}(2)\,\phi_{\alpha j}(1))/\sqrt{2}\,(1 \leqslant i \leqslant j \leqslant d_\alpha)$. The subduced representation of the group $O(3) \times P$ which is a subgroup of S leaving both spaces invariant is given by

$$D(g) = D_\alpha(g) \otimes D_\alpha(g) \quad \text{with respect to the first basis,}$$

$$D(P) = \begin{pmatrix} \mathbb{1}_n & 0 \\ 0 & -\mathbb{1}_{n'} \end{pmatrix} \quad \text{with respect to the second basis,}$$

$$(n = \tfrac{1}{2}d_\alpha(d_\alpha + 1),\, n' = \tfrac{1}{2}d_\alpha(d_\alpha - 1)\,.$$

The character of the subduced representation is given by $\chi(g) = [\chi_\alpha(g)]^2$ and $\chi(P) = d_\alpha$. An arbitrary representation of $O(3) \times C_2$ can be reduced into irreducible components $D_\beta^\pm$. However, we are more interested in the representation carried by the space of symmetric or antisymmetric functions. The representation carried by the (anti)symmetric functions is called the *(anti)-symmetrized Kronecker product*. The character of these representations are found from

$$T_g\{\phi_{\alpha i}(1)\,\phi_{\alpha j}(2) \pm \phi_{\alpha i}(2)\,\phi_{\alpha j}(1)\}$$

$$= \sum_{kl} (D_\alpha \otimes D_\alpha)(g)_{kl,\,ij}\,\{\phi_{\alpha k}(1)\,\phi_{\alpha l}(2) \pm \phi_{\alpha k}(2)\,\phi_{\alpha l}(1)\}\,.$$

Therefore, the character of the representation carried by the space of antisymmetric functions is

$$\chi^-(g) = \sum_{k<l} (D_\alpha \otimes D_\alpha)(g)_{kl,kl} = \tfrac{1}{2} \sum_{kl} (D_\alpha \otimes D_\alpha)(g)_{kl,kl} - \tfrac{1}{2} \sum_{k} D_\alpha^2(g)_{kk} ,$$

whereas the character for the space of symmetric functions is

$$\chi^+(g) = \sum_{k\leqslant l} (D_\alpha \otimes D_\alpha)(g)_{kl,kl} = \tfrac{1}{2} \sum_{kl} (D_\alpha \otimes D_\alpha)(g)_{kl,kl} + \tfrac{1}{2} \sum_{k} D_\alpha^2(g)_{kk} .$$

Hence

$$\chi^\pm(g) = \tfrac{1}{2}(\chi_\alpha(g)^2 \pm \chi_\alpha(g^2)) \tag{3.9}$$

$$\chi^\pm(P) = \tfrac{1}{2} d_\alpha (d_\alpha \pm 1) .$$

The level splitting of a system of two electrons is determined by the reduction of the representation in the space of antisymmetric functions, when the spin is neglected or when the spin of the system forms a triplet state (even in spin space). It is determined by the reduction of the representation in the symmetric space, when the spins form a singlet (odd) state.

As an example we consider again a system with 2 d-electrons. Now we assume that they are in states with the same principal quantum number. Also we again take the case of the atom placed in a crystal field of cubic symmetry. The Hamiltonian is $H(1,2) = H_0(1,2) + W(1,2) + V(1,2)$, where H_0 is the Hamiltonian of the two electrons in the spherically symmetric potential of the core, W is the (strong) crystal field potential and $V(1,2)$ is the interaction between the electrons. The invariance group of H_0 is the group generated by $O(3) \times O(3)$ and P. The eigenspace with eigenvalue $2E_\alpha$ carries a 25-dimensional representation of this group with basis functions $\phi_{\alpha i}(r_1)\,\phi_{\alpha j}(r_2)$ where $i, j = 1, ..., 5$. The space is irreducible under $O(3) \times O(3)$ and a fortiori under the whole invariance group. The representation restricted to the subgroup $K \times K$ is reducible: according to §3.3 it has four irreducible components: $\Gamma_{33}, \Gamma_{35}, \Gamma_{53}$, and Γ_{55}. The spaces carrying the representations Γ_{33} and Γ_{55} are also invariant under P, and consequently they carry irreducible representations of the invariance group of $H_0 + W$: the group generated by $K \times K$ and P. However, the spaces carrying the representations Γ_{53} and Γ_{35} are interchanged by P. Hence their direct sum carries an irreducible representation of the invariance group of $H_0 + W$. We draw the conclusion that by the crystal field the 25-fold degenerate level of H_0 splits up into three levels of degeneracy 4, 9 and 12 respectively. By the interaction V these levels are again split up.

The invariance group of H is the direct product $K \times C_2$, where C_2 is the group generated by P. Let us consider the three levels already found separately.
1) The space carrying the irreducible representation Γ_{33} carries a reducible representation of $K \times C_2$. The subduced representation is given by

$$D(g) = \Gamma_3(g) \otimes \Gamma_3(g)$$

$$D(P) = \begin{pmatrix} 1 & 0 & 0 & 0 \\ 0 & 0 & 1 & 0 \\ 0 & 1 & 0 & 0 \\ 0 & 0 & 0 & 1 \end{pmatrix} .$$

On the subspace of symmetric functions the subduced representation of $K \times C_2$ is given by the character (cf. eq. (3.9))

$$\chi(g) = \tfrac{1}{2}(\chi_3(g)^2 + \chi_3(g^2))$$

$$\chi(P) = \tfrac{1}{2} d_3(d_3+1) = 3 ,$$

whereas the representation in the subspace of antisymmetric functions is given by the character

$$\chi(g) = \tfrac{1}{2}(\chi_3(g)^2 - \chi_3(g^2))$$

$$\chi(P) = -1 .$$

For the group $K \cong 432 \cong O$ the character is $\chi(\epsilon, \beta, \alpha^2, \alpha, \alpha\beta) = (3, 0, 3, 1, 1)$ on the symmetric space and $(1, 1, 1, -1, -1)$ on the antisymmetric space. This means that the subduced representation of $K \times C_2$ on the space of symmetric functions reduces into $\Gamma_1^+ \oplus \Gamma_3^+$, whereas that on the space of antisymmetric functions is Γ_2^-. Hence

$$D(K \times C_2) = \Gamma_1^+ \oplus \Gamma_2^- \oplus \Gamma_3^+ .$$

2) In an analogous way one finds that the 9-dimensional representation subduced from Γ_{55} has four irreducible components

$$D(K \times C_2) = \Gamma_1^+ \oplus \Gamma_3^+ \oplus \Gamma_4^+ \oplus \Gamma_5^- .$$

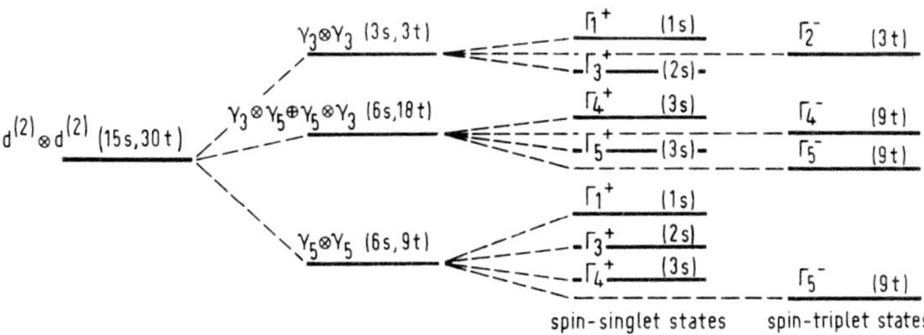

Fig. 3.5. Splitting of a level of 2 equivalent d-electrons in a strong cubic crystal field.

3) The case of the direct sum of the representations Γ_{35} and Γ_{53} corresponds to the case of nonequivalent electrons: one electron is in the state characterized by the representation Γ_{35}, the other in that of Γ_{53}. It splits into a 6-dimensional space of symmetric functions and a 6-dimensional space of antisymmetric functions. Hence the subduced representation of $K \times C_2$ reduces according to

$$D(K \times C_2) = \Gamma_4^+ + \Gamma_5^+ + \Gamma_4^- + \Gamma_5^- .$$

The level scheme of the subsequent splitting under the influence of W and V is given in fig. 3.5.

We considered in this section both symmetric and antisymmetric functions, although the particles involved, the electrons, are fermions. The reason for our interest in symmetric functions is the fact that the wave function of a pair of electrons is either the product of an antisymmetric spatial wave function and a symmetric spin wave function (triplet spin state), or the product of a symmetric spatial wave function and an antisymmetric spin wave function (singlet spin state). In this way the spin enters into our symmetry considerations, although we neglected spin-orbit interaction here. A proof for our assertion is the following. If $\mathcal{H}_r(i)$ is the Hilbert space of spatial wave functions of particle i, and $\mathcal{H}_s(i)$ its two-dimensional spin space ($i = 1, 2$), the Hilbert space of the two-electron system is the subspace of antisymmetric functions of the space $\mathcal{H}_r(1) \otimes \mathcal{H}_s(1) \otimes \mathcal{H}_r(2) \otimes \mathcal{H}_s(2)$. If $\mathcal{H}_r(1) \otimes \mathcal{H}_r(2)$ carries a representation D_r of the permutation group S_2, and if $\mathcal{H}_s(1) \otimes \mathcal{H}_s(2)$ carries a representation D_s, the product space carries a representation $D_r \otimes D_s$. Antisymmetric functions transform according to the representation Γ_2. Since $\Gamma_1 \otimes \Gamma_1 = \Gamma_2 \otimes \Gamma_2 = \Gamma_1$ and $\Gamma_1 \otimes \Gamma_2 = \Gamma_2$, an antisymmetric function has either a symmetric spatial part and an antisymmetric spin part or the inverse.

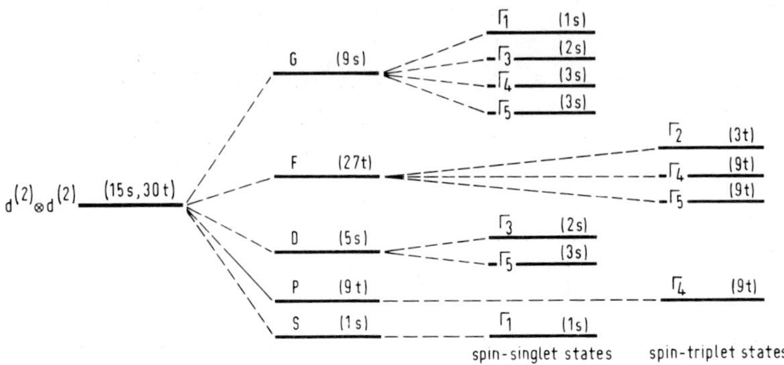

Fig. 3.6. Splitting of a level of 2 equivalent d-electrons in an intermediate crystal field of cubic symmetry.

Returning to our example and taking into account the spin degeneracy of the two-electron system with electrons in d-states, the total degeneracy is 45 for the d^2-level, 6 for the Γ_{33}-level, 15 for the Γ_{55}-level and 24 for the $\Gamma_{35} \oplus \Gamma_{53}$-level.

For an intermediate field, the invariance groups of H_o, $H_o + V$ and H are the group generated by $O(3) \times O(3)$ and P, the group $O(3) \times C_2$ and the group $K \times C_2$, respectively. The corresponding level scheme is given in fig. 3.6. It must be stressed that the figs. 3.5 and 3.6 only visualize the way the levels split, but that they are not realistic examples of level schemes. From other arguments, which involve not only group theory, one can derive general rules (e.g. Hunds rule) about the order of the energy levels. For more details we refer to books on crystal field theory, e.g. Griffith [1961].

The generalization to a system of n electrons is now straightforward. Consider n electrons in the same shell of an ion in a strong crystal field. The invariance group of $H(1,2,...,n) = H_o(1,2,...,n) + W(1,2,...,n) + V(1,2,...,n)$ is K. The symmetry group of H_o is generated by $O(3)^n$ and S_n, the permutation group of n elements. By the crystal field the symmetry of $H_o + W$ is lowered to the group generated by K^n and S_n. Because of the interaction V between the electrons, the symmetry of H is $K \times S_n$. Suppose the n electrons are in the same level of $H_o + W$ which belongs to an irreducible representation of its symmetry group. The eigenspace $\mathcal{H}_d$ carries a reducible representation of S_n. The component $\mathcal{H}_d^a$ which carries the fully antisymmetric representation of S_n (the representation with $+\mathbb{1}$ for even permutations and $-\mathbb{1}$ for odd ones) is invariant under K. The space $\mathcal{H}_d$ carries the n-fold Kronecker product of a representation $D_\alpha(K)$. The representation $D_\alpha(K) \otimes \ldots \otimes D_\alpha(K)$ (with n factors) restricted to $\mathcal{H}_d^a$ is called the antisymmetrized n-fold Kronecker product of $D_\alpha(K)$. In the same way the space which carries the trivial representation (the fully symmetric representation) of S_n carries also the symmetrized n-fold Kronecker product of $D_\alpha(K)$. However, if $n > 2$ one cannot assert that the spatial wave function belongs to either the symmetrized or the antisymmetrized Kronecker product, because in this case it is in general not possible to write the total wave function as a product of a symmetric spatial part and an antisymmetric spin part or the inverse. E.g. for a 3 particle system, the wave function in the product space must belong to the representation D_2 of S_3. As one sees from Ch. 1, §3.4 the representation D_2 occurs in the products $D_1 \otimes D_2, D_2 \otimes D_1$ and $D_3 \otimes D_3$. When the space $\mathcal{H}_r(i)$ is two-dimensional, as is the space $\mathcal{H}_s(i)$, neither $\mathcal{H}_r(1) \otimes \mathcal{H}_r(2) \otimes \mathcal{H}_r(3)$ nor $\mathcal{H}_s(1) \otimes \mathcal{H}_s(2) \otimes \mathcal{H}_s(3)$ can carry a representation D_2 of S_3. Hence the three-electron wave function must belong to $D_3 \otimes D_3$, i.e. both its spatial part and its spin part belong to the representation D_3. Of course, the n-electron wave function belongs to the antisymmetrized n-fold Kronecker product of the representation carried by $\mathcal{H}_r \otimes \mathcal{H}_s$.

3.3.5. Symmetry adapted functions

Consider again the problem of level splitting for one electron in an atom in a crystal field. For the unperturbed Hamiltonian H_0, the invariance group is $O(3)$. Then an energy level space carries a representation of $O(3)$, which is the representation $D^{(l)}$ in the case of natural degeneracy. Basis functions for this representation are the spherical harmonics Y_l^m. When the atom is placed in a crystal, the symmetry is lowered to the point group K. The subduced representation $D^{(l)}(K)$ is, in general, reducible, but this does not mean that the matrices are in reduced form with respect to this basis. A basis with respect to which the matrices are in reduced form is obtained by linear combinations of the spherical harmonics. This new basis will be a better zeroth order approximation to the eigenfunctions of H (according to Ch. 2, §2.3). When K is the cubic group $m3m$, the new basis functions are called *cubic harmonics*. In general, the functions which reduce $D^{(l)}(K)$ for a point group K are called *lattice harmonics* or *symmetry adapted functions*. They can be found by application of the projection operators from Ch. 2, §1.4 and they are tabulated for several crystallographic point groups and for values up to $l = 30$ in Bell [1954], Altmann and Bradley [1965], Altmann and Cracknell [1965], Puff [1970].

Table 3.5
Cubic harmonics for $l \leqslant 4$.

l	Irreducible components $D^l(K)$	Basis functions of the reduced representation
0	Γ_1	Y_0^0
1	Γ_4	Y_1^0, Y_1^1, Y_1^{-1}
2	Γ_3	$Y_2, \frac{1}{2}\sqrt{2}(Y_2^2 + Y_2^{-2})$
	Γ_5	$-i\sqrt{\pi}(Y_2^2 - Y_2^{-2}), \sqrt{\pi}(Y_2^1 + Y_2^{-1}), -i\sqrt{\pi}(Y_2^1 - Y_2^{-1})$
3	Γ_2	$-i\sqrt{\pi}(Y_3^2 - Y_3^{-2})$
	Γ_4	$\sqrt{\pi}(Y_3^2 + Y_3^{-2}), \sqrt{3\pi/8}(Y_3^3 + Y_3^{-3}) + \sqrt{5\pi/8}(Y_3^1 + Y_3^{-1}),$ $-i\sqrt{3\pi/8}(Y_3^3 - Y_3^{-3}) + i\sqrt{5\pi/8}(Y_3^1 - Y_3^{-1})$
	Γ_5	$\sqrt{2\pi}Y_3^0, \sqrt{5\pi/8}(Y_3^3 + Y_3^{-3}) - \sqrt{3\pi/8}(Y_3^1 + Y_3^{-1})$ $-i\sqrt{5\pi/8}(Y_3^3 - Y_3^{-3}) - i\sqrt{3\pi/8}(Y_3^1 - Y_3^{-1})$
4	Γ_1	$\sqrt{7\pi/6}\, Y_4^0 + \sqrt{5\pi/12}\,(Y_4^4 + Y_4^{-4})$
	Γ_3	$\sqrt{\pi}(Y_4^2 + Y_4^{-2}), \sqrt{5\pi/6}\, Y_4^0 - \sqrt{7\pi/12}\,(Y_4^4 + Y_4^{-4})$
	Γ_4	$-i\sqrt{\pi}(Y_4^2 - Y_4^{-2}), \sqrt{\pi/8}\,(Y_4^1 + Y_4^{-1}) + \sqrt{7\pi/8}(Y_4^3 + Y_4^{-3}),$ $-i\sqrt{\pi/8}\,(Y_4^1 - Y_4^{-1}) + i\sqrt{7\pi/8}\,(Y_4^3 - Y_4^{-3})$
	Γ_5	$\sqrt{\pi}(Y_4^4 + Y_4^{-4}), \sqrt{7\pi/8}\,(Y_4^1 + Y_4^{-1}) - \sqrt{\pi/8}\,(Y_4^3 + Y_4^{-3}),$ $-i\sqrt{7\pi/8}\,(Y_4^1 - Y_4^{-1}) - i\sqrt{\pi/8}\,(Y_4^3 - Y_4^{-3})$

As an example we give in table 3.5 the cubic harmonics for $l = 0, 1, 2, 3, 4$. The reduction of $D^{(l)}$ for the group $m3m$ has already been determined in §3.2.

CHAPTER IV

SPACE GROUPS

In this chapter we will treat another class of groups: groups of inhomogeneous spatial transformations which describe the symmetry of crystals. Their elements are translations, orthogonal transformations or combinations of both. The study of these so called space groups goes back to the 19th century when Schoenflies [1891] and von Fedorow [1892] determined the three-dimensional space groups. The general theory of n-dimensional space groups was given by Bieberbach [1911, 1912]. A nice derivation and a discussion of their properties can be found in Burckhardt [1966]. The first section of the present chapter is devoted to the general properties. The second section discusses the representations of space groups. Their determination was given for the first time in Bouckaert et al. [1936] and more general in Wintgen [1941]. A more recent treatment can be found in Koster [1957] and Bradley and Cracknell [1971]. The third section deals with some simple applications of the representations. A more complete discussion is postponed till Ch. 6. The representations of space groups are published in tabular form in Miller and Love [1967] and in Zak [1969]. Diagrams and many properties of the space groups are tabulated in the International Tables for X-ray Crystallography: Henry and Lonsdale [1965].

4.1. Properties of space groups

4.1.1. The Euclidean group $E(3)$

Consider a three-dimensional real vector space V. We define for any nonsingular linear transformation S and translation $\boldsymbol{t}$ a transformation of V by

$$\{S|\boldsymbol{t}\}\boldsymbol{x} = S\boldsymbol{x} + \boldsymbol{t} \qquad (\text{any } \boldsymbol{x} \in V) .$$

The product of two such transformations is defined by their successive action

$$(\{S_1|t_1\}\{S_2|t_2\})x = \{S_1|t_1\}(\{S_2|t_2\}x) = S_1S_2x + S_1t_2 + t_1$$

or

$$\{S_1|t_1\}\{S_2|t_2\} = \{S_1S_2|S_1t_2 + t_1\} . \tag{4.1}$$

It is easily verified that the set of elements $\{S|t\}$ forms a group under this multiplication rule. It is called the *affine group* $A(3)$. Notice that the elements of $A(3)$ are not linear transformations: for $x, y \in V$ one does not have $g(x+y) = gx + gy$! The unit element of $A(3)$ is $\{\mathbb{1}|O\}$. A subgroup of $A(3)$ is formed by the elements $\{\mathbb{1}|t\}$ (for any translation t). This subgroup, isomorphic with the three-dimensional group $T(3)$, is Abelian and invariant because for any $\{S|t\} \in A(3)$ and any $\{\mathbb{1}|t'\} \in T(3)$ one has

$$\{S|t\}\{\mathbb{1}|t'\}\{S|t\}^{-1} = \{S|t + St'\}\{S^{-1}|-S^{-1}t\} = \{\mathbb{1}|St'\} \in T(3) .$$

Another subgroup is the group of elements $\{S|O\}$ which is the group of non-singular transformations $GL(3,\mathbb{R})$. As $T(3)$ is an invariant subgroup, one can consider the factor group $A(3)/T(3)$. Two elements $\{S_1|t_1\}$ and $\{S_2|t_2\}$ belong to the same coset of $T(3)$, if there is an element $\{\mathbb{1}|u\} \in T(3)$ such that

$$\{S_1|t_1\} = \{\mathbb{1}|u\}\{S_2|t_2\} = \{S_2|u + t_2\} .$$

This means that they belong to the same coset if and only if $S_1 = S_2$. The multiplication in the factor group is defined by: the product of the coset of $\{S_1|t_1\}$ and the coset of $\{S_2|t_2\}$ is the coset of $\{S_1S_2|t_1 + S_1t_2\}$. Hence

$$A(3)/T(3) \cong GL(3,\mathbb{R}) . \tag{4.2}$$

A group with such a structure, partially recalling the structure of a direct product, is called a semidirect product. A group G is the *semidirect product* of two groups A and B if
1) for any $b \in B$ there is an automorphism $\varphi(b)$ of A such that

$$\varphi(b_1)\{\varphi(b_2)a\} = \varphi(b_1b_2)a \qquad (\text{any } a \in A) ,$$

2) there is an isomorphism between G and the group of pairs (a, b) with $a \in A$ and $b \in B$ having the multiplication law

$$(a_1, b_1)(a_2, b_2) = (a_1[\varphi(b_1)a_2], b_1b_2) .$$

When we use an additive notation for A (which does not mean that A is Abelian) and write ba for $\varphi(b)a$, the last formula reads

$$(a_1, b_1)(a_2, b_2) = (a_1 + b_1 a_2, b_1 b_2) .$$

The semidirect product of A and B is denoted by $A \boxtimes B$ or by $A \boxtimes_\varphi B$ when we wish to show explicitly the dependence on the action $\varphi(B)$. The *direct product* is a special case of the semidirect product, where the automorphism $\varphi(b)$ is the identity: $\varphi(b)a = a$ for any $a \in A$, $b \in B$. In the present case we have

$$A(3) = T(3) \boxtimes GL(3, \mathbb{R}) , \tag{4.3}$$

where $\varphi(S)\boldsymbol{t}$ is defined as $S\boldsymbol{t}$.

We now suppose that V is a three-dimensional *Euclidean vector space*, i.e. a real three-dimensional linear vector space with a positive definite inner product. One can define the distance between two points $\boldsymbol{x}$ and $\boldsymbol{y}$ as the norm of their difference: $d(\boldsymbol{x},\boldsymbol{y}) = (\boldsymbol{x}-\boldsymbol{y}, \boldsymbol{x}-\boldsymbol{y})^{1/2}$. The elements of the translation group $T(3)$ leave the distance between any pair of points invariant. One defines the *Euclidean group* $E(3)$ as the subgroup of elements of $A(3)$ keeping the distance of any two points invariant. It can be shown that all distance preserving mappings belong to $A(3)$. The element $g = \{S|\boldsymbol{t}\}$ does so if $d(g\boldsymbol{x}, g\boldsymbol{y}) = d(\boldsymbol{x},\boldsymbol{y})$ for any $\boldsymbol{x},\boldsymbol{y} \in V$. As this is equivalent to $(\boldsymbol{x}-\boldsymbol{y}, \boldsymbol{x}-\boldsymbol{y}) = (S\boldsymbol{x}-S\boldsymbol{y}, S\boldsymbol{x}-S\boldsymbol{y})$ this means that $\{S|\boldsymbol{t}\} \in E(3)$ if and only if $S \in O(3)$. Therefore, the Euclidean group $E(3)$ is the group of elements $\{R|\boldsymbol{t}\}$ with $R \in O(3)$ and $\boldsymbol{t} \in T(3)$. As for the affine group, one has the properties:

1) $T(3)$ is an Abelian invariant subgroup of $E(3)$,
2) $E(3)/T(3) \cong O(3)$,
3) $E(3)$ is the semidirect product of $T(3)$ and $O(3)$.

The elements of $E(3)$ which leave the origin (null vector) fixed form the subgroup $O(3)$. The elements which leave a point $\boldsymbol{x}$ fixed also form a subgroup. If $\{R|\boldsymbol{t}\}$ leaves $\boldsymbol{x}$ fixed and if $\boldsymbol{x}$ is obtained from the origin $\boldsymbol{x}_0$ by a translation $\{\mathbb{1}|\boldsymbol{u}\}$ one has

$$\{R|\boldsymbol{t}\}\{\mathbb{1}|\boldsymbol{u}\}\boldsymbol{x}_0 = \{\mathbb{1}|\boldsymbol{u}\}\boldsymbol{x}_0 .$$

Then $\{\mathbb{1}|\boldsymbol{u}\}^{-1}\{R|\boldsymbol{t}\}\{\mathbb{1}|\boldsymbol{u}\} = \{R|\boldsymbol{t}+R\boldsymbol{u}-\boldsymbol{u}\}$ is an element of $E(3)$ leaving the origin fixed and is therefore an element of $O(3)$. From this it follows that $\boldsymbol{t} = \boldsymbol{u} - R\boldsymbol{u}$. The elements $\{R|\boldsymbol{u}-R\boldsymbol{u}\}$, which leave $\boldsymbol{x} = \boldsymbol{x}_0 + \boldsymbol{u}$ fixed, form the *orthogonal group of* $\boldsymbol{x}$. It is denoted by $O_{\boldsymbol{x}}(3)$ and it is isomorphic to $O(3)$. Notice that each element $\{R\,|\,\boldsymbol{t}\} \in E(3)$ can be written as

$$\{R\,|\,\boldsymbol{t}\} = \{\mathbb{1}\,|\,\boldsymbol{t}-\boldsymbol{u}+R\boldsymbol{u}\}\{R\,|\,\boldsymbol{u}-R\boldsymbol{u}\}\,.$$

This means that the element $\{R\,|\,\boldsymbol{t}\}$ can be obtained by the successive action of the element $\{R\,|\,\boldsymbol{u}-R\boldsymbol{u}\} \in O_{\boldsymbol{x}}(3)$ and the translation $\{\mathbb{1}\,|\,\boldsymbol{t}-\boldsymbol{u}+R\boldsymbol{u}\}$. In fact, we have performed a change of origin.

The elements of $A(3)$, although not linear transformations, can be brought into one-to-one correspondence with the elements of a group of linear transformations in four dimensions. To the element $\{S\,|\,\boldsymbol{t}\}$ corresponds the transformation of the space $V \oplus \mathrm{R}$ (with elements $(\boldsymbol{x}, w)$) given by $\{S\,|\,\boldsymbol{t}\}(\boldsymbol{x}, w) = (S\boldsymbol{x}+w\boldsymbol{t}, w)$. A matrix for this linear transformation is

$$\begin{pmatrix} S & t \\ 0 & 1 \end{pmatrix}, \tag{4.4}$$

where $\boldsymbol{t}$ is a three-dimensional column vector. It is easily verified that this correspondence is in fact an isomorphism.

The elements of $E(3)$ are sometimes called *Euclidean motions*. As $E(3)$ is the semidirect product of $T(3)$ and $O(3)$ one could denote the Euclidean motions by $(\boldsymbol{t}, R)$. For crystallography the *Seitz notation* $\{R\,|\,\boldsymbol{t}\}$ is more customary. We will use it here.

4.1.2. Space groups

The atoms in an infinite crystal are arranged in a regular pattern. This means that there is a finite region in space which is repeated in all directions. In other words, there is a *lattice group U*, generated by three linearly independent basis vectors $\boldsymbol{a}_1, \boldsymbol{a}_2, \boldsymbol{a}_3$, which transforms the pattern into itself. Each element of such a lattice is an integral, linear combination of the three basis translation vectors. This means that the group U is isomorphic with $\mathbb{Z}^3$, the additive group of triples of integers. Any element of U is given by three integers, once the basis is chosen.

There may be other Euclidean motions which transform the pattern into itself. All these motions together form the space group of the pattern. We will now give a more rigorous definition of this very important notion.

A *space group G* is a subgroup of $E(3)$, such that its intersection with the translation group $T(3)$ is isomorphic to $\mathbb{Z}^3$ and is generated by three linearly independent basis vectors. We denote the translation subgroup $G \cap T(3)$ by U. Then the definition requires that U is a lattice group.

The *translation subgroup* U is an Abelian invariant subgroup of G. To prove this, take an element $g \in G$ and an element $u \in U$. Since $u \in G$, also $gug^{-1} \in G$. Furthermore, u is a translation and $g \in G \subset E(3)$ is an Euclidean motion. Hence $gug^{-1} \in G \cap T(3) = U$. This means that U is invariant.

The factor group G/U has as elements the cosets of U in G. Again it is easy to see that any two elements $\{R|\boldsymbol{t}\}$ and $\{S|\boldsymbol{v}\}$ of G belong to the same coset if and only if $S = R$. The elements R from all $\{R|\boldsymbol{t}\} \in G$ form the point group $K : K = \{R | \{R|\boldsymbol{t}\} \in G\}$. Therefore $G/U \cong K$.

Because U is an invariant subgroup, for any $\{\mathbb{1}|\boldsymbol{a}\} \in U$ and any $\{R|\boldsymbol{t}\} \in G$ one has

$$\{R|\boldsymbol{t}\}\{\mathbb{1}|\boldsymbol{a}\}\{R|\boldsymbol{t}\}^{-1} = \{\mathbb{1}|R\boldsymbol{a}\} \in U .$$

Consider now the lattice Λ obtained by the action of U on the origin $\boldsymbol{x}_o$. For any $\boldsymbol{x} \in \Lambda$ one has $R\boldsymbol{x} = R(\boldsymbol{x}_o + \boldsymbol{u}) = \boldsymbol{x}_o + R\boldsymbol{u}$ for any $R \in K$ and $\{\mathbb{1}|\boldsymbol{u}\} \in U$. Hence

$$K\Lambda = \Lambda . \tag{4.5}$$

Consequently the point group K is a crystallographic point group, which is of finite order. So any point group of a space group G is one of the groups discussed in Ch. 3. The definitions were given for the three-dimensional case. The generalization to n dimensions will be obvious.

4.1.3. Nonprimitive translations

Because the translation subgroup U is an invariant subgroup of the space group G, the latter can be decomposed into cosets

$$G = U + g_2 U + \ldots + g_N U .$$

The number of cosets is equal to the order N of the point group K. To any $R \in K$ belongs one coset with representative $g = \{R|\boldsymbol{t}_R\}$. The translation $\{\mathbb{1}|\boldsymbol{t}_R\}$ is determined up to an element of U. When $\{\mathbb{1}|\boldsymbol{t}_R\}$ is an element of U, it is called a *primitive translation*. Then one can choose a representative with $\boldsymbol{t}_R = \boldsymbol{O}$. If $\boldsymbol{t}_R$ does not belong to U, it is called a *nonprimitive translation*. A nonprimitive translation $\{\mathbb{1}|\boldsymbol{t}\}$ is never an element of the space group, as from $\{\mathbb{1}|\boldsymbol{t}\} \in G$ follows $\{\mathbb{1}|\boldsymbol{t}\} \in G \cap T(3) = U$ or $\{\mathbb{1}|\boldsymbol{t}\}$ is a primitive translation. The translation $\boldsymbol{t}$ is a real linear combination of the three basis vectors $\boldsymbol{a}_1, \boldsymbol{a}_2, \boldsymbol{a}_3$ of U : $\boldsymbol{t} = t_1\boldsymbol{a}_1 + t_2\boldsymbol{a}_2 + t_3\boldsymbol{a}_3$. When $\boldsymbol{t}$ is primitive, the

3 numbers t_i are integers, when $\boldsymbol{t}$ is nonprimitive, not all t_i are integers. However, one can write $t_i = n_i + \tau_i$ with n_i an integer and $0 \leqslant \tau_i < 1$. The coordinates of $\boldsymbol{t}$ with respect to the basis $\boldsymbol{a}_1, \boldsymbol{a}_2, \boldsymbol{a}_3$ are called *lattice coordinates.*

A set $\{\boldsymbol{t}_R | R \in K\}$ such that $\{R | \boldsymbol{t}_R\}$ are representatives of the cosets is called a *system of nonprimitive translations* (even when all the translations $\boldsymbol{t}_R$ are primitive). A space group is given by its translation subgroup U, determined by a basis $\boldsymbol{a}_1, \boldsymbol{a}_2, \boldsymbol{a}_3$, by its point group K and by a system of nonprimitive translations.

Once a basis for U is chosen one can express the orthogonal transformations $R \in K$ with respect to this basis. Because of eq. (4.5) these are matrices with integral entries: it is an integral three-dimensional representation of the point group K. Also the translations $\boldsymbol{t}_R$ can be expressed with respect to this basis in lattice coordinates, which can be chosen between 0 and 1 because $\boldsymbol{t}_R$ is determined only up to a primitive translation. When all $\boldsymbol{t}_R$ can be chosen to be $\boldsymbol{O}$ the space group is called symmorphic. A more general definition of symmorphic space group will be given in § 1.6.

When $\boldsymbol{t}_R = \boldsymbol{O}$ the orthogonal transformation $\{R | \boldsymbol{O}\}$ is an element of the space group G. However, in general the point group K is not a subgroup of G. The elements of K occur only in combination with nonprimitive translations (in general). When R is a rotation and $\boldsymbol{t}_R$ a translation in the direction of the axis of R, the element $\{R | \boldsymbol{t}_R\}$ represents a *screw motion*. When R is a reflection from a plane and $\boldsymbol{t}_R$ a translation in this plane, $\{R | \boldsymbol{t}_R\}$ is a *glide reflection*. A rotation R combined with a translation $\boldsymbol{t}_R$ perpendicular to the axis of rotation is a rotation around a point different from the origin. As R has eigenvalues 1 and $\exp(\pm i\varphi)$, the equation $(\mathbb{1} - R)\boldsymbol{t} = \boldsymbol{t}_R$ has a solution. Hence $\{R | \boldsymbol{t}_R\}$ is a rotation through an angle φ around the point $\boldsymbol{t}$. A roto-inversion R has eigenvalues -1, $\exp(i\pi \pm i\varphi)$. If $\varphi \neq \pi$ the equation $(\mathbb{1} - R)\boldsymbol{t} = \boldsymbol{t}_R$ has a solution. Hence $\{R | \boldsymbol{t}_R\}$ is then a roto-inversion around $\boldsymbol{t}$. Finally, if R is a reflection, it has 2 eigenvalues 1 and one eigenvalue -1. A nonprimitive translation $\boldsymbol{t}_R$ perpendicular to the mirror plane can be written as $\frac{1}{2}(\mathbb{1} - R)\boldsymbol{t}_R$. Therefore, in this case $\{R | \boldsymbol{t}_R\}$ is a reflection from a mirror plane through $\frac{1}{2}\boldsymbol{t}_R$. So any space group element is either a screw motion, a glide reflection or an orthogonal transformation which leaves invariant a point which can be different from the origin. As for point groups we can indicate the space group elements by diagrams. These can be found in Henry and Lonsdale [1965]. [1965].

4.1.4. Lattices

A lattice group U can be determined by its metric tensor, which is the matrix of the real symmetric bilinear form $g(\boldsymbol{x},\boldsymbol{y}) = \boldsymbol{x}\cdot\boldsymbol{y}$ with respect to a basis $\boldsymbol{a}_1, \boldsymbol{a}_2, \boldsymbol{a}_3$ of U: $g_{ij} = \boldsymbol{a}_i \cdot \boldsymbol{a}_j$ (cf. Ch. 1, §2.2). With respect to another basis, obtained from the first by a nonsingular transformation S, the metric tensor is given by $g' = \widetilde{S}gS$ according to eq. (1.6). When S is an orthogonal transformation, one has $g = \widetilde{S}gS$. Notice that, although S is an orthogonal transformation, it is in general not an orthogonal matrix with respect to the basis $\boldsymbol{a}_1, \boldsymbol{a}_2, \boldsymbol{a}_3$. So U is determined by its metric tensor up to an orthogonal transformation. This is as it should be, as the three vectors $\boldsymbol{a}_i$ have 9 components, whereas g has only 6. The remaining 3 parameters are the parameters of the orthogonal transformation: the orientation of the lattice in space is not determined by g.

When S is a basis transformation which transforms the basis $a = (\boldsymbol{a}_1, \boldsymbol{a}_2, \boldsymbol{a}_3)$ into another basis $a' = aS$ of U, the matrix of S with respect to the basis a must have integral entries. The matrices of dimension 3 with integral entries for which the inverse exists and has also integer entries, form the group $GL(3,\mathbb{Z})$ which is a subgroup of $GL(3,\mathbb{R})$. As the determinant of an element of $GL(3,\mathbb{Z})$ is integral and has an integral inverse, it must be ± 1. Any element of $GL(3,\mathbb{Z})$ transforms the lattice into itself. It is an orthogonal transformation, i.e. an element of the point group leaving the lattice invariant, if and only if $g = \widetilde{S}gS$.

A *lattice* Λ is determined by a lattice group U and a point $\boldsymbol{x}_0$. Choosing for $\boldsymbol{x}_0$ the origin one can identify Λ and U. Considering such a lattice one defines the *holohedry* of Λ as the subgroup of all elements of $O(3)$ which transform the lattice into itself. We call two lattices geometrically equivalent if their holohedries are geometrically equivalent, i.e. if there is a nonsingular transformation S such that the holohedries H and H' are related by $H' = S^{-1}HS$. This is the case when there are bases for the vector space such that the matrix groups corresponding to the holohedries are the same. The equivalence class of this relation is called a *system*. The lattices in 3 dimensions belong to one of 7 systems. For each system the crystal class of the holohedry is determined. As any crystallographic point group leaves a lattice invariant, any such group is a subgroup of the holohedry of that lattice. We say that a point group K belongs to a certain system, when the holohedry of that system is the smallest holohedry containing K. The 7 systems and their holohedries and point groups are given in table 4.1. A lattice left invariant by a point group K is at least left invariant by the holohedry of K.

The arithmetic holohedry of a lattice with metric tensor g is the subgroup of elements S of $GL(3,\mathbb{Z})$ such that $g = \widetilde{S}gS$. It is the group of matrices corresponding to the holohedry with respect to the basis of the lattice which gives g. We denote the arithmetic holohedry by $\phi(H)$. With respect to another

Table 4.1
Systems in 3 dimensions.

System	Holohedry	Point groups
Triclinic	$\bar{1}$	$1,\bar{1}$
Monoclinic	$2/m$	$2,m,2/m$
Orthorhombic	mmm	$222,2mm,mmm$
Tetragonal	$4/mmm$	$4,4/m,422, 4mm,\bar{4},\bar{4}2m,4/mmm$
Trigonal	$\bar{3}m$	$3,32,\bar{3},3m,\bar{3}m$
Hexagonal	$6/mmm$	$6,6mm,622,6/m,\bar{6},\bar{6}m2,6/mmm$
Cubic	$m3m$	$23,m3,432,\bar{4}3m,m3m$

basis of the lattice the arithmetic holohedry becomes $\phi'(H) = S^{-1}\phi(H)S$ according to eq. (1.4). We call two lattices *arithmetically equivalent* if there is a matrix $S \in GL(3, \mathbf{Z})$ such that their holohedries $\phi(H)$ and $\phi(H')$ are related by $\phi(H') = S^{-1}\phi(H)S$. The equivalence classes of this relation are called *Bravais classes*. Two lattices belong to the same Bravais class if and only if there are bases for them with respect to which the arithmetic holohedries are the same. As a basis for the lattice is also a basis for the space to which it belongs, two lattices from the same Bravais class belong also to the same system. Hence each system can be subdivided in a number of Bravais classes.

In several cases lattices from one Bravais class contain sublattices which belong to another Bravais class. A *sublattice* of U here is a subgroup of U which is again a lattice group. A basis for the sublattice is obtained by integral linear combinations of the basis of the original lattice. However, the matrix describing this basis transformation is not an element of $GL(3, \mathbf{Z})$. Otherwise the lattice group generated by the new basis would be the same as that generated by the old basis. On the other hand a basis of the original lattice can be obtained from a basis of the sublattice by a linear transformation described by the inverse of a matrix with integral entries. The sublattice is often chosen in such a way that the matrices of the holohedry are in a simple form. The sublattice is called the *primitive lattice*, the original lattice the *centered lattice*. The reason for this terminology will become clear in the following. In particular the different Bravais classes belonging to one system can be considered as centerings of each other. Usually one chooses the primitive lattice in such a way that the matrices of the holohedry get a simple form.

As an example consider a two-dimensional case. The system with holohedry $2mm$ contains two Bravais classes. A lattice Λ_1 of the first class is generated by $\boldsymbol{a}_1 = (a, 0)$ and $\boldsymbol{a}_2 = (0, b)$. The arithmetical holohedry of this Bravais class is given by the matrix group

$$\begin{pmatrix} 1 & 0 \\ 0 & 1 \end{pmatrix}, \quad \begin{pmatrix} 1 & 0 \\ 0 & -1 \end{pmatrix}, \quad \begin{pmatrix} -1 & 0 \\ 0 & 1 \end{pmatrix}, \quad \begin{pmatrix} -1 & 0 \\ 0 & -1 \end{pmatrix}.$$

A lattice Λ_2 of the second Bravais class is generated by $\boldsymbol{a}'_1 = (a, b)$ and $\boldsymbol{a}'_2 = (-a, b)$. The arithmetical holohedry for this lattice is the matrix group

$$\begin{pmatrix} 1 & 0 \\ 0 & 1 \end{pmatrix}, \quad \begin{pmatrix} 0 & 1 \\ 1 & 0 \end{pmatrix}, \quad \begin{pmatrix} 0 & -1 \\ -1 & 0 \end{pmatrix}, \quad \begin{pmatrix} -1 & 0 \\ 0 & -1 \end{pmatrix}.$$

It is easily verified that there is no matrix $S \in GL(2, \mathbb{Z})$ which links one group to the other. This means that they really belong to different Bravais classes. However, the sublattice Λ generated by $\boldsymbol{A}_2 = \boldsymbol{a}'_1 + \boldsymbol{a}'_2 = (0, 2b)$ and $\boldsymbol{A}_1 = \boldsymbol{a}'_1 - \boldsymbol{a}'_2 = (2a, 0)$ has the same holohedry and the same arithmetic holohedry as the first lattice Λ_1. The basis $\boldsymbol{A}_1, \boldsymbol{A}_2$ of the sublattice is obtained by the matrix S from the basis $\boldsymbol{a}'_1, \boldsymbol{a}'_2$ and conversely $\boldsymbol{a}'_1, \boldsymbol{a}'_2$ from $\boldsymbol{A}_1, \boldsymbol{A}_2$ by S^{-1} with

$$S = \begin{pmatrix} 1 & 1 \\ 1 & -1 \end{pmatrix}, \qquad S^{-1} = \tfrac{1}{2} \begin{pmatrix} 1 & 1 \\ 1 & -1 \end{pmatrix}.$$

The sublattice Λ belongs to the Bravais class of lattice Λ_1. The lattice Λ_2 is obtained from Λ by S^{-1}.

The different Bravais classes belonging to one system are given in table 4.2 by a primitive lattice (i.e. an conveniently chosen element of the system) and the matrices S^{-1} necessary to obtain a lattice in the Bravais class from the primitive lattice.

A *unit cell* for a lattice Λ is a region in the vector space such that for every $\boldsymbol{x} \in V$ there is exactly one point $\boldsymbol{y}$ in this region and one element $\boldsymbol{u}$ of the lattice group U such that $\boldsymbol{x} = \boldsymbol{y} + \boldsymbol{u}$. The unit cell of a lattice is not uniquely determined. A possible choice is the parallelopepid spanned by 3 basis vectors $\boldsymbol{a}_1, \boldsymbol{a}_2, \boldsymbol{a}_3$. Another possible choice is the so called *Wigner–Seitz cell* defined as the set of points which have a smaller distance to a fixed lattice point than to all other lattice points. An advantage of the latter choice is that this cell is transformed into itself by the holohedry of the lattice. A drawback is the fact that for lattices with a low-order holohedry it is often difficult to construct or to visualize.

Table 4.2
Bravais classes in 3 dimensions.

System	Metric tensor primitive lattice $a,b,c,d,e,f \in \mathbb{R}$	Bravais class	Centering matrix S^{-1}	Number of points per unit cell
Triclinic	$\begin{pmatrix} a & b & c \\ b & d & e \\ c & e & f \end{pmatrix}$	Primitive	–	1
Monoclinic	$\begin{pmatrix} a & b & 0 \\ b & c & 0 \\ 0 & 0 & d \end{pmatrix}$	Primitive	–	1
		Body-centered	$\frac{1}{2}\begin{pmatrix} -1 & 1 & 1 \\ 1 & -1 & 1 \\ 1 & 1 & -1 \end{pmatrix}$	2
Orthorhombic	$\begin{pmatrix} a & 0 & 0 \\ 0 & b & 0 \\ 0 & 0 & c \end{pmatrix}$	Primitive	–	1
		Body-centered	$\frac{1}{2}\begin{pmatrix} -1 & 1 & 1 \\ 1 & -1 & 1 \\ 1 & 1 & -1 \end{pmatrix}$	2
		Side-centered	$\frac{1}{2}\begin{pmatrix} -1 & 1 & 0 \\ 1 & 1 & 0 \\ 0 & 0 & 2 \end{pmatrix}$	2
		All-face-centered	$\frac{1}{2}\begin{pmatrix} 1 & 1 & 0 \\ 1 & 0 & 1 \\ 0 & 1 & 1 \end{pmatrix}$	4
Tetragonal	$\begin{pmatrix} a & 0 & 0 \\ 0 & a & 0 \\ 0 & 0 & b \end{pmatrix}$	Primitive	–	1
		Body-centered	$\frac{1}{2}\begin{pmatrix} -1 & 1 & 1 \\ 1 & -1 & 1 \\ 1 & 1 & -1 \end{pmatrix}$	2

Table 4.2 (continued)

System	Metric tensor primitive lattice $a,b,c,d,e,f \in \mathbb{R}$	Bravais class	Centering matrix S^{-1}	Number of points per unit cell
Trigonal *	$\begin{pmatrix} a & b & b \\ b & a & b \\ b & b & a \end{pmatrix}$	Primitive (h.c.p.)	–	1
Hexagonal	$\begin{pmatrix} a & \frac{1}{2}a & 0 \\ \frac{1}{2}a & a & 0 \\ 0 & 0 & b \end{pmatrix}$	Primitive	–	1
Cubic	$\begin{pmatrix} a & 0 & 0 \\ 0 & a & 0 \\ 0 & 0 & a \end{pmatrix}$	Primitive	–	1
		Body-centered (b.c.c.)	$\frac{1}{2}\begin{pmatrix} -1 & 1 & 1 \\ 1 & -1 & 1 \\ 1 & 1 & -1 \end{pmatrix}$	2
		Face-centered (f.c.c.)	$\frac{1}{2}\begin{pmatrix} 1 & 1 & 0 \\ 1 & 0 & 1 \\ 0 & 1 & 1 \end{pmatrix}$	4

* A trigonal lattice is a rhomboedric centering of a hexagonal lattice via the centering

$$\frac{1}{3}\begin{pmatrix} 1 & 1 & -2 \\ 1 & -2 & 1 \\ 1 & 1 & 1 \end{pmatrix}.$$

According to the definition, in each unit cell there is exactly one element of the lattice. When we have a primitive and a centered lattice, in each unit cell of the sublattice (the primitive lattice) the number of points of the lattice is equal to the determinant of the matrix S. Since $S \notin GL(3, \mathbb{Z})$, this number is an integer greater than one. This is the reason why the original lattice is called a centering of the sublattice. In table 4.2 the number of lattice points per unit cell of the primitive lattice is given. For one lattice from each Bravais class either a unit cell, or a unit cell of a sublattice and a centering is drawn in fig. 4.1.

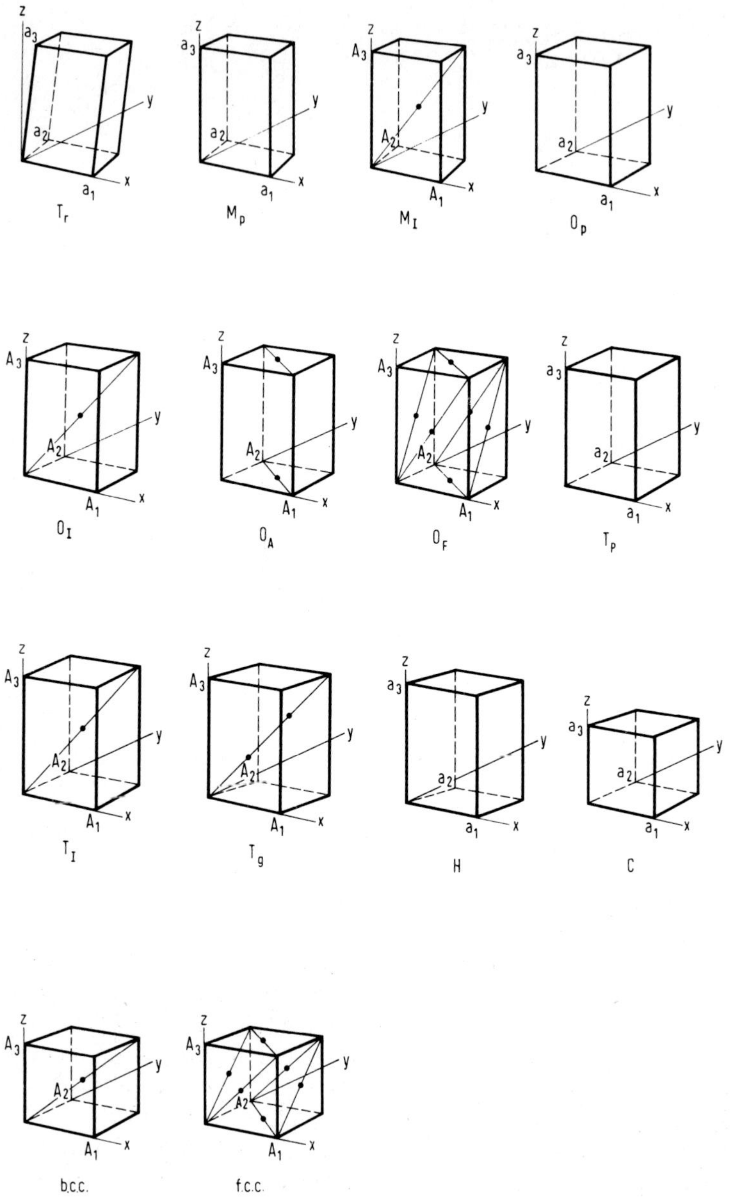

T_r
M_P
M_I
O_P
O_I
O_A
O_F
T_P
T_I
T_g
H
C
b.c.c.
f.c.c.

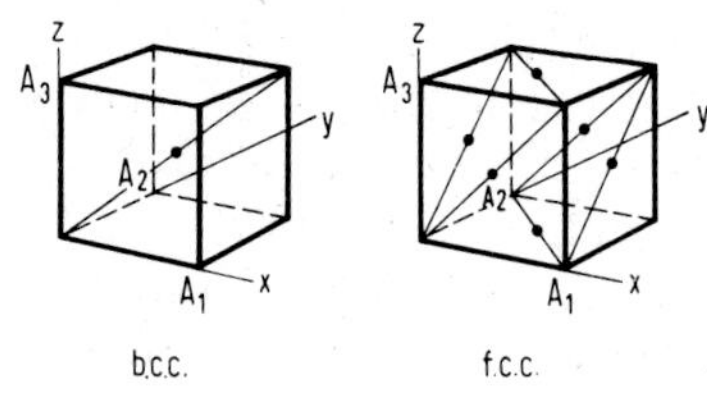

Unit cells of the centered sublattices.

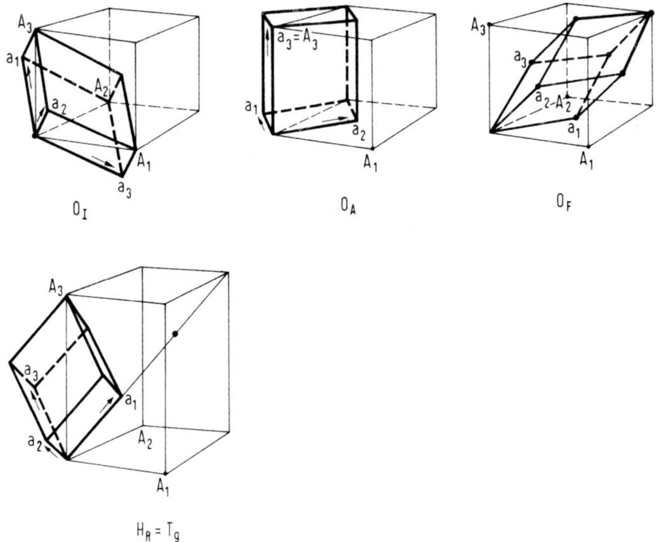

Bases of the lattices with respect to the centered sublattice

Fig. 4.1. The unit cells of the 14 Bravais classes.

4.1.5. *Examples of space groups*

In this section we will consider some space groups which occur as groups of Euclidean motions which transform regular patterns into itself. We choose our examples here in two dimensions in order to have a simple case. We consider three two-dimensional regular patterns (fig. 4.2).

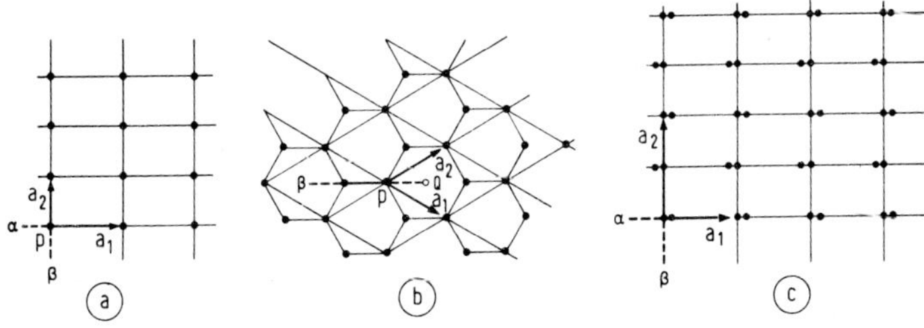

Fig. 4.2. Some two-dimensional space groups.

In the first example all points (fig. 4.2a) can be obtained from each other by elements of the lattice group U generated by $\boldsymbol{a}_1$ and $\boldsymbol{a}_2$. If we choose P as origin, the elements of the space group which leaves the pattern invariant are products of

$\{\epsilon | n_1\boldsymbol{a}_1 + n_2\boldsymbol{a}_2\} \in U$, where ϵ is the identity,

$\{\alpha | \boldsymbol{O}\}$ where α is the reflection leaving $\boldsymbol{a}_1$ fixed,

$\{\beta | \boldsymbol{O}\}$ where β is the reflection leaving $\boldsymbol{a}_2$ fixed,

$\{\alpha\beta | \boldsymbol{O}\}$ which is the 180° rotation around the origin P.

In the second example (fig. 4.2b) only half of the points can be obtained from one point by a lattice group. The lattice group U is generated by $\boldsymbol{a}_1$ and $\boldsymbol{a}_2$. If we choose P as origin, the pattern is transformed into itself by elements of U, by the Euclidean motions

$\{\beta | \boldsymbol{O}\}$ which is a reflection interchanging $\boldsymbol{a}_1$ and $\boldsymbol{a}_2$,

$\{\alpha^2 | \boldsymbol{O}\}$ which is a 120° rotation,

$\{\alpha | \frac{2}{3}\boldsymbol{a}_1 + \frac{2}{3}\boldsymbol{a}_2\}$ where α is a 60° rotation,

and products of these elements. The point group K has the elements $\epsilon, \alpha, \alpha^2, \alpha^3, \alpha^4, \alpha^5, \beta, \alpha\beta, \alpha^2\beta, \alpha^3\beta, \alpha^4\beta, \alpha^5\beta$. It is a point group from the geometrical crystal class $6mm$ with abstract point group D_6. Notice that U is invariant under each element of K.

When we choose the pattern with origin in Q, it is transformed into itself by all elements $\{\epsilon | \boldsymbol{u}\} \in U$ and by all elements $\{R | \boldsymbol{O}\}$ with $R \in K$. Hence in this case there are no nonprimitive translations. So the space group is *symmorphic*. This shows that a system of nonprimitive translations depends on the choice of the origin. As we will discuss in the next section the space groups of the same pattern are identified. We will call the space group of the pattern with origin in P also symmorphic, because the nonprimitive translations can be removed by another choice of origin.

In the example of fig. 4.2c the lattice group is generated by $\boldsymbol{a}_1$ and $\boldsymbol{a}_2$. The pattern is transformed into itself by the reflection $\{\alpha | \boldsymbol{O}\}$ and by the Euclidean motions $\{\beta | \frac{1}{2}\boldsymbol{a}_2\}$ and $\{\alpha\beta | \frac{1}{2}\boldsymbol{a}_2\}$. The nonprimitive translation $\frac{1}{2}\boldsymbol{a}_2$ cannot be removed by another choice of basis. The space group is then called *nonsymmorphic*.

4.1.6. *Identifications*

The number of space groups is infinite. However, many of these space groups describe patterns which are in many respects the same. In fig. 4.2a e.g. one can change the ratio of $\boldsymbol{a}_1$ and $\boldsymbol{a}_2$. The symmetry remains the same as long as $\boldsymbol{a}_1$ and $\boldsymbol{a}_2$ are of different length. Also, another choice of origin gives another space group for the same pattern which can be identified with the first space group.

In crystallography, it is customary to identify space groups which are conjugate as subgroups of the affine group $A(3)$. Two space groups G and G' are *equivalent* if there is an element $\{S|\boldsymbol{t}\} \in A(3)$ such that

$$\{S|\boldsymbol{t}\}G\{S|\boldsymbol{t}\}^{-1} = G' . \tag{4.6}$$

This implies that G and G' are isomorphic. Bieberbach has shown (Bieberbach [1911])

PROPOSITION 4.1. Two space groups are isomorphic if and only if they are conjugate subgroups of the affine group $A(3)$.

PROPOSITION 4.2. The number of isomorphism classes of space groups is finite.

In three dimensions *the number of nonisomorphic space groups* is 219. However, one sometimes identifies space groups only when they are conjugate subgroups of $A(3)$ with conjugating element $\{S|\boldsymbol{t}\}$ such that $\det(S) = +1$. With this identification there are 11 pairs of space groups which are pairwise isomorphic but not conjugate by an element with $\det(S) = +1$. Then there are 230 *different space groups*.

When we choose a basis for the translation subgroup U of a space group G, the elements of the point group K correspond to a group of matrices with integral entries. Such a group $\phi(K)$ with elements $\phi(\alpha)$ with α an element of the abstract point group K, is called an *arithmetic point group*. Choosing another basis for U related to the first by $S \in GL(3,\mathbb{Z})$ the point group corresponds to another arithmetic point group $\phi'(K) = S^{-1}\phi(K)S$. Two arithmetic point groups that are conjugate subgroups of $GL(3,\mathbb{Z})$ are called *arithmetically equivalent*. The equivalence classes of this relation are the *arithmetic crystal classes*. Notice that from arithmetic equivalence follows geometric equivalence. The converse is not true. Several arithmetic crystal classes may belong to the same geometric crystal class. So every space group determines an arithmetic point group up to arithmetic equivalence, i.e. it determines an arithmetic crystal class. Moreover, isomorphic space groups determine the same arithmetic crystal class. To prove this we first remark that by an element of $A(3)$ conjugating the space groups G and G' the translation subgroups U and U' are conjugated according to eq. (4.6). So an isomorphism $\chi : G \to G'$ maps U onto U'. Now consider for $g = \{R|\boldsymbol{t}\} \in G$ the relation $g a_i g^{-1} = \{1|(R\boldsymbol{a}_i)\} = \{1|\Sigma_j \phi(\alpha)_{ji}\boldsymbol{a}_j\}$ in G, where $\varphi(\alpha)$ is the element of $GL(3, \mathbb{Z})$ describing the element $R \in K$ with respect to the basis $\boldsymbol{a}_1, \boldsymbol{a}_2, \boldsymbol{a}_3$ of U. By the isomorphism χ, this relation is mapped on $\chi(g)\chi(a_i)\chi(g)^{-1} = \Sigma_j\, \varphi(\alpha)_{ji}\chi(a_j)$ in G'. As the translations $\chi(a_i)$ form a basis for U' one has $\chi(a_i) = \Sigma_j\, S_{ji}a_j'$. Therefore, one has $\chi(g)a_i'\chi(g)^{-1} = \Sigma_{j,k}\,(S^{-1})_{ji}\varphi(\alpha)_{kj}\chi(a_k) = \Sigma_l (S\varphi(\alpha)S^{-1})_{li}a_l'$. On the other hand $\chi(g)a_i'\chi(g)^{-1} = \Sigma_l \varphi'(\alpha)_{li}a_l'$. Consequently $\varphi'(K') = S\varphi(K)S^{-1}$ with $S \in GL(3, \mathbb{Z})$. This shows that the arithmetic point groups $\varphi'(K')$ and $\varphi(K)$ are in the same arithmetic crystal class. Hence *isomorphic space groups determine the same arithmetic crystal class.*

For every arithmetic crystal class there may be several nonisomorphic space groups. However, there is only one isomorphism class of space groups with the structure of a semidirect product. A space group with the arithmetic point group $\varphi(K)$ is given by the elements $\{R|\boldsymbol{u}\}$, with $\boldsymbol{u}$ a primitive translation from the lattice left invariant, and R an orthogonal transformation described by $\varphi(\alpha)$ with respect to the basis $\boldsymbol{a}_1, \boldsymbol{a}_2, \boldsymbol{a}_3$ of U. It is a space group without nonprimitive translations, so it is a symmorphic group. Every space group isomorphic with this group is also called *symmorphic*. As there is exactly one isomorphism class of symmorphic groups in every arithmetic crystal class, there is a one-to-one correspondence between the symmorphic space groups and the arithmetic crystal classes. In three dimensions there are 73 arithmetic crystal classes. Thus there are 73 symmorphic space groups. The groups which are not symmorphic are called *nonsymmorphic* space groups. There are 219–73 = 146 nonisomorphic nonsymmorphic space groups in three dimensions.

In summary one can classify the 219 nonisomorphic space groups as follows. For each of the 73 arithmetic crystal classes there is one symmorphic and possibly several nonsymmorphic space groups with an arithmetic point group in the crystal class. Each of the arithmetic crystal classes belongs to one of the 32 geometric crystal classes. The space groups belonging to such a crystal class have a point group from this class. Every geometric crystal class belongs to one of the 7 systems. The space groups belonging to a system have a translation subgroup with lattice belonging to this system. When we associate a space group to that Bravais class which has the smallest arithmetic holohedry containing the arithmetic point group of the space group as a subgroup, each space group belongs to one of the 14 Bravais classes. Finally each of the 14 Bravais classes belongs again to one of the 7 systems. In table 4.3 the 73

Table 4.3
The arithmetic crystal classes in 3 dimensions.

System	Bravais class	Holohedry	Arithmetic crystal class	Iso-morphism class	Generators point groups	Number space groups
Triclinic	Primitive	$P\bar{1}$	$P1$	C_1	1	1
			$P\bar{1}$	C_2	2	1
Monoclinic	Primitive	$P2/m$	Pm	C_2	7	2
			$P2$	C_2	3	2
			$P2/m$	D_2	3, 7	4
	Body-centered	$I2/m$	Im	C_2	$I7$	2
			$I2$	C_2	$I3$	1
			$I2/m$	D_2	$I3, I7$	2
Orthorhombic	Primitive	$Pmmm$	$P222$	D_2	3, 4	4
			$P2mm$	D_2	8, 9	10
			$Pmmm$	$D_2 \times C_2$	7, 8, 9	16
	Body-centered	$Immm$	$I222$	D_2	$I3, I4$	2
			$I2mm$	D_2	$I8, I9$	3
			$Immm$	$D_2 \times C_2$	$I7, I8, I9$	4
	Side-centered	$Ammm$	$A222$	D_2	$A3, A4$	2
			$1A2mm$	D_2	$A8, A9$	3
			$2A2mm$	D_2	$A7, A8$	4
			$Ammm$	$D_2 \times C_2$	$A7, A8, A9$	6
	All-face centered	$Fmmm$	$F222$	D_2	$F3, F4$	1
			$F2mm$	D_2	$F8, F9$	2
			$Fmmm$	$D_2 \times C_2$	$F7, F8, F9$	2
Tetragonal	Primitive	$P4/mmm$	$P4$	C_4	12	3(4)
			$P4/m$	$C_4 \times C_2$	12, 7	4
			$P422$	D_4	12, 5	6(8)
			$P4mm$	D_4	12, 8	8
			$P\bar{4}$	C_4	13	1
			$1P\bar{4}2m$	D_4	13, 5	4
			$2P\bar{4}2m$	D_4	13, 4	4
			$P4/mmm$	$D_4 \times C_2$	12, 8, 7	16

Table 4.3 (continued)

System	Bravais class	Holohedry	Arithmetic crystal class	Iso-morphism class	Generators point groups	Numbe space groups
Tetragonal	Body-centered	$I4/mmm$	$I4$	C_4	$I12$	2
			$I4/m$	$C_4 \times C_2$	$I12, I7$	2
			$I422$	D_4	$I12, I5$	2
			$I4mm$	D_4	$I12, I8$	4
			$I\bar{4}$	C_4	$I13$	1
			$1I\bar{4}2m$	D_4	$I13, I5$	2
			$2I\bar{4}2m$	D_4	$I13, I4$	2
			$I4/mmm$	$D_4 \times C_2$	$I12, I8, I7$	4
Trigonal	Primitive	$R\bar{3}m$	$R3$	C_3	$R14$	1
			$R32$	D_3	$R14, R6$	1
			$R\bar{3}$	C_6	$R15$	1
			$R3m$	D_3	$R14, R10$	2
			$R\bar{3}m$	D_6	$R15, R10$	2
Hexagonal	Primitive	$P6/mmm$	$P3$	C_3	14	2(3)
			$1P32$	D_3	14, 5	2(3)
			$2P32$	D_3	14, 6	2(3)
			$P\bar{3}$	C_6	15	1
			$1P3m$	D_3	14, 11	2
			$2P3m$	D_3	14, 10	2
			$1P\bar{3}m$	D_6	15, 11	2
			$2P\bar{3}m$	D_6	15, 10	2
			$P6$	C_6	16	4(6)
			$P6mm$	D_6	16, 10	4
			$P622$	D_6	16, 5	4(6)
			$P6/m$	$C_6 \times C_2$	16, 7	2
			$P\bar{6}$	C_6	17	1
			$1P\bar{6}m2$	D_6	17, 5	2
			$2P\bar{6}m2$	D_6	17, 6	2
			$P6/mmm$	$D_6 \times C_2$	16, 10, 7	4
Cubic	Primitive	$Pm3m$	$P23$	T	18, 3	2
			$Pm3$	$T \times C_2$	18, 3, 2	3
			$P432$	O	12, 19	3(4)
			$P\bar{4}3m$	O	13, 19	2
			$Pm3m$	$O \times C_2$	12, 19, 2	4

Table 4.3 (continued)

System	Bravais class	Holohedry	Arithmetic crystal class	Iso-morphism class	Generators point groups	Number space groups
Cubic	B.c.c.	$Im3m$	$I23$	T	$I18, I3$	2
			$Im3$	$T \times C_2$	$I18, I3, 2$	2
			$I432$	O	$I12, I19$	2
			$I\bar{4}3m$	O	$I13, I19$	2
			$Im3m$	$O \times C_2$	$I12, I19, 2$	2
	F.c.c.	$Fm3m$	$F23$	T	$F18, F3$	1
			$Fm3$	$T \times C_2$	$F18, F3, 2$	2
			$F432$	O	$F12, F19$	2
			$F\bar{4}3m$	O	$F13, F19$	2
			$Fm3m$	$O \times C_2$	$F12, F19, 2$	4
						219(230)

1) The arithmetic crystal classes are denoted by the symbol for the corresponding symmorphic space group.

2) The generators of the point groups are given in the following table

$$1 = \begin{pmatrix} 1 & 0 & 0 \\ 0 & 1 & 0 \\ 0 & 0 & 1 \end{pmatrix} \quad 2 = \begin{pmatrix} -1 & 0 & 0 \\ 0 & -1 & 0 \\ 0 & 0 & -1 \end{pmatrix} \quad 3 = \begin{pmatrix} -1 & 0 & 0 \\ 0 & -1 & 0 \\ 0 & 0 & 1 \end{pmatrix} \quad 4 = \begin{pmatrix} 1 & 0 & 0 \\ 0 & -1 & 0 \\ 0 & 0 & -1 \end{pmatrix} \quad 5 = \begin{pmatrix} 0 & 1 & 0 \\ 1 & 0 & 0 \\ 0 & 0 & -1 \end{pmatrix}$$

$$6 = \begin{pmatrix} 1 & 1 & 0 \\ 0 & -1 & 0 \\ 0 & 0 & -1 \end{pmatrix} \quad 7 = \begin{pmatrix} 1 & 0 & 0 \\ 0 & 1 & 0 \\ 0 & 0 & -1 \end{pmatrix} \quad 8 = \begin{pmatrix} 1 & 0 & 0 \\ 0 & -1 & 0 \\ 0 & 0 & 1 \end{pmatrix} \quad 9 = \begin{pmatrix} -1 & 0 & 0 \\ 0 & 1 & 0 \\ 0 & 0 & 1 \end{pmatrix} \quad 10 = \begin{pmatrix} 0 & 1 & 0 \\ 1 & 0 & 0 \\ 0 & 0 & 1 \end{pmatrix}$$

$$11 = \begin{pmatrix} 1 & 1 & 0 \\ 0 & -1 & 0 \\ 0 & 0 & 1 \end{pmatrix} \quad 12 = \begin{pmatrix} 0 & -1 & 0 \\ 1 & 0 & 0 \\ 0 & 0 & 1 \end{pmatrix} \quad 13 = \begin{pmatrix} 0 & 1 & 0 \\ -1 & 0 & 0 \\ 0 & 0 & -1 \end{pmatrix} \quad 14 = \begin{pmatrix} -1 & -1 & 0 \\ 1 & 0 & 0 \\ 0 & 0 & 1 \end{pmatrix} \quad 15 = \begin{pmatrix} 1 & 1 & 0 \\ -1 & 0 & 0 \\ 0 & 0 & -1 \end{pmatrix}$$

$$16 = \begin{pmatrix} 0 & -1 & 0 \\ 1 & 1 & 0 \\ 0 & 0 & 1 \end{pmatrix} \quad 17 = \begin{pmatrix} 0 & 1 & 0 \\ -1 & -1 & 0 \\ 0 & 0 & -1 \end{pmatrix} \quad 18 = \begin{pmatrix} 0 & 1 & 0 \\ 0 & 0 & 1 \\ 1 & 0 & 0 \end{pmatrix} \quad 19 = \begin{pmatrix} 0 & 0 & 1 \\ 1 & 0 & 0 \\ 0 & 1 & 0 \end{pmatrix}$$

$$An = \tfrac{1}{2}\begin{pmatrix} -1 & 1 & 0 \\ 1 & 1 & 0 \\ 0 & 0 & 1 \end{pmatrix} n \begin{pmatrix} -1 & 1 & 0 \\ 1 & 1 & 0 \\ 0 & 0 & 2 \end{pmatrix} \qquad Fn = \tfrac{1}{2}\begin{pmatrix} 1 & 1 & -1 \\ 1 & -1 & 1 \\ -1 & 1 & 1 \end{pmatrix} n \begin{pmatrix} 1 & 1 & 0 \\ 1 & 0 & 1 \\ 0 & 1 & 1 \end{pmatrix}$$

$$In = \tfrac{1}{2}\begin{pmatrix} 0 & 1 & 1 \\ 1 & 0 & 1 \\ 1 & 1 & 0 \end{pmatrix} n \begin{pmatrix} -1 & 1 & 1 \\ 1 & -1 & 1 \\ 1 & 1 & -1 \end{pmatrix} \qquad Rn = \tfrac{1}{3}\begin{pmatrix} 1 & 1 & 1 \\ 0 & -1 & 1 \\ -1 & 0 & 1 \end{pmatrix} n \begin{pmatrix} 1 & 1 & -2 \\ 1 & -2 & 1 \\ 1 & 1 & 1 \end{pmatrix}$$

arithmetic crystal classes and the 219 associated space groups are given as well as the Bravais class and system to which they belong.

A space group can be given by the underlying lattice (determined by its metric tensor), generators for the arithmetic point group $\varphi(K)$ and the lattice coordinates of the nonprimitive translations associated to these generators. It is sufficient to give only the nonprimitive translations associated to the generators, because from $\{R\,|\,\boldsymbol{t}_R\}\{S\,|\,\boldsymbol{t}_S\} = \{RS\,|\,\boldsymbol{t}_R + R\boldsymbol{t}_S\}$ it follows that the nonprimitive translation associated to RS is equal to $\boldsymbol{t}_R + R\boldsymbol{t}_S$ up to a primitive translation (a system of nonprimitive translations is determined up to primitive translations). The 219 space groups are given in the appendix. More extensive information about the three-dimensional space groups can be found in the International Tables for X-ray Crystallography (Henry and Lonsdale [1965]). See also Burckhardt [1966], Ascher and Janner [1965, 1968] for mathematical aspects, and Buerger [1963] for a crystallographer's point of view.

4.1.7. Crystal structure, in particular diamond structure

The particles constituting an infinite crystal form a regular pattern. This means that when we consider the particles as point particles their equilibrium positions form such a regular pattern. The group of Euclidean motions which transform the crystal into itself (the *symmetry group of the crystal*) is a space group. When the crystal vibrates, when the particles are described by quantum mechanical wave functions, or when the crystal has imperfections, the symmetry is broken. The space group of the pattern formed by the equilibrium positions does not, in general, transform the real crystal into itself. However, the Hamiltonian of a perfect crystal will again show the symmetry of the space group.

When the space group of the crystal is G, one chooses a unit cell of the translation subgroup U to describe the positions of the particles. Because any point in space is the sum of a primitive translation and a vector in the unit cell, a point is given by $\boldsymbol{x} = \boldsymbol{a} + \boldsymbol{\tau}$ with $\boldsymbol{a} \in U$ and $\boldsymbol{\tau}$ in the unit cell. For a crystal with the particles at their equilibrium positions, the pattern is left invariant by U. The s particles in the unit cell can be described by the vector $\boldsymbol{\tau}_j$ with $j = 1, ..., s$. The position of any particle in the crystal can then be described by $\boldsymbol{x}\binom{\boldsymbol{n}}{j} = \boldsymbol{a}(\boldsymbol{n}) + \boldsymbol{\tau}_j$, where $\boldsymbol{a}(\boldsymbol{n})$ is the element of U given by the triple of integers $\boldsymbol{n} = (n_1, n_2, n_3)$: $\boldsymbol{a}(\boldsymbol{n}) = n_1\boldsymbol{a}_1 + n_2\boldsymbol{a}_2 + n_3\boldsymbol{a}_3$.

The crystal is transformed into itself by the Euclidean motion $\{R\,|\,\boldsymbol{t}\}$ if for any $\boldsymbol{n}$ and j at the position $R\boldsymbol{x}\binom{\boldsymbol{n}}{j} + \boldsymbol{t}$ one has a particle of the same kind as at the position $\boldsymbol{x}\binom{\boldsymbol{n}}{j}$. The expression "of the same kind" must be understood in connection with the problem one wants to consider. It can happen that they are different in some sense. For example, in considering lattice vibrations one is usually not interested in excited states of the constituting atoms. In that case atoms in different states may be considered as "of the same kind".

As an example of crystal structure we will consider a crystal with the *structure of diamond.* It is transformed into itself by a lattice group from the Bravais class of face-centered cubic lattices. This lattice has a sublattice from the Bravais class of primitive cubic lattices. We will give the point group elements and the nonprimitive translations with respect to a basis $\boldsymbol{A}_1, \boldsymbol{A}_2, \boldsymbol{A}_3$ of this lattice. The unit cell of the f.c.c. lattice may be chosen as the parallelopiped spanned by basis vectors $\boldsymbol{a}_1, \boldsymbol{a}_2, \boldsymbol{a}_3$ (fig. 4.3). In each unit cell there are two carbon atoms. When we put one of the atoms at the origin, the second atom has a position in the unit cell given by its lattice coordinates $[\frac{1}{4}\,\frac{1}{4}\,\frac{1}{4}]$ with respect to $\boldsymbol{A}_1, \boldsymbol{A}_2, \boldsymbol{A}_3$. The points with integer lattice coordinates form a lattice Λ_1, the other atoms a lattice Λ_2 obtained by a translation $[\frac{1}{4}\,\frac{1}{4}\,\frac{1}{4}]$ from Λ_1.

The point group of the space group is of order 48. It is a group from the geometric crystal class $m3m$ and from the isomorphism class $O \times C_2$. Its elements with respect to $\boldsymbol{A}_1, \boldsymbol{A}_2, \boldsymbol{A}_3$ are given in table 4.4. The point group contains as a subgroup the group 23 (the tetrahedral group). Another subgroup is formed by the subgroup 23 and the products of the rotations $\notin 23$ with the central inversion. This is the group $\bar{4}3m$.

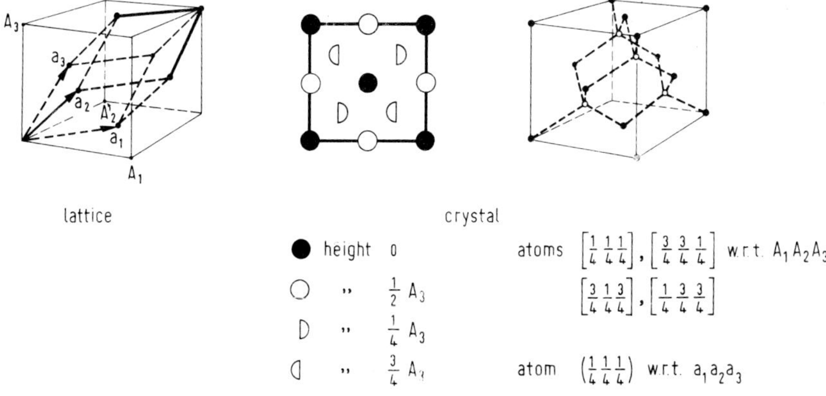

Fig. 4.3. Diamond structure.

Table 4.4
Matrices of the point group of $F d3m$.
Elements of the rotation subgroup 432 (the numbers correspond with those in table 3.2).

1 $1 = \begin{pmatrix} 1 & 0 & 0 \\ 0 & 1 & 0 \\ 0 & 0 & 1 \end{pmatrix}$	10 $2_z = \begin{pmatrix} -1 & 0 & 0 \\ 0 & -1 & 0 \\ 0 & 0 & 1 \end{pmatrix}$	11 $2_x = \begin{pmatrix} 1 & 0 & 0 \\ 0 & -1 & 0 \\ 0 & 0 & -1 \end{pmatrix}$	12 $2_y = \begin{pmatrix} -1 & 0 & 0 \\ 0 & 1 & 0 \\ 0 & 0 & -1 \end{pmatrix}$
2 $3_{xyz} = \begin{pmatrix} 0 & 0 & 1 \\ 1 & 0 & 0 \\ 0 & 1 & 0 \end{pmatrix}$	3 $3_{xy\bar{z}} = \begin{pmatrix} 0 & 0 & -1 \\ 1 & 0 & 0 \\ 0 & -1 & 0 \end{pmatrix}$	4 $3_{x\bar{z}\bar{y}} = \begin{pmatrix} 0 & 0 & -1 \\ -1 & 0 & 0 \\ 0 & 1 & 0 \end{pmatrix}$	5 $3_{xz\bar{y}} = \begin{pmatrix} 0 & 0 & 1 \\ -1 & 0 & 0 \\ 0 & -1 & 0 \end{pmatrix}$
6 $3^2_{xyz} = \begin{pmatrix} 0 & 1 & 0 \\ 0 & 0 & 1 \\ 1 & 0 & 0 \end{pmatrix}$	7 $3^2_{xy\bar{z}} = \begin{pmatrix} 0 & 1 & 0 \\ 0 & 0 & -1 \\ -1 & 0 & 0 \end{pmatrix}$	8 $3^2_{x\bar{z}\bar{y}} = \begin{pmatrix} 0 & -1 & 0 \\ 0 & 0 & 1 \\ -1 & 0 & 0 \end{pmatrix}$	9 $3^2_{xz\bar{y}} = \begin{pmatrix} 0 & -1 & 0 \\ 0 & 0 & -1 \\ 1 & 0 & 0 \end{pmatrix}$
13 $4_z = \begin{pmatrix} 0 & -1 & 0 \\ 1 & 0 & 0 \\ 0 & 0 & 1 \end{pmatrix}$	14 $4_z^{-1} = \begin{pmatrix} 0 & 1 & 0 \\ -1 & 0 & 0 \\ 0 & 0 & 1 \end{pmatrix}$	15 $4_x = \begin{pmatrix} 1 & 0 & 0 \\ 0 & 0 & -1 \\ 0 & 1 & 0 \end{pmatrix}$	16 $4_y = \begin{pmatrix} 0 & 0 & 1 \\ 0 & 1 & 0 \\ -1 & 0 & 0 \end{pmatrix}$
17 $4_x^{-1} = \begin{pmatrix} 1 & 0 & 0 \\ 0 & 0 & 1 \\ 0 & -1 & 0 \end{pmatrix}$	18 $4_y^{-1} = \begin{pmatrix} 0 & 0 & -1 \\ 0 & 1 & 0 \\ 1 & 0 & 0 \end{pmatrix}$	19 $2_{y\bar{z}} = \begin{pmatrix} -1 & 0 & 0 \\ 0 & 0 & -1 \\ 0 & -1 & 0 \end{pmatrix}$	20 $2_{xz} = \begin{pmatrix} 0 & 0 & 1 \\ 0 & -1 & 0 \\ 1 & 0 & 0 \end{pmatrix}$
21 $2_{yz} = \begin{pmatrix} -1 & 0 & 0 \\ 0 & 0 & 1 \\ 0 & 1 & 0 \end{pmatrix}$	22 $2_{x\bar{z}} = \begin{pmatrix} 0 & 0 & -1 \\ 0 & -1 & 0 \\ -1 & 0 & 0 \end{pmatrix}$	23 $2_{x\bar{y}} = \begin{pmatrix} 0 & -1 & 0 \\ -1 & 0 & 0 \\ 0 & 0 & -1 \end{pmatrix}$	24 $2_{xy} = \begin{pmatrix} 0 & 1 & 0 \\ 1 & 0 & 0 \\ 0 & 0 & -1 \end{pmatrix}$

Elements of $m3m$–432 can be obtained by putting a – sign in front of each of the matrices of 432.

Nonprimitive translations associated with the point group elements are:

$$t_R = \begin{cases} [0\,0\,0] & \text{for the elements of } \bar{4}3m\,, \\ [\tfrac{1}{4}\,\tfrac{1}{4}\,\tfrac{1}{4}] & \text{for the other elements}\,. \end{cases}$$

The space group is denoted by $F d3m$ or by O_h^7. It is generated by the translations $a_1 = \frac{1}{2}(A_1 + A_2)$, $a_2 = \frac{1}{2}(A_1 + A_3)$, $a_3 = \frac{1}{2}(A_2 + A_3)$ and the elements

$$\left\{\begin{bmatrix} 0 & 1 & 0 \\ -1 & 0 & 0 \\ 0 & 0 & 1 \end{bmatrix} \middle| \begin{matrix} \frac{1}{4} \\ \frac{1}{4} \\ \frac{1}{4} \end{matrix}\right\}, \quad \left\{\begin{bmatrix} 0 & 1 & 0 \\ 0 & 0 & 1 \\ 1 & 0 & 0 \end{bmatrix} \middle| \begin{matrix} 0 \\ 0 \\ 0 \end{matrix}\right\} \quad \left\{\begin{bmatrix} -1 & 0 & 0 \\ 0 & -1 & 0 \\ 0 & 0 & -1 \end{bmatrix} \middle| \begin{matrix} \frac{1}{4} \\ \frac{1}{4} \\ \frac{1}{4} \end{matrix}\right\}$$

The point group $\bar{4}3m$ combined with the primitive translations forms a symmorphic subgroup $F\bar{4}3m$ or T_d^2. This subgroup of $Fd3m$ does not interchange the atoms in the unit cell. The coset $Fd3m - F\bar{4}3m$ does interchange the atoms. Hence, if the two atoms in the unit cell were different the space group of the crystal would be only the subgroup $F\bar{4}3m$.

4.1.8. Space groups as extensions

We will discuss here once more some of the algebraic properties of space groups, as these properties can be seen in a formulation which is more abstract than the one used in the preceding sections. We recapitulate some results. A space group has an invariant subgroup U which is isomorphic to the group of triples of integers $\mathbb{Z}^3$ and the factor group G/U is isomorphic to the point group K. This point group acts as a group of automorphisms on U. This situation is well known in group theory. We call a group G *an extension of a group A by a group B* if G has an invariant subgroup isomorphic to A (we denote this subgroup also by A) such that the factor group G/A is isomorphic to B. So a space group is an extension of $\mathbb{Z}^3$ by a finite group K. In the following we assume that A is Abelian and that B is finite.

The mapping σ which assigns to each element $g \in G$ the coset to which it belongs is an epimorphism of G onto the factor group G/A and consequently onto B (fig. 4.4). We write the products in A and G with a + sign, although this does not imply that G is Abelian. The product in B is written as multiplication. The coset which is mapped by σ on $\alpha \in B$ can be written as $A + r(\alpha)$, where $r(\alpha)$ is a representative of the coset. Notice that $\sigma r(\alpha) = \alpha$. Any element $g \in G$ can be written in a unique way as $g = a + r(\alpha)$ for some $a \in A$ and some $\alpha \in B$.

Because A is an invariant subgroup, it is invariant under an inner automorphism of G: for any $g \in G$ and any $a \in A$ one has $g + a - g \in A$. In particular $g = r(\alpha)$ gives an automorphism of A, denoted by $\varphi(\alpha)$ and defined by

$$\varphi(\alpha)\, a = r(\alpha) + a - r(\alpha) \tag{4.7}$$

for any $a \in A$ and any $\alpha \in B$. We also write $\varphi(\alpha)\, a = \alpha a$.

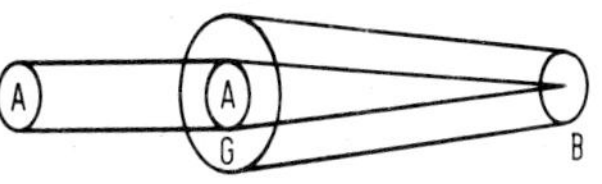

Fig. 4.4. Extension of A by B.

For two representatives $r(\alpha)$ and $r(\beta)$ their product $r(\alpha) + r(\beta)$ is in general not a representative, but it belongs to the coset with representative $r(\alpha\beta)$. Then there is an element $m(\alpha,\beta) \in A$ such that

$$r(\alpha) + r(\beta) = m(\alpha,\beta) + r(\alpha\beta) \,. \tag{4.8}$$

The set $m(\alpha,\beta)$ with $\alpha,\beta \in B$ is called a *factor system*.

For a fixed choice of the representatives $r(\alpha)$ we denote the element $a + r(\alpha)$ by (a,α). The product of two such pairs is given by

$$\begin{aligned}(a,\alpha) + (b,\beta) &= a + r(\alpha) + b + r(\beta)\\ &= a + r(\alpha) + b - r(\alpha) + r(\alpha) + r(\beta)\\ &= (a + \alpha b + m(\alpha,\beta), \alpha\beta) \,. \end{aligned} \tag{4.9}$$

On the other hand one can ask when the pairs (a,α) with $a \in A$ and $\alpha \in B$ with the product rule (4.9) will form a group. When one chooses $m(\alpha,\epsilon) = m(\epsilon,\alpha) = e \in A$, which can always be done by the choice $r(\epsilon) = e$, the unit element is (e,ϵ) and the inverse of (a,α) is $(-\alpha^{-1}a - \alpha^{-1}m(\alpha,\alpha^{-1}),\alpha^{-1})$. The associativity postulate requires

$$[(a,\alpha)+(b,\beta)] + (c,\gamma) = (a,\alpha) + [(b,\beta)+(c,\gamma)]$$

or

$$m(\alpha,\beta) + m(\alpha\beta,\gamma) = \alpha m(\beta,\gamma) + m(\alpha,\beta\gamma) \tag{4.10}$$

for any $a,b,c \in A$ and any $\alpha,\beta,\gamma \in B$. It can be shown (see e.g. Hall [1959]) that the pairs (a,α) with product (4.9) form a group if and only if the set $m(\alpha,\beta)$ satisfies eq. (4.10).

When two sets $m(\alpha,\beta)$ and $m'(\alpha,\beta)$ satisfy eq. (4.10) also their sum $(m+m')(\alpha,\beta) = m(\alpha,\beta) + m'(\alpha,\beta)$, the set $m(\alpha,\beta) = 0$ and $-m(\alpha,\beta)$ satisfy this relation. Therefore, the factor systems $m(\alpha,\beta)$ form an Abelian group, denoted by $Z^2_\varphi(B,A)$. Any factor system determines a group which has as invariant subgroup the group of elements (a,ϵ) which is isomorphic to A, with factor group isomorphic to B. Therefore, each element $m \in Z^2_\varphi(B,A)$ determines an extension of A by B.

For another choice of the representatives of the cosets of A in G one obtains another factor system m'. When $r'(\alpha)$ is another representative of the coset to which $r(\alpha)$ belongs, one has an element $u(\alpha)$ such that $r'(\alpha) =$

$u(\alpha)+r(\alpha)$. Then the factor system m' is determined by

$$m'(\alpha,\beta)+r'(\alpha\beta)=r'(\alpha)+r'(\beta)=u(\alpha)+r(\alpha)+u(\beta)+r(\beta)$$

$$=u(\alpha)+\alpha u(\beta)+m(\alpha,\beta)+r(\alpha\beta)$$

or

$$m'(\alpha,\beta)=u(\alpha)+\alpha u(\beta)-u(\alpha\beta)+m(\alpha,\beta)\,. \tag{4.11}$$

Two *factor systems* m and m' satisfying eq. (4.11) are called *equivalent*. The *extensions* of A by B with these factor systems are also called *equivalent*. It is easy to show that equivalent extensions are isomorphic.

One extension always exists, because the factor system $m(\alpha,\beta)=0$ for any $\alpha,\beta\in B$ satisfies eq. (4.10). For this extension the product rule is

$$(a,\alpha)+(b,\beta)=(a+\alpha b,\alpha\beta)\,.$$

This is the semidirect product $A \boxtimes_\varphi B$. As remarked already in § 1.1 the semidirect product is a direct product if the automorphism $\varphi(\alpha)$ is the identical one for any $\alpha\in B$. An extension which is equivalent to the semidirect product has a factor system given by eq. (4.11)

$$m(\alpha,\beta)=u(\alpha)+\alpha u(\beta)-u(\alpha\beta) \tag{4.12}$$

for some $u(\alpha), u(\beta), u(\alpha\beta)\in A$. The factor systems satisfying eq. (4.12) form an Abelian subgroup of $Z^2_\varphi(B,A)$. It is denoted by $B^2_\varphi(B,A)$. The cosets of $B^2_\varphi(B,A)$ in $Z^2_\varphi(B,A)$ are formed by the equivalence classes of factor systems. Hence the factor group

$$H^2_\varphi(B,A)\cong Z^2_\varphi(B,A)/B^2_\varphi(B,A) \tag{4.13}$$

corresponds to the set of nonequivalent extensions. Notice the similarity with the case of factor systems of projective representations. There the automorphism $\varphi(\alpha)$ is always the identical one.

As we have seen, each space group is an extension of $\mathbb{Z}^3$ by a finite group K. This group K, the point group, acts as a group of automorphisms on the Abelian group $\mathbb{Z}^3$, because (with respect to a basis of the translation subgroup U) the group K corresponds to a group $\varphi(K)$ of integral 3×3 matrices. It has been shown (Ascher and Janner [1965]):

PROPOSITION 4.3. Any extension of $\mathbb{Z}^3$ with a finite group K such that its action on $\mathbb{Z}^3$ is faithful (the matrices $\varphi(\alpha)$ for $\alpha \in K$ are all different) is isomorphic to a space group.

PROPOSITION 4.4. When the arithmetic point groups $\varphi(K)$ and $\varphi'(K)$ are arithmetically equivalent, for each extension with $\varphi(K)$ there is an isomorphic extension with $\varphi'(K)$.

To find all space groups one can take one arithmetic point group from each arithmetic crystal class and determine all nonequivalent extensions of $\mathbb{Z}^3$ with these point groups. The number of nonequivalent extensions of $\mathbb{Z}^3$ obtained in this way is 305. Two equivalent extensions are isomorphic, but the converse is not true. Therefore, one has to determine the nonisomorphic groups among the 305 nonequivalent extensions. Their number is 219 as we have mentioned already. Among the 219 space groups there are 73 which are equivalent to a semidirect product. These are the 73 symmorphic space groups. Among the 73 symmorphic groups, none is a direct product, except the one with point group of order one, which is isomorphic to $\mathbb{Z}^3$. Algorithms for obtaining the nonisomorphic space groups for arbitrary dimension, once the arithmetic crystal classes are known, are given in Zassenhaus [1947], Janssen et al. [1969b] and Fast and Janssen [1971].

4.2. Representations of space groups

4.2.1. Representations of the translation subgroup U

The translation subgroup U of a space group G is an Abelian group generated by three basis vectors $\boldsymbol{a}_1, \boldsymbol{a}_2, \boldsymbol{a}_3$. Its elements can be written as $\boldsymbol{a} = n_1\boldsymbol{a}_1 + n_2\boldsymbol{a}_2 + n_3\boldsymbol{a}_3$ for some triple of integers n_1, n_2, n_3. As Euclidean motion one has

$$a = \{\mathbb{1} \mid n_1\boldsymbol{a}_1 + n_2\boldsymbol{a}_2 + n_3\boldsymbol{a}_3\} = \{\mathbb{1} \mid \boldsymbol{a}_1\}^{n_1}\{\mathbb{1} \mid \boldsymbol{a}_2\}^{n_2}\{\mathbb{1} \mid \boldsymbol{a}_3\}^{n_3} .$$

The group U is an infinite group. Although the unitary representations of this group can be found, one usually applies a trick which makes the group finite. It consists of imposing periodic boundary conditions, i.e. we put $\{\mathbb{1} \mid \boldsymbol{a}\}^N = \{\mathbb{1} \mid N\boldsymbol{a}\} = \{\mathbb{1} \mid \boldsymbol{O}\}$ for any $a \in U$ and for some large, but finite number N. This means that we consider a space filled with exactly the same crystals, each with N^3 unit cells. As one always has to do with finite crystals there never is an infinite group of transformations leaving the crystal invariant. So the space group of a crystal is only an approximate symmetry. However, when N is large enough the boundary effects will be unimportant. One can put the situation in a more precise form. The group U has an invariant, Abelian subgroup NU generated by Na_1, Na_2, Na_3. The factor group U/NU is a finite group which is isomorphic to the direct product of three cyclic groups of order N. When $D(U/NU)$ is a representation of this finite group and σ the epimorphism $U \to U/NU$, the composition $D(\sigma U)$ is a representation of U with the property that all elements of the subgroup NU are represented by the unit matrix. Identification of the elements Na with the identity gives just the definition: $Na = O$ for any $a \in U$.

The irreducible representations of the group $C_N \times C_N \times C_N$ of order N^3 are all one-dimensional. According to Ch. 3, §2.1 and §2.3, they can be characterized by a triple p_1, p_2, p_3 of integers which give

$$D^{(p_1,p_2,p_3)}(\{\mathbb{1} \mid \sum_{i=1}^{3} n_i a_i\}) = \exp\{2\pi i(n_1p_1 + n_2p_2 + n_3p_3)/N\}\,, \tag{4.14}$$

because a representation of C_N is characterized by the N-th root $\exp\{2\pi ip/N\}$ with

$$D^{(p)}(\{\mathbb{1} \mid a_i\}) = \exp\{2\pi ip/N\}\,.$$

Because $0 \leqslant p_i < N$ there are exactly N^3 irreducible representations of the group. All these representations are representations of U. We recall that we consider only those representations of U which give $D(a)^N = \mathbb{1}$ for any $a \in U$. To label the representations given in eq. (4.14) in a simpler way one introduces the reciprocal lattice.

When a_1, a_2, a_3 are basis vectors of the group U, a dual basis is defined by (the dot denotes the inner product)

$$b_i \cdot a_j = 2\pi\delta_{ij}\,. \tag{4.15}$$

The integral linear combinations of b_1, b_2, b_3 form a lattice called the *reciprocal lattice* Λ^*. Thus $K = n_1b_1 + n_2b_2 + n_3b_3$ is a vector of Λ^*. Any vector of Λ^* has the following inner product with a vector $a = m_1a_1 + m_2a_2 + m_3a_3$ of the lattice Λ, which is called *direct lattice* to make the distinction:

$$K \cdot a = 2\pi(n_1m_1 + n_2m_2 + n_3m_3)\,. \tag{4.16}$$

When an orthogonal transformation R transforms Λ into itself one has $(Rb_i) \cdot a_j = b_i \cdot R^{-1}a_j = 2\pi$ (integer). This means that Rb_i is an element of the reciprocal lattice. So the reciprocal lattice is transformed into itself by every orthogonal transformation which does so for the direct lattice. The holohedries of direct and reciprocal lattice are the same.

When $D^{(p_1,p_2,p_3)}(U)$ is an irreducible representation of U one defines a vector $\boldsymbol{k}$ by

$$\boldsymbol{k} = (p_1\boldsymbol{b}_1 + p_2\boldsymbol{b}_2 + p_3\boldsymbol{b}_3)/N\,, \tag{4.17}$$

which gives another notation for the representation using $\boldsymbol{k}$

$$D^{(p_1,p_2,p_3)}(a) = D^{(\boldsymbol{k})}(a) = \mathrm{e}^{i\boldsymbol{k}\cdot\boldsymbol{a}}\,. \tag{4.18}$$

Representations characterized by vectors $\boldsymbol{k}$ and $\boldsymbol{k}+\boldsymbol{K}$ [written as $\boldsymbol{k}+\boldsymbol{K} \equiv \boldsymbol{k}(\mathrm{mod}\,\Lambda^*)$] with $\boldsymbol{K} \in \Lambda^*$ are the same because $\exp\{i(\boldsymbol{k}+\boldsymbol{K})\cdot\boldsymbol{a}\} = \exp\{i\boldsymbol{k}\cdot\boldsymbol{a}\}\exp\{2\pi i(\text{integer})\} = \exp\{i\boldsymbol{k}\cdot\boldsymbol{a}\}$. Therefore, the nonequivalent irreducible representations of U are characterized by vectors from a unit cell of the reciprocal lattice. Usually one takes the unit cell as the Wigner–Seitz cell: the volume around a reciprocal lattice point consisting of all points which have a smaller distance to this point than to any other lattice point. It is bounded by the perpendicular bisecting planes of the lines connecting a lattice point with all other lattice points. This unit cell is called the (*first*) *Brillouin zone*. The vectors in the Brillouin zone characterizing representations of U form a lattice with generators $\boldsymbol{a}_1/N$, $\boldsymbol{a}_2/N$, $\boldsymbol{a}_3/N$. When $N \to \infty$ the density of these $\boldsymbol{k}$-vectors becomes infinite.

4.2.2. Irreducible representations of space groups

The fact that every space group has an Abelian invariant subgroup, for which we know the irreducible representations, will be of great help in finding all nonequivalent irreducible representations of the space group. As for the translation subgroup, we will be interested only in those representations which map the elements $N\boldsymbol{a}$ of U onto the identity. The translations $N\boldsymbol{a}$ form also an invariant subgroup (NU) of the space group (see exercise 4.5). Therefore, we will consider the representations of the factor group G/NU. In the following by G we mean G/NU, by U the group U/NU.

Consider an n-dimensional representation $D(G)$ of the space group G. The matrices $D(\{\mathbb{1}\,|\,\boldsymbol{a}\})$ with $\{\mathbb{1}\,|\,\boldsymbol{a}\} \in U$ form a reducible representation of U, unless $n = 1$. This means that there is a basis for the representation space such that the matrices of $D(U)$ are diagonal. The irreducible components of the representation $D(U)$ are characterized by vectors $\boldsymbol{k}_1, ..., \boldsymbol{k}_n$.

$$D(\{\mathbb{1}\,|\,\boldsymbol{a}\}) = \begin{pmatrix} e^{i\boldsymbol{k}_1\cdot\boldsymbol{a}} & 0 & \ldots & \ldots & 0 \\ 0 & e^{i\boldsymbol{k}_2\cdot\boldsymbol{a}} & 0 & \ldots & 0 \\ 0 & 0 & e^{i\boldsymbol{k}_3\cdot\boldsymbol{a}} & 0 & \ldots \\ \ldots & \ldots & \ldots & \ldots & \ldots \\ 0 & \ldots & \ldots & 0 & e^{i\boldsymbol{k}_n\cdot\boldsymbol{a}} \end{pmatrix} . \tag{4.19}$$

Among the vectors $\boldsymbol{k}_1, ..., \boldsymbol{k}_n$ equal ones may occur. When ψ is a basis function for the irreducible representation of U that is characterized by the vector $\boldsymbol{k}$ in the Brillouin zone, we will say that ψ transforms with $\boldsymbol{k}$, or according to the vector $\boldsymbol{k}$. If ψ is a basis function of the irreducible space transforming with $\boldsymbol{k}_1$ and if $T_{\{R|\boldsymbol{t}\}}$ is the linear operator on the representation space corresponding to the space group element $\{R\,|\,\boldsymbol{t}\}$, the function $\psi' = T_{\{R|\boldsymbol{t}\}}\psi$ transforms under a translation $\{\mathbb{1}\,|\,\boldsymbol{a}\} \in U$ according to

$$T_{\{\mathbb{1}|\boldsymbol{a}\}}T_{\{R|\boldsymbol{t}\}}\psi = T_{\{R|\boldsymbol{t}\}}T_{\{\mathbb{1}|R^{-1}\boldsymbol{a}\}}\psi = e^{i\boldsymbol{k}_1\cdot R^{-1}\boldsymbol{a}}\,\psi' .$$

Because $\boldsymbol{k}_1 \cdot R^{-1}a = R\boldsymbol{k}_1 \cdot a$ the function ψ' transforms under U according to the vector $R\boldsymbol{k}$. As ψ' belongs to the representation space, the vector $R\boldsymbol{k}_1$ must occur among $\boldsymbol{k}_1, ..., \boldsymbol{k}_n$. This is the case for any R from the point group K. We call the set of $\boldsymbol{k}$-vectors $\{R\boldsymbol{k}\,|\,R\in K\}$ the *star of* $\boldsymbol{k}$. As $\boldsymbol{k}$ and $\boldsymbol{k} + \boldsymbol{K}$ characterize the same representation, two vectors in the star, $R_1\boldsymbol{k}$ and $R_2\boldsymbol{k}$, are the same if there is a $\boldsymbol{K} \in \Lambda^*$ such that $R_1\boldsymbol{k} = R_2\boldsymbol{k} + \boldsymbol{K}$. Then, when $\boldsymbol{k}$ occurs in the matrix (4.19), all the members of the star of $\boldsymbol{k}$ occur also. It can be shown (details are given in §2.7), that if $D(G)$ is an irreducible representation all occurring $\boldsymbol{k}$-vectors belong to one star, and that the multiplicities of the irreducible components characterized by the vectors of this star are the same. So when a vector $\boldsymbol{k}$ occurs d times, the other vectors of the star of $\boldsymbol{k}$ occur also d times. When $R_1 = \mathbb{1}, R_2, ..., R_s$ are elements of K such that $\boldsymbol{k}, R_2\boldsymbol{k}, ..., R_s\boldsymbol{k}$ are the vectors of the star of $\boldsymbol{k}$, denoted by $\boldsymbol{k}_1, ..., \boldsymbol{k}_s$, respectively, the matrix (4.19) can be written

$$D(\{\mathbb{1}\,|\,\boldsymbol{a}\}) = \begin{pmatrix} e^{i\boldsymbol{k}_1\cdot\boldsymbol{a}}\,\mathbb{I} & & 0 \\ & e^{i\boldsymbol{k}_2\cdot\boldsymbol{a}}\,\mathbb{I} & \\ \ldots & \ldots & \ldots \\ 0 & & e^{i\boldsymbol{k}_s\cdot\boldsymbol{a}}\,\mathbb{I} \end{pmatrix} . \tag{4.20}$$

Each block in the diagonal is of the same dimension d. For the dimension n of $D(G)$, one has $n = sd$.

For a given vector $\boldsymbol{k}$ one defines a group $G_{\boldsymbol{k}}$ called the *group of* $\boldsymbol{k}$. It is the subgroup of G of all elements with a homogeneous part R which leaves $\boldsymbol{k}$ invariant (up to a reciprocal lattice vector):

$$G_{\boldsymbol{k}} = \{\{R\,|\,\boldsymbol{t}\} \in G \,|\, R\boldsymbol{k} = \boldsymbol{k} + \boldsymbol{K} \text{ for some } \boldsymbol{K} \in \Lambda^*\}\,. \tag{4.21}$$

Notice that $U \subseteq G_{\boldsymbol{k}}$. The group G can be decomposed into cosets of the subgroup $G_{\boldsymbol{k}}$. As representatives of the cosets one can choose elements $\{\mathbb{1}\,|\,\boldsymbol{t}_1\}$, $\{R_2\,|\,\boldsymbol{t}_2\}$, ..., $\{R_s\,|\,\boldsymbol{t}_s\}$ such that $\boldsymbol{k}_i = R_i\boldsymbol{k}_1$. Denoting $\{R_i\,|\,\boldsymbol{t}_i\}$ by g_i one has

$$G = G_{\boldsymbol{k}} + g_2 G_{\boldsymbol{k}} + \ldots + g_s G_{\boldsymbol{k}}\,.$$

The subspace of the space which carries $D(G)$ that transforms under U according to the vector $\boldsymbol{k}$, is invariant under $G_{\boldsymbol{k}}$: when $\{R\,|\,\boldsymbol{t}\} \in G_{\boldsymbol{k}}$ and ψ transforms as $T_{\{\mathbb{1}\,|\,\boldsymbol{a}\}}\psi = \mathrm{e}^{i\boldsymbol{k}\circ\boldsymbol{a}}\psi$, the function $T_{\{R\,|\,\boldsymbol{t}\}}\psi$ transforms with $R\boldsymbol{k} \equiv \boldsymbol{k}$ (up to $\boldsymbol{K} \in \Lambda^*$), so also belongs to the same space. Choose a basis $\psi_{1\mu}$ (with $\mu = 1, ..., d$) of this space. Then the functions

$$\psi_{i\mu} = T_{g_i}\psi_{1\mu} \qquad (i = 1, ..., s\,;\, \mu = 1, ..., d) \tag{4.22}$$

form a basis for the space carrying $D(G)$ because for fixed i the functions $\psi_{i\mu}$ carry a space of functions belonging to $\boldsymbol{k}_i$. The matrix representation of T_G with respect to the basis (4.22) is constructed as follows.

Consider $\{R\,|\,\boldsymbol{t}\} \in G$. The element $\{R\,|\,\boldsymbol{t}\}g_i$ belongs to a coset $g_j G_{\boldsymbol{k}}$. So there is an element $\{S\,|\,\boldsymbol{u}\} \in G_{\boldsymbol{k}}$ such that $\{R\,|\,\boldsymbol{t}\}g_i = g_j\{S\,|\,\boldsymbol{u}\}$. The index j and the element $\{S\,|\,\boldsymbol{u}\}$ are uniquely determined by $\{R\,|\,\boldsymbol{t}\}$ and i. With this relation one can now write

$$T_{\{R\,|\,\boldsymbol{t}\}}\psi_{i\mu} = T_{g_j}T_{\{S\,|\,\boldsymbol{u}\}}T_{g_i}^{-1}\psi_{i\mu} = T_{g_j}T_{\{S\,|\,\boldsymbol{u}\}}\psi_{1\mu}$$

$$= T_{g_j}\sum_{\nu=1}^{d} D_{\boldsymbol{k}}(\{S\,|\,\boldsymbol{u}\})_{\nu\mu}\psi_{1\nu} = \sum_{\nu=1}^{d} D_{\boldsymbol{k}}(\{S\,|\,\boldsymbol{u}\})_{\nu\mu}\psi_{j\nu}\,.$$

Here $D_{\boldsymbol{k}}(G_{\boldsymbol{k}})$ denotes the representation of $G_{\boldsymbol{k}}$ carried by the space of the vector $\boldsymbol{k}$. One can divide the matrix $D(\{R\,|\,\boldsymbol{t}\})$ into s^2 blocks of dimension d, according to the division of the matrix (4.20). In the i-th column of these blocks the matrix $D(\{R\,|\,\boldsymbol{t}\})$ is different from zero only in the j-th row, where j is determined by $\{R\,|\,\boldsymbol{t}\}g_i \in g_j G_{\boldsymbol{k}}$. The block on the intersection of

i-th column and j-th row is the matrix $D_k(\{S|u\})$. When we define an $s \times s$ matrix with exactly one nonzero element in each row and each column given by $M_{ji} = 1$ for i, j such that $\{R|t\}g_i \in g_j G_k$, the matrix $D(\{R|t\})$ is

$$D(\{R|t\})_{j\nu,i\mu} = M(\{R|t\})_{ji} D_k(g_j^{-1}\{R|t\}g_i)_{\nu\mu} \, . \tag{4.23}$$

So $D(G)$ is determined by the representations $D_k(G_k)$. It can be shown that $D(G)$ is irreducible and unitary if and only if $D_k(G_k)$ is irreducible and unitary.

To find the irreducible representations $D_k(G_k)$ one proceeds as follows. First we remark that $U \subseteq G_k$ and that under elements of U basis functions of $D_k(G_k)$ transform according to the vector $\boldsymbol{k}$, so $D_k(\{\mathbb{1}|\boldsymbol{a}\}) = \exp\{i\boldsymbol{k}\cdot\boldsymbol{a}\}\,\mathbb{1}$. As any element $\{S|u\} \in G_k$ can be written as $\{\mathbb{1}|\boldsymbol{a}\}\{S|\boldsymbol{t}_S\}$ for some nonprimitive translation $\boldsymbol{t}_S$ one can write

$$D_k(\{S|u\}) = \mathrm{e}^{i\boldsymbol{k}\cdot\boldsymbol{a}} D_k(\{S|\boldsymbol{t}_S\}) = \mathrm{e}^{i\boldsymbol{k}\cdot(\boldsymbol{a}+\boldsymbol{t}_S)}\Gamma(S) \, . \tag{4.24}$$

Here the matrix $\Gamma(S)$ is defined by $\Gamma(S) = \exp\{-i\boldsymbol{k}\cdot\boldsymbol{t}_S\} D_k(\{S|\boldsymbol{t}_S\})$. The matrix $\Gamma(S)$ is independent of the choice of $\boldsymbol{t}_S$, because for $\boldsymbol{t}_S' = \boldsymbol{t}_S + \boldsymbol{a}$ one has $\Gamma'(S) = \exp(-i\boldsymbol{k}\cdot(\boldsymbol{a}+\boldsymbol{t}_S)) D_k(\{1|\boldsymbol{a}\}\{S|\boldsymbol{t}_S\}) = \Gamma(S)$. The product of two matrices $\Gamma(S)$ and $\Gamma(S')$ is

$$\begin{aligned} \Gamma(S)\,\Gamma(S') &= \exp\{-i\boldsymbol{k}\cdot(\boldsymbol{t}_S + \boldsymbol{t}_{S'})\}\, D_k(\{SS'|\boldsymbol{t}_S + S\boldsymbol{t}_{S'}\}) \\ &= \exp\{-i\boldsymbol{k}\cdot(\boldsymbol{t}_{S'} - S\boldsymbol{t}_{S'})\}\,\Gamma(SS') \, . \end{aligned} \tag{4.25}$$

Thus the matrices $\Gamma(S)$ form a projective representation of the point group K_k of the space group G_k. In two cases the projective representation is an ordinary representation: when G_k is symmorphic or when $\boldsymbol{k}$ is inside the first Brillouin zone. When G_k is symmorphic, both $\boldsymbol{t}_S$ and $\boldsymbol{t}_{S'}$ in eq. (4.25) can be chosen to be zero. When $\boldsymbol{k}$ is inside the Brillouin zone, one has $\exp\{-i\boldsymbol{k}\cdot(\boldsymbol{t}_{S'} - S\boldsymbol{t}_{S'})\} = \exp\{-i(\boldsymbol{k} - S^{-1}\boldsymbol{k})\cdot\boldsymbol{t}_{S'}\} = 1$ because $S\boldsymbol{k} = \boldsymbol{k} + \boldsymbol{K}$ but this is possible only with $\boldsymbol{K} = \boldsymbol{O}$. So when G_k is symmorphic or when $\boldsymbol{k}$ is inside the Brillouin zone (not on the border) the irreducible representations of G_k are given by eq. (4.24) with $\Gamma(K_k)$ an irreducible representation of K_k. The nonequivalent irreducible representations of K_k give the nonequivalent irreducible representations of G_k. When G_k is a nonsymmorphic space group and $\boldsymbol{k}$ is on the border of the Brillouin zone, the nonequivalent irreducible representations of G_k are given by the nonequivalent irreducible projective representations with factor system

$$\omega(S,S') = e^{-i(\boldsymbol{k}-S^{-1}\boldsymbol{k})\cdot \boldsymbol{t}_{S'}} . \tag{4.26}$$

This method of finding the irreducible representations of space groups is in fact the method of induction of representations. This method will be treated in more detail in §2.7, as it is also of importance for other problems in group theory. In §2.7 we will also be a bit more precise in our statements.

The nonequivalent representations of G are characterized by the star of a vector $\boldsymbol{k}$ and by an irreducible (projective) representation of the point group $K_{\boldsymbol{k}}$. One can define a domain in the Brillouin zone with the property that in this domain lies exactly one vector from each star. This domain is called a *fundamental region, reduced zone*, or *representation domain*. Then the non-equivalent irreducible representations of G are characterized by a vector $\boldsymbol{k}$ from such a fundamental region and an irreducible representation of $K_{\boldsymbol{k}}$.

4.2.3. Example

Consider the two-dimensional symmorphic space group G for the pattern of fig. 4.2b. Its translation subgroup U is generated by two translations of equal length making an angle of 60°. The point group K is the group $6mm$ of order 12, which is isomorphic to $D_6 \cong D_3 \times C_2$. The reciprocal lattice is again a hexagonal lattice generated by $\boldsymbol{b}_1$ and $\boldsymbol{b}_2$ (fig. 4.5a). The first Brillouin zone is a regular hexagon. Each vector in this zone belongs to a star of a vector in the triangle PQR, which is a fundamental region in the Brillouin zone. The group of a vector $\boldsymbol{k}$ is a symmorphic space group with point group $K_{\boldsymbol{k}}$. This point group and the number s of vectors in the star is

$$\begin{array}{llll}
K_{\boldsymbol{k}} \cong 6mm \cong D_3 \times C_2 , & s=1 & \text{for} & \boldsymbol{k}=P=\boldsymbol{O} , \\
K_{\boldsymbol{k}} \cong 3m \quad \cong D_3 , & s=2 & \text{for} & \boldsymbol{k}=Q \qquad \text{(fig. 4.5b)} , \\
K_{\boldsymbol{k}} \cong 2mm \cong D_2 , & s=3 & \text{for} & \boldsymbol{k}=R , \\
K_{\boldsymbol{k}} \cong m \quad \cong C_2 , & s=6 & \text{for} & \boldsymbol{k}=S, V \text{ or } T \quad \text{(fig. 4.5b)} , \\
K_{\boldsymbol{k}} \cong 1 \quad \cong C_1 , & s=12 & \text{for} & \boldsymbol{k} \text{ inside } PQR \quad \text{(fig. 4.5c)} .
\end{array}$$

The nonequivalent irreducible representations of G are found as follows for each $\boldsymbol{k}$ from PQR.

1) For $\boldsymbol{k} = P$. Here $G_{\boldsymbol{k}} = G$ and the point group $K_{\boldsymbol{k}}$ has 6 nonequivalent irreducible representations $\Gamma_1, ..., \Gamma_6$. The corresponding representations $D_{Pi}(G)$ are given by

$$D_{Pi}(\{R \mid \boldsymbol{t}\}) = \Gamma_i(\alpha) ,$$

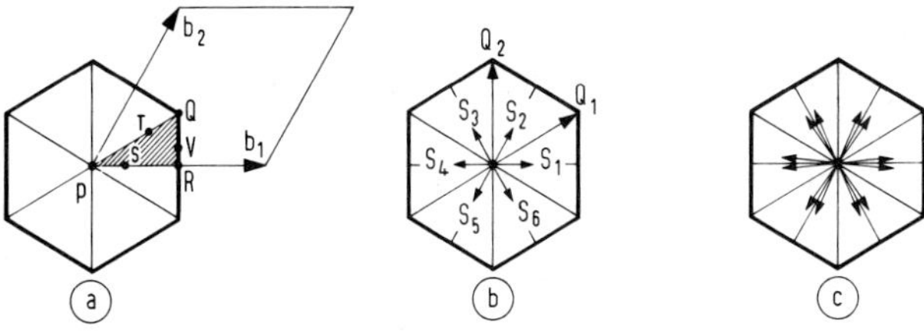

Fig. 4.5. $\boldsymbol{k}$ vectors and their stars in the Brillouin zone of $P6mm$.

where α is the element of the abstract point group corresponding to R.
2) For $\boldsymbol{k} = Q$. Here $G = G_{\boldsymbol{k}} + g_2 G_{\boldsymbol{k}}$, where one can choose $g_2 = \{R\,|\,\boldsymbol{O}\}$ with R a 60° rotation. The point group is generated by α (corresponding to R) and β (which corresponds to the reflection which interchanges $\boldsymbol{b}_1$ and $\boldsymbol{b}_2$). The group D_3 has 3 nonequivalent irreducible representations. When $\{S\,|\,\boldsymbol{t}\} \in G_{\boldsymbol{k}}$, one has $\{S\,|\,\boldsymbol{t}\}g_1 = \{S\,|\,\boldsymbol{t}\}\{\mathbb{1}\,|\,\boldsymbol{O}\} = \{\mathbb{1}\,|\,\boldsymbol{O}\}\{S\,|\,\boldsymbol{t}\}$ and $\{S\,|\,\boldsymbol{t}\}g_2 = \{S\,|\,\boldsymbol{t}\}\{R\,|\,\boldsymbol{O}\} = \{R, O\}\{R^{-1}SR\,|\,R^{-1}\boldsymbol{t}\} = g_2\{R^{-1}SR\,|\,R^{-1}\boldsymbol{t}\}$. The matrix $D_{Qi}(\{S\,|\,\boldsymbol{t}\})$ has the form

$$D_{Qi}(\{S\,|\,\boldsymbol{t}\}) = \begin{bmatrix} e^{i\boldsymbol{k}\cdot\boldsymbol{t}}\Gamma_i(\gamma) & 0 \\ 0 & e^{iR\boldsymbol{k}\cdot\boldsymbol{t}}\Gamma_i(\alpha^{-1}\gamma\alpha) \end{bmatrix}$$

when γ in the abstract point group corresponds to S and when $\{S\,|\,\boldsymbol{t}\} \in G_{\boldsymbol{k}}$. The matrices of D_{Qi} for the elements of the coset $g_2 G_{\boldsymbol{k}}$ are the products of the matrix $D_{Qi}(g_2)$ and the matrices representing the elements of $G_{\boldsymbol{k}}$. Thus the representation is known if one gives $D_{Qi}(g_2)$. This matrix follows from the relations $g_2 g_1 = g_2\{\mathbb{1}\,|\,\boldsymbol{O}\}$ and $g_2 g_2 = g_1\{R^2\,|\,\boldsymbol{O}\}$. Therefore,

$$D_{Qi}(\{R\,|\,\boldsymbol{O}\}) = \begin{bmatrix} 0 & \Gamma_i(\alpha^2) \\ \mathbb{1} & 0 \end{bmatrix} \qquad (i = 1, 2, 3)\,.$$

3) For $\boldsymbol{k}$ in general position one has the decomposition $G = U + g_2 U + \ldots$ $\ldots + g_N U$, where $g_i = \{R_i\,|\,\boldsymbol{O}\}$ are the elements of the point group K. The representation for this $\boldsymbol{k}$ is 12-dimensional. There is only one representation for $K_{\boldsymbol{k}} \cong C_1$. The matrix $D(\{R\,|\,\boldsymbol{t}\})$ is obtained as the product of the matrices $D(\{R\,|\,\boldsymbol{O}\})$ and $D(\{\mathbb{1}\,|\,\boldsymbol{t}\})$. The matrix for g_l is obtained from the relations $g_l g_i = g_j = g_j\{\mathbb{1}\,|\,\boldsymbol{O}\}$, where g_j is the product of g_l and g_i. The matrix $D(g_l)$ is the matrix in the regular representation. In each row and column there is

exactly one 1 and the other elements are 0. For an element $\{\mathbb{1}\,|\,\boldsymbol{u}\} \in U = G_{\boldsymbol{k}}$ the matrix $D(\{\mathbb{1}\,|\,\boldsymbol{u}\})$ for this $\boldsymbol{k}$ is

$$D(\{\mathbb{1}\,|\,\boldsymbol{u}\}) = \begin{pmatrix} e^{i\boldsymbol{k}\cdot\boldsymbol{u}} & 0 \ldots\ldots\ldots\ldots 0 \\ 0 & e^{iR_2\boldsymbol{k}\cdot\boldsymbol{u}} \quad 0 \ldots 0 \\ \ldots\ldots\ldots\ldots & \ldots\ldots\ldots\ldots \\ 0 \ldots\ldots\ldots\ldots & e^{iR_{12}\boldsymbol{k}\cdot\boldsymbol{u}} \end{pmatrix} .$$

One can proceed in an analogous way for the other points in the fundamental region.

4.2.4. Subduced representations

When H is a subgroup of a group G and $D(G)$ a representation of G, the matrices $D(h)$ with $h \in H$ form a representation of H. The representation property follows in a trivial way from that of $D(G)$. The representation $D(H)$, also denoted by $(D \downarrow H)$, is called a *representation subduced* from $D(G)$. When the subduced representation is irreducible, the representation $D(G)$ is also irreducible, but the converse is in general not true. The subduced representations can be decomposed into irreducible components.

In the case that H is an invariant subgroup of G one can define representations $D_g(H)$ for an arbitrary representation $D(H)$ by

$$D_g(h) = D(ghg^{-1}) \qquad (g \in G,\ h \in H) . \tag{4.27}$$

The representations $D(H)$ and $D_g(H)$ are called *conjugate representations* with respect to G. The representation property is easily verified from

$$D_g(h_1)\,D_g(h_2) = D(gh_1g^{-1}gh_2g^{-1}) = D_g(h_1h_2) .$$

Notice that the matrices of $D(H)$ and $D_g(H)$ are the same. Only the correspondence between matrices and elements of H has been changed.

The dimensions of $D(H)$ and $D_g(H)$ are the same. The representation $D_g(H)$ is irreducible if and only if $D(H)$ is irreducible, as follows from the fact that the matrices are the same. The representations $D(H)$ and $D_g(H)$ are not necessarily equivalent. They are equivalent when g is an element of H, because then

$$D_g(h) = D(ghg^{-1}) = D(g)\,D(h)\,D(g)^{-1} .$$

For the same reason the representations $(D\downarrow H)_g(H)$ and $(D\downarrow H)(H)$ are equivalent for $g\in G$. When a representation $D(H)$ of $H\subseteq G$ is equivalent to $D_g(H)$ for any $g\in G$, the representation is *self-conjugate* with respect to G. All elements of G such that $D(H)$ and $D_g(H)$ are equivalent are called elements of the *little group* of $D(H)$ in G. The little group of a subduced representation or of a selfconjugate representation is always the group G itself. In general the little group L satisfies $H\subseteq L\subseteq G$. The first inclusion is a consequence of the fact that for any $h\in H$ the representations $D(H)$ and $D_h(H)$ are equivalent, the second inclusion follows from the definition. The little group is indeed a group. To prove the group property one takes $l_1, l_2\in L$. Then $D_{l_1}(H)=U_1D(H)U_1^{-1}$ and $D_{l_2}(H)=U_2D(H)U_2^{-1}$ for certain nonsingular transformations U_1 and U_2. Then $D_{l_1l_2}(h)=D(l_1l_2hl_2^{-1}l_1^{-1})=D_{l_1}(l_2hl_2^{-1})=U_1D(l_2hl_2^{-1})U_1^{-1}=U_1U_2D(h)U_2^{-1}U_1^{-1}$. Thus $D_{l_1l_2}(H)$ is equivalent to $D(H)$ and l_1l_2 is an element of the little group. Moreover, $e\in H\subseteq L$ and if $l\in L$ also $l^{-1}\in L$.

In general for a subduced representation one has

$$(D\downarrow H)(H)=\sum_{\alpha} m_\alpha D_\alpha(H)\,, \tag{4.28}$$

a decomposition into irreducible components. In particular one can consider the subduced representation $D(L)\downarrow H$, where L is the little group of $D'(H)$. When the irreducible components of this subduced representation are all equivalent to $D'(H)$, the representation $D(L)$ is called an *allowable representation*. The meaning of several of the concepts introduced here will become clear in the next section.

4.2.5. Induced representations

A process which is in some sense inverse to subducing is the induction of a representation of a group G from a representation of a subgroup H. The classical theory of induced representations of finite groups was given by Frobenius (1898), Weyl (1920) and Clifford (1937). See Clifford [1937]. The next important contribution was given by Mackey (Mackey [1955]) who discussed induced representations for a large class of infinite groups. The method has become of great importance in physics. Examples are the papers by Wigner (Wigner [1939]) on the representations of the Poincaré group and by Koster (Koster [1957]) on the representations of space groups. A nice review is given in Coleman [1968]. The results of this section will be a repetition of those of §2.2. However, we will treat induced representations here in view of their importance. We will not give full proofs of all the propositions, but state only the most important results.

An induced representation is a generalization of the concept of regular representation. Let H be a subgroup of a group G. Suppose $D(H)$ is a representation of H in a linear vector space V. In Ch. 1, § 1.7 we considered functions on a group. Here we consider functions on G with values in V. When f_1 and f_2 are such functions, so that $f_1(g)$ and $f_2(g)$ are elements of V for each $g \in G$, the function $(\alpha f_1 + \beta f_2)(g) = \alpha f_1(g) + \beta f_2(g)$ is also a function on G. Thus these functions form a linear vector space V^G. A subset of this space is formed by those functions f which satisfy

$$f(gh) = D(h^{-1})\, f(g) \qquad \text{(any } g \in G, \text{ any } h \in H) .$$

On this linear vector space W we define linear operators T_a for any $a \in G$ by

$$T_a f(g) = f(a^{-1}g) . \tag{4.29}$$

Since $(T_a f)(gh) = f(a^{-1}gh) = D(h^{-1}) f(a^{-1}g) = D(h^{-1})(T_a f)(g)$ it is an operator on W. As $T_a(\alpha f_1 + \beta f_2)(g) = \alpha f_1(a^{-1}g) + \beta f_2(a^{-1}g)$, it is a linear operator. As $T_a T_b f(g) = f(b^{-1}a^{-1}g) = T_{ab} f(g)$, T_G is a representation of G. This representation is called the *representation of G induced from D(H)*. It is denoted by $D \uparrow G$ or $D(H) \uparrow G$.

To obtain a matrix representation for $D \uparrow G$ one has to choose a basis in W. Suppose that $\boldsymbol{e}_1, ..., \boldsymbol{e}_d$ forms a basis for the vector space V. Then a basis for V^G is formed by the functions f_{ai} $(a \in G, i = 1, ..., d)$ defined by

$$f_{ai}(g) = \begin{cases} 0 & \text{when } g \neq a \\ \boldsymbol{e}_i & \text{when } g = a . \end{cases}$$

The functions belonging to W are already determined by their values on representatives g_i of the cosets of H in G, where

$$G = H + g_2 H + ... + g_s H . \tag{4.30}$$

Therefore, a basis of W is given by the functions $f_{\mu i}$ $(\mu = 1, ..., s;\ i = 1, ..., d)$ defined by

$$f_{\mu i}(g_\nu) = \boldsymbol{e}_i \delta_{\mu\nu} .$$

These functions form a basis because an arbitrary function f can be written as a linear combination

$$f = \sum_{\mu i} f(g_\mu)_i f_{\mu i} ,$$

where $\Sigma_i f(g)_i e_i = f(g)$. Then one obtains for the function $T_a f_{\mu i}$

$$T_a f_{\mu i}(g_\nu) = f_{\mu i}(a^{-1} g_\nu) = f_{\mu i}(g_\sigma h^{-1}) = \sum_j D(h)_{ji} e_j \delta_{\mu\sigma} ,$$

where $a^{-1} g_\nu$ belongs to the coset $g_\sigma H$ and $a g_\sigma = g_\nu h$ for some $h \in H$. Finally one has

$$T_a f_{\mu i} = \sum_{\nu j} (T_a f_{\mu i})(g_\nu)_j f_{\nu j} = \sum_{\nu j} D(h)_{ji} \delta_{\sigma\mu} f_{\nu j} .$$

So the matrix $(D \uparrow G)(a)$ is given by

$$(D \uparrow G)(a)_{\nu j, \mu i} = D(h)_{ji} \delta_{\sigma\mu} . \tag{4.31}$$

It is an $sd \times sd$ matrix which can be divided in $d \times d$ blocks. There is only one nonzero block in each row. In the ν-th row it is the block in the σ-th column. This block is given by $D(h) = D(g_\nu^{-1} a g_\sigma)$. Compare this with eq. (4.23). Introducing

$$M(a,h)_{\nu\mu} = \begin{cases} 1 & \text{if} \quad a g_\mu = g_\nu h \\ 0 & \text{otherwise} \end{cases} \tag{4.32}$$

one can write

$$(D \uparrow G)(a) = \sum_{h \in H} \{M(a,h) \otimes D(h)\} .$$

From the construction it is easily seen that an equivalent representation $D(H)$ or another choice of representatives g_μ give an equivalent induced representation. A special case of an induced representation is the regular representation of G. It is obtained by choosing the unit element as the subgroup H. The only irreducible representation of this group is given by the complex number $D(e) = 1$. Then $D \uparrow G(g) = M(g)$ with the definition of eq. (4.32). For the character of $D \uparrow G$ one finds from eq. (4.31)

$$\chi(g) = \sum_\nu{}' \operatorname{tr} D(g_\nu^{-1} g g_\nu) , \tag{4.33}$$

where the prime means summation over all ν such that $gg_\nu \in g_\nu H$.

We now mention three propositions without proof. The proofs can be found, e.g. in Coleman [1968], or in Jansen and Boon [1967].

PROPOSITION 4.5. Induction is transitive:

$$(D(H) \uparrow L) \uparrow G = D(H) \uparrow G \quad \text{when} \quad H \subseteq L \subseteq G\,.$$

PROPOSITION 4.6. (Frobenius reciprocity theorem). When D is an irreducible representation of a subgroup H of G, and Δ an irreducible representation of G, the multiplicity of Δ in $D \uparrow G$ is equal to the multiplicity of D in $\Delta \downarrow H$.

PROPOSITION 4.7. Let D be an irreducible representation of $H \subseteq G$. Let H_g be the intersection $gHg^{-1} \cap H$ for any $g \in G$. The induced representation $D \uparrow G$ is irreducible if and only if the representations $D(g^{-1}H_g g)$ and $D(H_g)$ have no irreducible component in common, unless $g \in H$.

The two last propositions imply that in general the representation induced from an irreducible representation is not irreducible. In the following we will suppose that H is an invariant subgroup of G. Even then, as proposition 4.7 shows, a representation of G induced from an irreducible $D(H)$ is in general not irreducible. To find the irreducible representations of G one proceeds as follows. When $D(H)$ is a representation, its *orbit* is the set of equivalence classes of the conjugate representations $D_g(H)$ for all $g \in G$. Thus G maps the elements of the orbit onto each other. The little group of $D(H)$ maps the equivalence class of $D(H)$ onto itself. For $G = L + g_2L + \dots + g_sL$ the elements of the orbit are the s equivalence classes of $D_{g_i}(H)$ with $i = 1, \dots, s$. If $D(H)$ is irreducible, all elements of the orbit of $D(H)$ are irreducible, because $D(H)$ and $D_g(H)$ are the same sets of matrices. Now we can formulate

PROPOSITION 4.8. When H is an invariant subgroup of the finite group G, $D(H)$ an irreducible representation of H and Δ an allowable irreducible representation of the little group L of $D(H)$ with respect to G, then

a) the representation $\Delta \uparrow G$ of G is irreducible,

b) all irreducible representations of G may be obtained in this way.

PROPOSITION 4.9. All irreducible nonequivalent representations of G can be found by taking one representative $D(H)$ from each orbit of irreducible representations of H, constructing all nonequivalent allowable representations of the little groups of the representations $D(H)$ and forming the induced representations of G from the representations of the little groups.

4.2.6. Representations of space groups by the method of induction

To obtain the nonequivalent irreducible representations of a space group

G one uses the fact that G has an invariant subgroup U for which the irreducible representations are known. The method used in this section is essentially the same as that used in §2.2, but we now apply the more precise propositions 4.8 and 4.9 which have a great generality. Later on we will see other examples of the method of induction.

Consider a representation $D^{(\boldsymbol{k})}$ of the translation subgroup U. The orbit of this representation is obtained from

$$D_g^{(\boldsymbol{k})}(\{\mathbb{1}\,|\,\boldsymbol{a}\}) = D^{(\boldsymbol{k})}(g\{\mathbb{1}\,|\,\boldsymbol{a}\}g^{-1}) = D^{(\boldsymbol{k})}(\{\mathbb{1}\,|\,R\boldsymbol{a}\})$$

$$= \mathrm{e}^{i\boldsymbol{k}\cdot R\boldsymbol{a}} = \mathrm{e}^{iR^{-1}\boldsymbol{k}\cdot\boldsymbol{a}} = D^{(R^{-1}\boldsymbol{k}+\boldsymbol{K})}(\{\mathbb{1}\,|\,\boldsymbol{a}\})\,,$$

where $g = \{R\,|\,\boldsymbol{t}\}$, $\{\mathbb{1}\,|\,\boldsymbol{a}\} \in U$ and $\boldsymbol{K}$ a reciprocal lattice vector such that $R^{-1}\boldsymbol{k} + \boldsymbol{K}$ in the first Brillouin zone. Thus the orbit of $D^{(\boldsymbol{k})}$ has elements characterized by the $\boldsymbol{k}$-vectors $R\boldsymbol{k}$ for all point group elements R, i.e. by the $\boldsymbol{k}$-vectors of the star of $\boldsymbol{k}$. The orbit is characterized by the star of $\boldsymbol{k}$. The number of points of the star is the number of elements of the orbit.

The second step is the determination of the little group of one representation from each orbit, i.e. for one point from each star. An element $g = \{R\,|\,\boldsymbol{t}\}$ belongs to the little group of $D^{(\boldsymbol{k})}$ if $D_g^{(\boldsymbol{k})}$ is equivalent to $D^{(\boldsymbol{k})}$, i.e. if $R^{-1}\boldsymbol{k} = \boldsymbol{k} + \boldsymbol{K}$ for some $\boldsymbol{K} \in \Lambda^*$. This means that the little group of $D^{(\boldsymbol{k})}$ is the group of $\boldsymbol{k}$ denoted by $G_{\boldsymbol{k}}$. Among the representations of $G_{\boldsymbol{k}}$ those are allowable which subduce a multiple of the representation $D^{(\boldsymbol{k})}$, i.e. those $D_{\boldsymbol{k}}(G_{\boldsymbol{k}})$ for which $D_{\boldsymbol{k}}(\{1\,|\,\boldsymbol{a}\}) = \mathrm{e}^{i\boldsymbol{k}\cdot\boldsymbol{a}_1}\,\mathbb{1}$ for $a \in U$. The allowable representations of $G_{\boldsymbol{k}}$ are given by

$$D_{\boldsymbol{k}}(\{R\,|\,\boldsymbol{t}\}) = \mathrm{e}^{i\boldsymbol{k}\cdot\boldsymbol{t}}\,\Gamma_\alpha(R) \qquad \text{for} \qquad \{R\,|\,\boldsymbol{t}\} \in G_{\boldsymbol{k}}\;,$$

as was derived in §2.2. Here $\Gamma_\alpha(K_{\boldsymbol{k}})$ are nonequivalent irreducible projective representations of the point group $K_{\boldsymbol{k}}$ of the group $G_{\boldsymbol{k}}$.

According to eq. (4.31) the representations induced from $D_{\boldsymbol{k}}(G_{\boldsymbol{k}})$ are found using the decomposition

$$G = G_{\boldsymbol{k}} + g_2 G_{\boldsymbol{k}} + \ldots + g_s G_{\boldsymbol{k}}\;.$$

The matrices of $D_{\boldsymbol{k}} \uparrow G$ have a block structure. In the i-th row of $d \times d$ blocks of the matrix $(D_{\boldsymbol{k}} \uparrow G)(g)$, only the block in the j-th column is different from zero, where $gg_j \in g_i G_{\boldsymbol{k}}$. The nonzero block in the i-th row is the matrix $D_{\boldsymbol{k}}(g_i^{-1} g g_j)$, as we found already in §2.2. However, because of the propositions 4.8 and 4.9 we are now sure that we have found all representations of G.

To be more precise, we have constructed all nonequivalent irreducible representations of the factor group G/NU where NU is the subgroup of U generated by $N\boldsymbol{a}_1, N\boldsymbol{a}_2, N\boldsymbol{a}_3$. A discussion of the projective representations of space groups can be found in Backhouse [1970, 1971] and Backhouse and Bradley [1970].

4.2.7. Irreducible representations of the group $F\,d3m$

In section 2.3 we considered the irreducible representations of a two-dimensional space group. In order to show the technique once more in an example in three dimensions, and in view of later applications, we will consider the irreducible representations of the space group of § 1.7, the space group of diamond. The reciprocal lattice of Λ^s generated by $\boldsymbol{A}_1, \boldsymbol{A}_2, \boldsymbol{A}_3$ is again a simple cubic lattice with basis elements $\boldsymbol{B}_i = 2\pi \boldsymbol{A}_i / |\boldsymbol{A}_i|^2$. The reciprocal lattice of Λ generated by $\boldsymbol{a}_1, \boldsymbol{a}_2, \boldsymbol{a}_3$ is then generated by

$$\boldsymbol{b}_1 = \boldsymbol{B}_1 + \boldsymbol{B}_2 - \boldsymbol{B}_3$$

$$\boldsymbol{b}_2 = \boldsymbol{B}_1 - \boldsymbol{B}_2 + \boldsymbol{B}_3$$

$$\boldsymbol{b}_3 = -\boldsymbol{B}_1 + \boldsymbol{B}_2 + \boldsymbol{B}_3 .$$

The reciprocal lattice on $\boldsymbol{b}_1, \boldsymbol{b}_2, \boldsymbol{b}_3$ is a b.c.c. (body-centered cubic) lattice. The Brillouin zones of the lattices Λ and Λ^s are drawn in fig. 4.6. A possible choice for the fundamental region, i.e. a region in the BZ in which one point of each star lies, is $\Gamma LUXWK$. Any vector from this region characterizes a star. One obtains all nonequivalent irreducible representations of $F\,d3m$ by taking all vectors $\boldsymbol{k}$ from the fundamental region, determining the associated groups $G_{\boldsymbol{k}}$ and considering the allowable irreducible representations of the

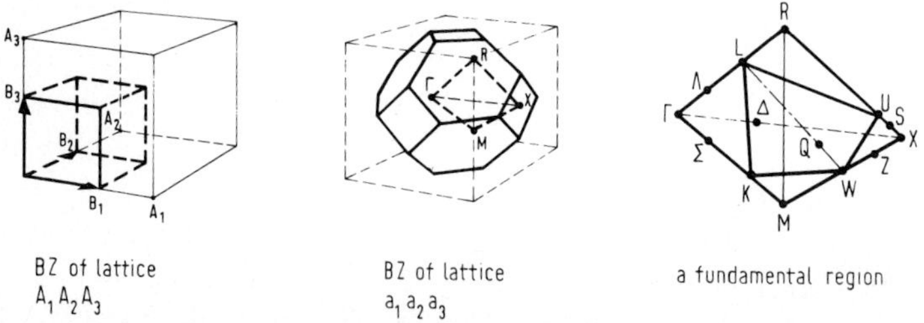

Fig. 4.6. The Brillouin zone and a fundamental region for the space group of diamond.

Table 4.5.
Groups of $\boldsymbol{k}$ for $\boldsymbol{k}$ in a fundamental region of the diamond Brillouin zone.

$\boldsymbol{k}$-vector w.r.t. B_1, B_2, B_3	point group $K_{\boldsymbol{k}}$	elements of $K_{\boldsymbol{k}}$	s	number of irreducible representations of dimension 1	2	3
$\Gamma = [000]$	$m3m \cong O \times C_2$	all elements of K	1	4	2	4
$X = [100]$	$4/mmm \cong D_4 \times C_2$	$\pm(1, 4_x, 2_x, 4_x^2, 2_y, 2_z, 2_{yz}, 2_{y\bar{z}})$	3	0	4	0 proj.
$L = [\frac{1}{2}\ -\frac{1}{2}\ \frac{1}{2}]$	$\bar{3}m \cong D_6$	$\pm(1, 3_{xyz}, 3_{xyz}^2, 2_{x\bar{y}}, 2_{x\bar{z}}, 2_{yz})$	4	4	2	0
$W = [1\ -\frac{1}{2}\ 0]$	$\bar{4}2m \cong D_4$	$1, -2_x, 2_y, -2_z, -4_y, -4_y^3, 2_{xz}, 2_{x\bar{z}}$	6	0	2	0 proj.
$\Lambda = \alpha L\ (0 < \alpha < 1)$	$3m \cong D_3$	$1, 3_{xyz}, 3_{xyz}^2, -2_{x\bar{y}}, -2_{x\bar{z}}, -2_{y\bar{z}}$	8	2	1	0
$\Delta = \alpha X\ (0 < \alpha < 1)$	$4mm \cong D_4$	$1, 4_x, 2_x, 4_x^3, -2_y, -2_z, -2_{yz}, -2_{y\bar{z}}$	6	4	1	0
$\Sigma = \alpha K\ (0 < \alpha < 1)$	$2mm \cong D_2$	$1, -2_z, 2_{xy}, -2_{x\bar{y}}$	12	4	0	0
$Z = \alpha W + (1-\alpha) X\ (0 < \alpha < 1)$	$2mm \cong D_2$	$1, 2_y, -2_z, -2_x$	12	0	1	0 proj.
$S = \alpha U + (1-\alpha) X\ (0 < \alpha < 1)$	$2mm \cong D_2$	$1, -2_x, 2_{yz}, -2_{y\bar{z}}$	12	4	0	0
$Q = \alpha L + (1-\alpha) W (0 < \alpha < 1)$	$2 \cong C_2$	$1, 2_{x\bar{z}}$	24	2	0	0
$\Gamma\Lambda\Delta$-plane	$m \cong C_2$	$1, -2_{y\bar{z}}$	24	2	0	0
$\Gamma\Sigma\Lambda$-plane	$m \cong C_2$	$1, -2_{x\bar{y}}$	24	2	0	0
$\Gamma\Delta\Sigma$-plane	$m \cong C_2$	$1, -2_z$	24	2	0	0
WUX-plane	$m \cong C_2$	$1, -2_x$	24	2	0	0
LKW-plane, LWU-plane, inside	1	1	48	1	0	0

groups $G_{\boldsymbol{k}}$. For the various points in the fundamental region the groups $G_{\boldsymbol{k}}$ and $K_{\boldsymbol{k}}$ are given in table 4.5. For some points we will discuss the corresponding representations in more detail.

1) The point Γ : $\boldsymbol{k} = [000]$. Here $G_{\boldsymbol{k}} = G$. The irreducible representations of G with the star of $\boldsymbol{k} = 0$ are the allowable representations of $G_{\boldsymbol{k}}$, which in turn are, according to eq. (4.24), the irreducible representations of $K_{\boldsymbol{k}} \cong O \times C_2$. The group $O \times C_2$ has 10 nonequivalent irreducible representations: $\Gamma_1^{\pm}, ..., \Gamma_5^{\pm}$. Hence there are four one-dimensional representations of G with $\boldsymbol{k} = \boldsymbol{O}$, two two-dimensional ones and four three-dimensional ones.

2) The point X : $\boldsymbol{k} = [100]$. The group $G_{\boldsymbol{k}}$ is of index 3 in G. One can choose for the coset representatives the elements $g_1 = \{\mathbb{1} \mid \boldsymbol{O}\}$, $g_2 = \{3_{xyz} | \boldsymbol{O}\}$, $g_3 = g_2^2$. The allowable irreducible representations of $G_{\boldsymbol{k}}$ are found from the irreducible projective representations of the point group $K_{\boldsymbol{k}} \cong D_4 \times C_2$ with factor system given by eq. (4.26). The group $D_4 \times C_2$ has generators α, β, γ with defining relations $\alpha^4 = \beta^2 = (\alpha\beta)^2 = \gamma^2 = \alpha\gamma\alpha^{-1}\gamma = \beta\gamma\beta\gamma = \epsilon$. One can choose for the orthogonal transformations of the group K the elements $R(\alpha) = 4_x$, $R(\beta) = -2_{yz}$, $R(\gamma) = -\mathbb{1} = I$. The allowable representations of $G_{\boldsymbol{k}}$ are obtained from projective representations $\Gamma(D_4 \times C_2)$ with factor system determined by eq. (4.26). As $R(\alpha)\boldsymbol{k} = R(\beta)\boldsymbol{k} = \boldsymbol{k}$ one has $\Gamma(\alpha)^4 = \Gamma(\beta)^2 = [\Gamma(\alpha)\,\Gamma(\beta)]^2 = \mathbb{1}$. Because $\boldsymbol{k} - R(\gamma)\boldsymbol{k} = 2\boldsymbol{k} = [200] = 2\boldsymbol{B}_1$ one has $\Gamma(\gamma)^2 = \omega(\gamma,\gamma)\,\mathbb{1} = \exp(-2i\boldsymbol{B}_1 \cdot \boldsymbol{\tau})\,\mathbb{1} = -\mathbb{1}$ with $\boldsymbol{\tau} = [\frac{1}{4}\,\frac{1}{4}\,\frac{1}{4}] = \boldsymbol{t}_{R(\gamma)}$. Furthermore, one has $\Gamma(\alpha)\,\Gamma(\gamma) = \omega(\alpha,\gamma)\,\Gamma(\alpha\gamma) = \Gamma(\alpha\gamma)$ because $R(\alpha)\boldsymbol{k} = \boldsymbol{k}$, whereas $\Gamma(\gamma)\,\Gamma(\alpha) = \omega(\gamma,\alpha)\,\Gamma(\alpha\gamma) = \exp(-2i\boldsymbol{B}_1 \cdot \boldsymbol{\tau})\,\Gamma(\alpha\gamma) = -\Gamma(\alpha\gamma)$. Finally, $\Gamma(\beta)\,\Gamma(\gamma) = \omega(\beta,\gamma)\,\Gamma(\beta\gamma) = \Gamma(\beta\gamma)$, and $\Gamma(\gamma)\,\Gamma(\beta) = \omega(\gamma,\beta)\,\Gamma(\beta\gamma) = \Gamma(\beta\gamma)$, because $\boldsymbol{t}_{R(\beta)} = \boldsymbol{O}$. This means that the representation $\Gamma(D_4 \times C_2)$ has relations

$$\Gamma(\alpha)^4 = \Gamma(\beta)^2 = [\Gamma(\alpha)\,\Gamma(\beta)]^2 = \Gamma(\beta)\,\Gamma(\gamma)\,\Gamma(\beta)^{-1}\Gamma(\gamma)^{-1} = \mathbb{1}\ ,$$

$$\Gamma(\gamma)^2 = \Gamma(\alpha)\,\Gamma(\gamma)\,\Gamma(\alpha)^{-1}\Gamma(\gamma)^{-1} = -\mathbb{1}\ .$$

To compare this with the relations for the representations given in the appendix we notice that $\Gamma(D_4 \times C_2)$ is associated to the representation $\Gamma'(\alpha) = \Gamma(\alpha)$, $\Gamma'(\beta) = \Gamma(\beta)$, $\Gamma'(\gamma) = \exp(i\pi/2)\,\Gamma(\gamma)$ with relation $\Gamma'(\gamma)^2 = \mathbb{1}$. In the appendix the factor system is characterized by the parameters $\lambda_1 = \lambda_3 = 1$, $\lambda_2 = -1$. There are four nonequivalent irreducible representations with this factor system. They are all of dimension 2 (notice that indeed $4 \times 2^2 = 16 =$ order of K). The induced representations are 6-dimensional.

3) The point Δ : $\boldsymbol{k} = [c00]$ with $0 < c < \frac{1}{2}$. The group $G_{\boldsymbol{k}}$ is a symmorphic space group with point group $K_{\boldsymbol{k}} \cong D_4$. There are 5 nonequivalent

irreducible ordinary representations of this group. For the decomposition of G into cosets of $G_{\boldsymbol{k}}$ one can choose coset representatives $g_1 = \{\mathbb{1} \mid O\}$, $g_2 = \{3_{xyz} \mid O\}$, $g_3 = g_2^2$, $g_4 = I$, $g_5 = Ig_2$, $g_6 = Ig_3$. For this $\boldsymbol{k}$ there are five irreducible representations of G: four of dimension 6, one of dimension 12.

For the other points we can proceed in an analogous way. For X, L and W projective representations occur, for the other points the irreducible representations are determined by ordinary representations of $K_{\boldsymbol{k}}$. The groups $G_{\boldsymbol{k}}$, $K_{\boldsymbol{k}}$ and the information for the representations of the 219 space groups are given in Miller and Love [1967] and Zak [1969].

4.3. Periodic potentials

4.3.1. Bloch electrons

The Hamiltonian of a crystal can be split up in a natural way in parts describing the cores, consisting of nuclei and electrons of the inner shells, the outer electrons, and their interaction, respectively. This can be done, because the interactions between the different cores are in general small compared with the binding energies of the inner electrons to the cores. The outer electron system can be described, neglecting spin effects, by

$$H = \sum_i \frac{p_i^2}{2m} + \sum_i V(r_i) + \sum_{i<j} V_{ij} ,$$

where $V(\boldsymbol{r})$ is the potential created by the cores and V_{ij} is the interaction between the i-th and j-th electrons. In the way which is usual in many-body quantum mechanics, the system without mutual interaction can be described by Slater determinants of one particle solutions of the Schrödinger problem with Hamiltonian

$$H = \frac{p^2}{2m} + V(r) , \tag{4.34}$$

where $V(\boldsymbol{r})$ is the crystal potential. This potential has the symmetry of the space group of the crystal: $V(g\boldsymbol{r}) = V(\boldsymbol{r})$ for any $g \in G$. In particular $V(\boldsymbol{r})$ is a periodic potential with the periodicity of the lattice of the space group. Then the eigenfunctions of H (eq. (4.34)) may be chosen as basis functions of irreducible representations of the translation subgroup U. These representations are characterized by a vector $\boldsymbol{k}$ from the first Brillouin zone. A basis function for the representation $D^{(\boldsymbol{k})}(U)$ transforms under $a \in U$ according to

$$T_{\boldsymbol{a}}\psi_{\boldsymbol{k}}(\boldsymbol{r}) = \mathrm{e}^{i\boldsymbol{k}\cdot\boldsymbol{a}}\psi_{\boldsymbol{k}}(\boldsymbol{r}) .$$

On the other hand for spinless particles the Hamiltonian H commutes with the substitution operators $P_{\boldsymbol{a}}$. So the representation of U is given by these operators. This leads to

$$P_{\boldsymbol{a}}\psi_{\boldsymbol{k}}(\boldsymbol{r}) = \psi_{\boldsymbol{k}}(\boldsymbol{r}-\boldsymbol{a}) = \mathrm{e}^{i\boldsymbol{k}\cdot\boldsymbol{a}}\psi_{\boldsymbol{k}}(\boldsymbol{r}) .$$

Then $\psi_{\boldsymbol{k}}(\boldsymbol{r})$ can be written as $\exp(-i\boldsymbol{k}\cdot\boldsymbol{a})\,\psi_{\boldsymbol{k}}(\boldsymbol{r}-\boldsymbol{a})$, so $\exp(i\boldsymbol{k}\cdot\boldsymbol{r})\,\psi_{\boldsymbol{k}}(\boldsymbol{r}) = \exp(i\boldsymbol{k}\cdot\boldsymbol{r}-i\boldsymbol{k}\cdot\boldsymbol{a})\,\psi_{\boldsymbol{k}}(\boldsymbol{r}-\boldsymbol{a}) = u_{\boldsymbol{k}}(\boldsymbol{r})$ for a function $u_{\boldsymbol{k}}(\boldsymbol{r})$ which has the property

$$u_{\boldsymbol{k}}(\boldsymbol{r}-\boldsymbol{a}) = u_{\boldsymbol{k}}(\boldsymbol{r}) \qquad \text{for any } a \in U . \tag{4.35}$$

Thus the basis functions of the eigenspaces of H (eq. (4.34)) can be chosen to be

$$\psi_{\boldsymbol{k}}(\boldsymbol{r}) = \mathrm{e}^{-i\boldsymbol{k}\cdot\boldsymbol{r}}u_{\boldsymbol{k}}(\boldsymbol{r}) . \tag{4.36}$$

This is the form of a *Bloch function.*

4.3.2. Eigenfunctions of an electron in a crystal

The Hamiltonian H (4.34) commutes with the substitution operators P_g for g an element of the space group G. This implies that the eigenfunctions of H transform according to irreducible representations of G, barring accidental degeneracy. The levels of H can be characterized by irreducible representations and a label distinguishing between levels with equivalent representations. The irreducible representations of a space group are characterized by a $\boldsymbol{k}$-vector from a fundamental region and a (projective) representation of the point group of $G_{\boldsymbol{k}}$. Therefore, the eigenvalues of H can be characterized by a $\boldsymbol{k}$-vector, a representation of $K_{\boldsymbol{k}}$ and an integer numbering the levels with the same representation. This is not the customary way of labelling the levels. Usually one gives a $\boldsymbol{k}$-vector for the level and one numbers the different levels for each $\boldsymbol{k}$ by their order on the energy scale starting from below. Then a level is denoted by $E_{\boldsymbol{k}\nu}$. The integer ν is called the *band index*. As the vector is only one point from the star of the representation for the level one has $E_{\boldsymbol{k}\nu} = E_{R\boldsymbol{k}\nu}$ for any R from the point group. This means that the eigenvalue function, which is a many-valued function on the first Brillouin zone, has the symmetry of the point group. One can describe $E_{\boldsymbol{k}\nu}$ by a hypersurface in the four-dimensional $(E,\boldsymbol{k})$-space. To each point $\boldsymbol{k}$ in the first Brillouin zone correspond a number of such hypersurfaces distinguished by their band index. For each value of ν there is a representation of $G_{\boldsymbol{k}}$ such that this representation together with $\boldsymbol{k}$ gives a representation of the space group.

The eigenfunctions carrying the eigenspace of H with eigenvalue $E_{\boldsymbol{k}\nu}$ are characterized by their transformation property under the translation subgroup U, i.e. by a vector from the star of $\boldsymbol{k}$, and an index labelling the rows of the corresponding representation of $K_{\boldsymbol{k}}$. When $\psi^{(\nu)}_{\boldsymbol{k}1}, ..., \psi^{(\nu)}_{\boldsymbol{k}d_\nu}$ denote the basis functions of the representation of $K_{\boldsymbol{k}}$, a basis for the eigenspace $E_{\boldsymbol{k}\nu}$ is given by (cf. eq; (4.22))

$$\psi^{(\nu)}_{i\mu} = P_{g_i}\psi^{(\nu)}_{\boldsymbol{k}\mu} \qquad (i=1,...,s\,;\,\mu=1,...,d_\nu)\,.$$

The transformation properties of these functions under space group elements are given by eq. (4.23).

4.3.3. Selection rules for space groups

Selection rules for systems with space group symmetry are determined by the decomposition of Kronecker products of representations into irreducible components according to Ch. 2. These decompositions in turn may be determined from the characters of the representations. Let us first consider the translation subgroup U. The character of the representation $D^{(\boldsymbol{k})}$ is given by $e^{i\boldsymbol{k}\cdot\boldsymbol{a}}$. We have seen that vectors $\boldsymbol{k}$ and $\boldsymbol{k}+\boldsymbol{K}$ with $\boldsymbol{K}\in\Lambda^*$ give the same representations. This means that a representation with $\boldsymbol{k}=\boldsymbol{K}$ is equivalent with the trivial one. Therefore

$$\frac{1}{N^3}\sum_{a\in U} e^{i\boldsymbol{k}\cdot\boldsymbol{a}} = \begin{cases} 1 & \text{when } \boldsymbol{k}\in\Lambda^*\,, \\ 0 & \text{otherwise}\,. \end{cases}$$

The product representation of $D^{(\boldsymbol{k})}\otimes D^{(\boldsymbol{k}')}(U)$ is "decomposed" according to

$$D^{(\boldsymbol{k})}\otimes D^{(\boldsymbol{k}')} = D^{(\boldsymbol{k}+\boldsymbol{k}'+\boldsymbol{K})}\,, \qquad \text{where } \boldsymbol{k}+\boldsymbol{k}'+\boldsymbol{K}\in \text{Brillouin zone}\,.$$

The matrix element of a tensor operator transforming according to the representation $D^{(\boldsymbol{k})}$ between states belonging to $D^{(\boldsymbol{k}')}$ and $D^{(\boldsymbol{k}'')}$ satisfies

$$\langle \psi_{k'} | O_k \phi_{k''} \rangle = \sum_K \delta_{-k'+k+k''+K} \langle \psi_{k'} | O_k \phi_{k''} \rangle \, . \tag{4.37}$$

This can only be different from zero when $\boldsymbol{k}, \boldsymbol{k}', \boldsymbol{k}''$ satisfy the relation called *quasi-momentum conservation*

$$\boldsymbol{k}' = \boldsymbol{k} + \boldsymbol{k}'' + \boldsymbol{K} \qquad \text{for some } \boldsymbol{K} \in \Lambda^* \, . \tag{4.38}$$

Selection rules using the *full space group* symmetry can be found from decompositions

$$D^{(\boldsymbol{k}\mu)} \otimes D^{(\boldsymbol{k}'\nu)} = \sum_{\boldsymbol{k}''\sigma} (\boldsymbol{k}\mu\boldsymbol{k}'\nu | \boldsymbol{k}''\sigma) D^{(\boldsymbol{k}''\sigma)} \, , \tag{4.39}$$

where $(\boldsymbol{k}\mu\boldsymbol{k}'\nu|\boldsymbol{k}''\sigma)$ denotes the multiplicity of the irreducible component characterized by a vector $\boldsymbol{k}''$ from a fundamental region and a representation $\Gamma_\sigma(K_{\boldsymbol{k}''})$ of the group $K_{\boldsymbol{k}''}$. The multiplicity is found using eq. (1.18).

First one can determine which $\boldsymbol{k}''$-vectors may occur in the decomposition (4.39). When the star of $\boldsymbol{k}$ has s points, and that of $\boldsymbol{k}'$ has s' points, when the dimension of $\Gamma_\mu(K_{\boldsymbol{k}})$ is d_μ and that of $\Gamma_\nu(K_{\boldsymbol{k}'})$ is d_ν, the $ss'd_\mu d_\nu$ basis functions of the product representation carry a representation of U which can be reduced into irreducible components characterized by the vectors $\boldsymbol{k}''_1, ..., \boldsymbol{k}''_{s''}$ of the stars of the $\boldsymbol{k}''$-vectors occurring in (4.39). Because of eq. (4.38) the points $\boldsymbol{k}''_l$ must satisfy the relation

$$\boldsymbol{k}_i + \boldsymbol{k}'_j \equiv \boldsymbol{k}''_l \quad (\text{mod}\, \Lambda^*) \, . \tag{4.40}$$

Thus there are ss' triples of vectors $\boldsymbol{k}_i, \boldsymbol{k}'_j, \boldsymbol{k}''_l$ from the stars of $\boldsymbol{k}, \boldsymbol{k}', \boldsymbol{k}''$ respectively, which satisfy eq. (4.40). As eq. (4.40) implies $R(\boldsymbol{k}_i + \boldsymbol{k}'_j) \equiv R\boldsymbol{k}''_l$ (mod Λ^*) for any $R \in K$, the triples can be taken together in such a way that the vectors $\boldsymbol{k}''_l$ form the star of a vector $\boldsymbol{k}''$. Often the triples belonging to one star $\boldsymbol{k}''$ can be obtained from each other by the action of elements of the point group, but this is not always true. Anyhow, in the decomposition (4.39) the multiplicity $(\boldsymbol{k}\mu\boldsymbol{k}'\nu|\boldsymbol{k}''\sigma)$ can only be different from zero, if there are vectors $\boldsymbol{k}_i$ and $\boldsymbol{k}'_j$ in the stars of $\boldsymbol{k}$ and $\boldsymbol{k}'$ such that the vector $\boldsymbol{k}_i + \boldsymbol{k}'_j$ occurs (up to a reciprocal lattice vector) in the star of $\boldsymbol{k}''$.

As an example we consider the two-dimensional space group treated in §2.3. Consider two irreducible representations of the space group both characterized by the same vector S in the fundamental region (fig. 4.5): $\boldsymbol{k} = \boldsymbol{k}' = [k0]$. From the 12 vectors in the stars of $\boldsymbol{k}$ and $\boldsymbol{k}'$ one can form the following triples: denoting $\boldsymbol{k}_i + \boldsymbol{k}'_j$ by (ij) one has

1) (14),(25),(36),(41),(52),(63) giving the star of the point $P = [00]$ 6 times,

2) (11),(22),(33),(44),(55),(66) giving the star of the point $S' = [2k\,0]$,

3) (12),(23),(34),(45),(56),(61),(16),(65),(54),(43),(32),(21) giving the star of $T = [kk]$ twice,

4) (26),(13),(24),(35),(46),(15),(62),(31),(42),(53),(64),(51) giving the star of the point $S = [k0]$ twice. Thus in the decomposition of $D^{(\boldsymbol{k}\mu)} \otimes D^{(\boldsymbol{k}'\nu)}$, only representations with the stars of $\boldsymbol{k}'' = [00], [2k\,0]$, $[kk]$ or $[k\,0]$ can occur.

The character of the irreducible representation $D^{(\boldsymbol{k}\mu)}$ of the space group is given by eq. (4.33)

$$\chi_{\boldsymbol{k}\mu}(g) = \sum_{i=1}^{s}{}' \operatorname{tr} D_{\boldsymbol{k}\mu}(g_i^{-1} g\, g_i)\,.$$

In the following we denote $\operatorname{tr} D_{\boldsymbol{k}\mu}(g_i^{-1} g\, g_i)$ by $\varphi_{\boldsymbol{k}_i\mu}(R)$ when $g = \{R \mid \boldsymbol{t}_R\}$. The prime on the summation means summation over those values of i for which $gg_i \in g_i G_{\boldsymbol{k}}$. Choosing for any $R \in K$ an element $g_R = \{R \mid \boldsymbol{t}_R\}$ one has (using eq. (1.18))

$$(\boldsymbol{k}\mu \boldsymbol{k}'\nu \mid \boldsymbol{k}''\sigma) = \frac{1}{nN^3} \sum_{g \in G} \chi_{\boldsymbol{k}\mu}(g)\, \chi_{\boldsymbol{k}'\nu}(g)\, \chi^*_{\boldsymbol{k}''\sigma}(g)\,,$$

where n is the order of the point group K and N^3 that of U. It is equal to

$$\frac{1}{nN^3} \sum_{\boldsymbol{a} \in U} \sum_{R \in K} \sum_{ijl}{}' \mathrm{e}^{i(\boldsymbol{k}_i + \boldsymbol{k}'_j - \boldsymbol{k}''_l)\cdot \boldsymbol{a}}\, \varphi_{\boldsymbol{k}_i\mu}(R)\, \varphi_{\boldsymbol{k}'_j\nu}(R)\, \varphi^*_{\boldsymbol{k}''_l\sigma}(R)$$

$$= \frac{1}{n} \sum_{R \in K} \sum_{ijl}{}' \delta^*(\boldsymbol{k}_i + \boldsymbol{k}'_j - \boldsymbol{k}''_l)\, \varphi_{\boldsymbol{k}_i\mu}(R)\, \varphi_{\boldsymbol{k}'_j\nu}(R)\, \varphi^*_{\boldsymbol{k}''_l\sigma}(R)$$

with

$$\delta^*(\boldsymbol{k}) = \sum_{\boldsymbol{K} \in \Lambda^*} \delta(\boldsymbol{k} - \boldsymbol{K})\,,$$

because of the quasi-momentum conservation. Now we define a subgroup K_{ij} of the point group $K_{\boldsymbol{k}_i}$ of the group of $\boldsymbol{k}_i$ by

$$K_{ij} = \{R \in K \mid R\boldsymbol{k}_i \equiv \boldsymbol{k}_i\,;\, R\boldsymbol{k}'_j \equiv \boldsymbol{k}'_j\}\,.$$

Because of the δ-function one has $K_{ij} = K_{\boldsymbol{k}_i} \cap K_{\boldsymbol{k}'_j} \cap K_{\boldsymbol{k}''_l}$. Summation over all $R \in K$ and the restriction on ijl is equivalent to summation over all triples ijl and all $R \in K_{ij}$. Thus the multiplicity becomes

$$(\boldsymbol{k}\mu\boldsymbol{k}'\nu|\boldsymbol{k}''\sigma) = (1/n) \sum_{ij} \sum_{R\in K_{ij}} \delta^*(\boldsymbol{k}_i + \boldsymbol{k}'_j - \boldsymbol{k}''_l)\, \varphi_{\boldsymbol{k}_i\mu}(R)\, \varphi_{\boldsymbol{k}'_j\nu}(R)\, \varphi^*_{\boldsymbol{k}''_l\sigma}(R)\,. \tag{4.41}$$

Suppose now that the triple $\boldsymbol{k}_{i'}, \boldsymbol{k}'_{j'}, \boldsymbol{k}''_{l'}$ is found from the triple $\boldsymbol{k}_i, \boldsymbol{k}'_j, \boldsymbol{k}''_l$ by an element $S_r \in K$. Choose an element $g_r = \{S_r|\boldsymbol{t}_r\} \in G$. The following statements are easily seen to be true:
1) $\boldsymbol{k}_{i'} + \boldsymbol{k}'_{j'} - \boldsymbol{k}''_{l'} \equiv \boldsymbol{O}$ if and only if $\boldsymbol{k}_i + \boldsymbol{k}'_j - \boldsymbol{k}''_l \equiv \boldsymbol{O}$;
2) since $\boldsymbol{k}_i = R_i\boldsymbol{k}_1$ and $\boldsymbol{k}_{i'} = S_r\boldsymbol{k}_i$, one can choose $g_{i'} = g_r g_i$;
3) when $R \in K_{ij}$, then $R' = S_r R S_r^{-1} \in K_{i'j'}$;
4) $\varphi_{\boldsymbol{k}_{i'}\mu}(R') = \mathrm{tr}\, D_{\boldsymbol{k}\mu}(g_i^{-1} g_r^{-1} g_r g_R g_r^{-1} g_r g_i) = \varphi_{\boldsymbol{k}_i\mu}(R)$.
These statements imply that the terms in eq. (4.41) are equal for triples obtained from each other by an element $S_r \in K$. When we call triples equivalent if there is such a S_r, the set of ss' triples in the product representation can be divided into equivalence classes (called stars, because often an equivalence class is the set of triples with a $\boldsymbol{k}''_l$ such that the $\boldsymbol{k}''_l$ form one or more stars of $\boldsymbol{k}''$). If w_s is the number of triples in a "star" one has finally

$$(\boldsymbol{k}\mu\boldsymbol{k}'\nu|\boldsymbol{k}''\sigma) = \sum_{\substack{\text{stars of}\\ \text{triples}}} \frac{w_s}{n} \sum_{R\in K_{ij}} \delta^*(\boldsymbol{k}_i + \boldsymbol{k}'_j - \boldsymbol{k}''_l)\varphi_{\boldsymbol{k}_i\mu}(R)\varphi_{\boldsymbol{k}'_j\nu}(R)\varphi^*_{\boldsymbol{k}''_l\sigma}(R)\,, \tag{4.42}$$

where $(\boldsymbol{k}_i, \boldsymbol{k}'_j, \boldsymbol{k}''_l)$ are representatives of the stars of triples. For symmorphic space groups eq. (4.42) can be simplified using

$$\varphi_{\boldsymbol{k}_i\mu}(R) = \chi_\mu(R_i^{-1} R R_i)\,.$$

Hence for symmorphic space groups one has the expression

$$(\boldsymbol{k}\mu\boldsymbol{k}'\nu|\boldsymbol{k}''\sigma) = \sum_{\substack{\text{stars of}\\ \text{triples}}} \frac{w_s}{n} \sum_{R\in K_{ij}} \delta^*(\boldsymbol{k}_i + \boldsymbol{k}'_j - \boldsymbol{k}''_l)\chi_\mu(R)\chi_\nu(R)\chi^*_\sigma(R)\,. \tag{4.43}$$

Let us illustrate the procedure with the example of $D^{(\boldsymbol{k}\mu)} \otimes D^{(\boldsymbol{k}'\nu)}$ with $\mu = 1, \nu = 2$ for the two-dimensional space group from the beginning of this section. The 36 basis functions can be collected into four stars of triples.
1) $\boldsymbol{k}'' = [00]$; $i=1, j=4, l=1$; $K_{ij} = m = C_2$; $n = 12$, $w_s = 6$; $K_{\boldsymbol{k}''} = 6mm = D_6$;
2) $\boldsymbol{k}'' = [2k0]$; $i=1, j=1, l=1$; $K_{ij} = m = C_2$; $n = 12$, $w_s = 6$; $K_{\boldsymbol{k}''} = m$;
3) $\boldsymbol{k}'' = [kk]$; $i=1, j=2, l=1$; $K_{ij} = e$; $n = 12$, $w_s = 12$; $K_{\boldsymbol{k}''} = m$;
4) $\boldsymbol{k}'' = [k0]$; $i=2, j=6, l=1$; $K_{ij} = e$; $n = 12$, $w_s = 12$; $K_{\boldsymbol{k}''} = m$.
For $\boldsymbol{k}'' = [00]$ the eq. (4.43) gives

$$(\boldsymbol{k}\mu\boldsymbol{k}'\nu|\boldsymbol{k}''\sigma) = \tfrac{6}{12} \sum_{R\in K_{ij}} \chi_\mu(R)\chi_\nu(R)\chi_\sigma^*(R) = \tfrac{1}{2}(d_\mu d_\nu d_\sigma + \chi_\mu(\beta)\chi_\nu(\beta)\chi_\sigma^*(\beta))\,.$$

From the representations of D_6 given in Ch. 3, §2.3, one finds

$$(\boldsymbol{k}\mu\boldsymbol{k}'\nu|\boldsymbol{k}''\sigma) = \begin{cases} 1 & \text{for } \mu=1,\ \nu=2,\ \sigma=2,3,5, \text{ or } 6 \\ 0 & \text{for } \mu=1,\ \nu=2,\ \sigma=1 \text{ or } 4\,. \end{cases}$$

For $\boldsymbol{k}'' = [2k\ 0]$ one finds with $\mu=1, \nu=2$

$$(\boldsymbol{k}\mu\boldsymbol{k}'\nu|\boldsymbol{k}''\sigma) = \begin{cases} 1 & \text{for } \sigma=2 \\ 0 & \text{for } \sigma=1\,, \end{cases}$$

as there are two irreducible representations of $K_{\boldsymbol{k}''}$ in this case. For $\boldsymbol{k}'' = [kk]$ and for $\boldsymbol{k}'' = [k\,0]$ only $R = 1$ occurs in the summation. Therefore, the multiplicity in this case is

$$(\boldsymbol{k}\mu\boldsymbol{k}'\nu|\boldsymbol{k}''\sigma) = d_\mu d_\nu d_\sigma\,.$$

In summary, for the decomposition of the product representation with $\mu = 1$ and $\nu = 2$ and with the notation $P = [00], S = [k0], S' = [2k\ 0], T = [kk]$ one has

$$D^{(S1)} \otimes D^{(S2)} = D^{(P2)} \oplus D^{(P3)} \oplus D^{(P5)} \oplus D^{(P6)} \oplus D^{(S'2)}$$

$$\oplus\, D^{(T1)} \oplus D^{(T2)} \oplus D^{(S1)} \oplus D^{(S2)}\,,$$

with the dimensions

$$6 \times 6 = 1 + 2 + 1 + 2 + 6 + 6 + 6 + 6 + 6\,.$$

To determine selection rules here we have made use of the *full space group*. This method gives an answer to the question: is the matrix element of a tensor operator associated with vector $\boldsymbol{k}'$ between states belonging to representations with stars $\boldsymbol{k}$ and $\boldsymbol{k}''$ different from zero? The answer is: it can be different from zero only when in the combinations of vectors from the stars of $\boldsymbol{k}'$ and $\boldsymbol{k}''$ vectors from the star of $\boldsymbol{k}$ occur. Another question one can ask is: is the matrix element between states belonging to specified points $\boldsymbol{k}'$ and $\boldsymbol{k}''$ of a tensor operator belonging to a given $\boldsymbol{k}'$-vector different from zero?

To answer this question one has to consider representations of the subgroup $G_{\boldsymbol{k}} \cap G_{\boldsymbol{k}'} \cap G_{\boldsymbol{k}''}$. The reduction of product representations of groups $G_{\boldsymbol{k}}$ will not be treated here, but is useful in answering the last question. The former method is called the *full-group method*, the latter is called the *subgroup method*. A discussion of these problems has been given by several authors. We can refer to Zak [1962], Birman [1962, 1963, 1966], Bradley [1966], Litvin and Zak [1968] and Bradley and Cracknell [1970].

CHAPTER V

SPIN AND TIME REVERSAL

In the physical systems we considered in the preceding chapters, we neglected the spin of the electrons completely. Since we know that spin effects are very important, this is rather unrealistic. However, in this way we could keep the presentation simpler. In this chapter we will discuss the changes we have to introduce in order to take the spin into account. The general procedure remains the same, but the symmetry groups become slightly more involved. They are known as the crystallographic double groups.

Another refinement of the theory is obtained by taking account, not only of space symmetries, but also of symmetries involving the time. Apart from time translations, the time reversal operation is the operation involving the time which is most frequently encountered in physical systems. This time reversal will be treated here in some detail. Still more general space-time symmetries will be discussed in the last chapter.

5.1. Double groups

5.1.1. Spin representations of the rotation group

In chapter II we found that, for physical systems, not only the ordinary representations of the symmetry group are important, but also the projective representations. In chapter III we discussed the ordinary representations of the orthogonal group, and in particular those of the rotation group $SO(3)$. We now want to discuss also the projective representations of this group. As in chapter III, we will only give some results needed in the following sections. For more details we refer to books treating the representations of this continuous group, like Boerner [1967].

It turns out that the multiplicator of $SO(3)$ consists of two elements. This means that there are two classes of representations: the ordinary ones, and those with a nontrivial factor system. These are called *spin representations*.

The factor system ω can be chosen in such a way that $\omega(R_1,R_2) = \pm 1$ for any pair of rotations R_1 and R_2. This means that for a projective representation D of $SO(3)$ one has $D(R_1)D(R_2) = \omega(R_1,R_2)D(R_1R_2) = \pm D(R_1R_2)$. For an ordinary representation, ω is identically equal to unity, for a spin representation there are elements R_1 and R_2 such that $\omega(R_1,R_2) = -1$. Therefore, the former are also called *single valued representations*, and the latter *double valued representations*.

A representation group for $SO(3)$ is the group $SU(2)$ of two-dimensional unitary matrices. It is easily found from the definition of unitary matrix, that an element $u \in SU(2)$ has the form

$$u = \begin{pmatrix} a & b \\ -b^* & a^* \end{pmatrix} \quad \text{with} \quad |a|^2 + |b|^2 = 1 . \tag{5.1}$$

A homomorphism $\varphi : SU(2) \to SO(3)$ is found as follows. It is easily seen that the set of two-dimensional hermitian, traceless matrices

$$L = \{\Lambda \in SL(2,\mathbb{C}) \mid \Lambda^\dagger = \Lambda, \operatorname{tr} \Lambda = 0\}$$

forms a real three-dimensional vector space. A basis for this space is formed by the 3 Pauli matrices. An isomorphism of $\mathbb{R}^3$ and L is determined by

$$\boldsymbol{r} = (x,y,z) \to \boldsymbol{r} \cdot \boldsymbol{\sigma} = \begin{pmatrix} z & x-iy \\ x+iy & -z \end{pmatrix} .$$

The Euclidean norm in $\mathbb{R}^3$ induces a norm in L:

$$|\boldsymbol{r} \cdot \boldsymbol{\sigma}|^2 = r^2 = -\operatorname{Det}(\boldsymbol{r} \cdot \boldsymbol{\sigma}) .$$

Any $u \in SU(2)$ determines a linear transformation of L, if one defines

$$\Lambda' = u \Lambda u^{-1} .$$

Moreover, this transformation conserves the norm:

$$(r')^2 = -\operatorname{Det}(\boldsymbol{r}' \cdot \boldsymbol{\sigma}) = -\operatorname{Det}(\boldsymbol{r} \cdot \boldsymbol{\sigma}) = r^2 \qquad (\forall \boldsymbol{r} \in \mathbb{R}^3) .$$

This means that for any $u \in SU(2)$ there exists a unique $R \in SO(3)$ such that

$$u(\boldsymbol{r} \cdot \boldsymbol{\sigma})u^{-1} = (R\boldsymbol{r}) \cdot \boldsymbol{\sigma} , \qquad (\forall \boldsymbol{r} \in \mathbb{R}^3) .$$

We denote the mapping $SU(2) \to SO(3)$ defined in this way by $\varphi : R = \varphi u$. Actually the mapping φ is a homomorphism, because

$$u_2 u_1 (\boldsymbol{r} \cdot \boldsymbol{\sigma}) u_1^{-1} u_2^{-1} = u_2 (R_1 \boldsymbol{r} \cdot \boldsymbol{\sigma}) u_2^{-1} = (R_2 R_1 \boldsymbol{r} \cdot \boldsymbol{\sigma}) \qquad (\forall \boldsymbol{r} \in \mathbb{R}^3) .$$

The kernel of φ is the group of matrices u satisfying

$$u \Lambda u^{-1} = \Lambda \qquad (\forall \Lambda \in L) .$$

Hence Ker_φ consists of the matrices $\pm \mathbb{1}$. Therefore, to each $R \in SO(3)$ correspond two elements $\pm u(R)$ of $SU(2)$. Both $\pm u(R)$ satisfy the equation

$$\sum_j R_{ij} \sigma_j = u(R)^{-1} \sigma_i u(R) . \tag{5.2}$$

An explicit expression for $u(R)$ with R given by its axis $\boldsymbol{n}$ and its rotation angle φ is

$$u(R) = \pm \left[\cos(\varphi/2)\, \mathbb{1} + i \sin(\varphi/2)\, \boldsymbol{\sigma} \cdot \boldsymbol{n}\right] . \tag{5.3}$$

We recall that the statement that $SU(2)$ is a representation group for $SO(3)$ means that any projective representation of $SO(3)$ can be obtained from an ordinary representation of $SU(2)$. When D' is an ordinary representation of $SU(2)$, one chooses for any element R of $SO(3)$ one of the two elements $u(R)$. The definition $P(R) = D'(u(R))$ gives a projective representation of $SO(3)$, because $P(R_1)P(R_2) = D'(u(R_1)u(R_2)) = D'(\pm u(R_1 R_2)) = \pm P(R_1 R_2)$. The ordinary representations are found from those $SU(2)$ representations for which $D(\mathbb{1}) = D(-\mathbb{1})$, whereas the others (spin representations) come from $D(SU(2))$ with $D(-\mathbb{1}) = -D(\mathbb{1})$.

According to the representation theory of the group $SU(2)$ the irreducible representations are characterized by $j = 0, \frac{1}{2}, 1, \frac{3}{2}, 2, \ldots$. The dimension of representation $D^{(j)}$ is $2j + 1$. The integral j representations (odd dimensions) give the ordinary representations of $SO(3)$ discussed in Ch. 3, § 2.4. The half-integral j representations (even dimensions) are faithful and give the spin representations of $SO(3)$. For an element u of $SU(2)$, which is mapped by φ on the rotation R through an angle ψ the character in the representation $D^{(j)}$ is given by

$$\chi_j(u) = (\pm)^{2j} \frac{\sin(j+\frac{1}{2})\psi}{\sin \frac{1}{2}\psi} . \tag{5.4}$$

In a spherical symmetric potential the eigenspaces with orbital angular momentum l carry a representation $D^{(l)}$ with integer l. For spin $\frac{1}{2}$ particles the eigenspaces with total angular momentum J carry a representation $D^{(J)}$ with half-integral J.

5.1.2. *Double groups*

To each rotation R correspond two elements $\pm u(R)$ of $SU(2)$. Therefore, to a crystallographic point group of the first kind, consisting only of rotations, and of order N, corresponds a set of $2N$ elements of $SU(2)$. As $u(R)u(R') = \pm u(RR')$ for any R, R' in $SO(3)$, it is easily verified that this set is a subgroup of $SU(2)$. It is denoted by K^{d}, and is called the *double point group* for K. As we will see in the next section, these crystallographic double groups will play the same role for spin $\frac{1}{2}$ particles as the ordinary point groups for spinless particles. When K^{d} is the double group for K, to the unit element of K correspond the two elements $\pm \mathbb{1}_2$ in K^{d}. As these two elements commute with all the elements of K^{d}, they form an invariant subgroup of order two in K^{d}. The factor group $K^{\mathrm{d}}/\{\pm \mathbb{1}_2\}$ is isomorphic to the group K. Therefore, from any representation Γ of K one can obtain a representation of K^{d} by assigning $\Gamma(R)$ to both elements $\pm u(R)$ of K^{d}. However, in general K^{d} has also representations for which one does not have $\Gamma(+\mathbb{1}_2) = \Gamma(-\mathbb{1}_2)$. These representations of K^{d} are called *extra representations*.

To find these extra representations we study the classes of the double group. One has the following proposition.

PROPOSITION 5.1. The elements of K^{d} which are mapped on one class in K, form either one or two classes in K^{d}. When the class in K consists of 180° rotations and if there exists in K a 180° rotation axis perpendicular to the 180° rotation axis of one of the elements of this class, the elements of K^{d} mapped onto this class form just one class in K^{d}. In all other cases, the elements of K^{d} mapped onto the class in K form two classes in K^{d}.

Proof. Suppose that S and R are elements of the same class in K. Then there is a rotation T such that $R = TST^{-1}$. Now choose for each element of K a fixed element $u(R)$ in K^{d}. Then one has $u(R) = \pm u(T)u(S)u(T)^{-1}$, which means that either $u(R)$ and $u(S)$, or $u(R)$ and $-u(S)$ are in the same class in K^{d}. So for each class in K one finds at most two classes in K^{d}. One finds only one class when $u(R)$ and $-u(R)$ are in the same class. To investigate this, we recall that all elements of a class in K describe rotations through the same angle. If this angle is φ the element $u(R)$ can, by a suitable choice of basis, be written in the form

$$u(R) = \begin{pmatrix} e^{i\varphi/2} & 0 \\ 0 & e^{-i\varphi/2} \end{pmatrix}.$$

However, there is no element U in $SU(2)$ such that $Uu(R) = -u(R)\,U$, unless $\varphi = \pi$. Hence only for classes of 180° rotations it is possible to find only one class in K^{d}, classes of another kind always lead to two classes. In the case of $\varphi = \pi$ the $SU(2)$ matrix U relating $u(R)$ and $-u(R)$ is given by

$$U = \begin{pmatrix} 0 & e^{i\psi} \\ -e^{-i\psi} & 0 \end{pmatrix} \qquad (\psi \text{ arbitrary real number})$$

which corresponds to a rotation through 180° with axis perpendicular to the axis of R. This proves that the class of the 180° rotation R gives rise to only one class in K^{d}, if U belongs to K^{d}, or in other words if the corresponding rotation belongs to K. This proves the proposition.

Consider as an example the point group $K = 32$, which is of order six. It is generated by a three-fold axis taken to be the z-axis, and a perpendicular two-fold axis, taken here along the y-axis. Then the corresponding double group is generated by

$$\alpha = u(R) = \begin{pmatrix} \frac{1}{2} + \frac{1}{2}i\sqrt{3} & 0 \\ 0 & \frac{1}{2} - \frac{1}{2}i\sqrt{3} \end{pmatrix} \quad \text{and} \quad \beta = u(S) = \begin{pmatrix} 0 & 1 \\ -1 & 0 \end{pmatrix}.$$

The twelve elements of K^{d} are

$$\alpha, \alpha^2, \alpha^3 = -\mathbb{1}_2, \alpha^4, \alpha^5, \alpha^6 = \mathbb{1}_2 = \epsilon$$

$$\beta, \alpha\beta, \alpha^2\beta, \alpha^3\beta = -\beta, \alpha^4\beta, \alpha^5\beta\,.$$

Each of the three classes of K gives rise to two classes of K^{d}. The classes of K^{d} are: $\{\epsilon\}$, $\{-\epsilon\}$, $\{\alpha, \alpha^5\}$, $\{\alpha^2, \alpha^4\}$, $\{\beta, \alpha^2\beta, \alpha^4\beta\}$, $\{\alpha\beta, \alpha^3\beta, \alpha^5\beta\}$. From this it follows that there are six nonequivalent irreducible representations for K^{d}: 4 one-dimensional ones, and 2 two-dimensional ones. ($12 = 1 + 1 + 1 + 1 + 2^2 + 2^2$). Three of them are already determined by the representations of K. The extra representations (denoted here by a prime) can be found using the techniques of Ch. 1. The character table for 32^{d} is given in table 5.1. Compare this with the character table of 32 discussed in Ch. 1, §3.

Table 5.1
The character table for the double group of 32.

32^{d}	ϵ	$-\epsilon$	α	α^2	β	$\alpha\beta$
Γ_1	1	1	1	1	1	1
Γ_2	1	1	1	1	−1	−1
Γ_3'	1	−1	−1	1	i	$-i$
Γ_4'	1	−1	−1	1	$-i$	i
Γ_5	2	2	−1	−1	0	0
Γ_6'	2	−2	1	−1	0	0

The extra representations are tabulated, e.g. in Koster [1957] and in appendix B. In fact the representations of the double group K^{d} are projective representations of the group K. Choosing one element $u(R) \in K^{\mathrm{d}}$ for any $R \in K$, a representation Γ of K^{d} gives a projective representation D of K on putting $D(R) = \Gamma(u(R))$. If Γ is not an extra representation, D is an ordinary representation of K, as we have seen. However, it is quite possible that even an extra representation is not a projective representation of K with nontrivial factor system. That means that the factor system may be associated to the trivial one. As an example, consider the representation Γ_6' of the group 32^{d}. This gives a projective representation of $K = 32$ by $D(R) = \Gamma_6'(\alpha)$, $D(R^2) = \Gamma_6'(\alpha)^2$, $D(S) = \Gamma_6'(\beta)$, $D(RS) = \Gamma_6'(\alpha\beta)$, $D(R^2S) = \Gamma_6'(\alpha^2\beta)$. As one easily checks, the factor system for this representation is not trivial. E.g. one has for a certain choice of the representation Γ_6'

$$D(R) = \begin{pmatrix} \frac{1}{2}+\frac{1}{2}i\sqrt{3} & 0 \\ 0 & \frac{1}{2}-\frac{1}{2}i\sqrt{3} \end{pmatrix}, \qquad D(S) = \begin{pmatrix} 0 & 1 \\ -1 & 0 \end{pmatrix}$$

and consequently $D(R^2)D(R) = -D(\mathbb{1})$, $D(S)^2 = -D(\mathbb{1})$, and $D(RS)^2 = -D(\mathbb{1})$. If one now takes the associated representation $D'(R) = -D(R)$, $D'(S) = iD(S)$, one obtains an ordinary representation of 32. This proves that every extra representation of a double group K^{d} gives a projective representation of the group K, but it is possible that the factor system of this projective representation is associated to the trivial factor system. The projective representations of the point groups corresponding to the extra representations of their double groups are given in the appendix.

5.1.3. Crystal field theory with spin-orbit coupling

In chapter III we considered the level splitting by a crystal field with spin effects neglected. Let us once again consider the same problem, now taking into account the spin terms in the Hamiltonian. Here we restrict ourselves to the one-electron problem. The extension to cases with more electrons can be made in the same way as was done in chapter III. We consider an atom or ion in a crystal field. Suppose that this atom contains one electron outside closed shells. Then the Hamiltonian for such an electron can be written

$$H = \frac{p^2}{2m} + V(\boldsymbol{r}) + W(\boldsymbol{r}) + V_s \tag{5.5}$$

where the first term gives the kinetic energy, $V(\boldsymbol{r})$ is the spherical symmetric potential of the atom core, $W(\boldsymbol{r})$ is the crystal field potential created by charges in the neighbourhood, and V_s is the spin-orbit coupling given by

$$V_s = v(r)\, \boldsymbol{S} \cdot \boldsymbol{L} \,.$$

In this expression $v(r)$ is some function of the radius r, $\boldsymbol{S}$ is the spin operator with components $S_i = \frac{1}{2}\hbar\, \sigma_i$, and $\boldsymbol{L}$ is the angular momentum operator. The Hamiltonian H (5.5) is an operator in the space of two-component spinors. Each spinor $\psi(\boldsymbol{r})$ has components $\psi_+(\boldsymbol{r})$ and $\psi_-(\boldsymbol{r})$. The first 3 terms of H act trivially in spin space, i.e. they can be written as $(p^2/2m + V(\boldsymbol{r}) + W(\boldsymbol{r}))\, \mathbb{1}$. The term V_s acts both on space and spin variables. We proceed in the same way as in chapter III. Thus we have to determine the symmetry group of H and of the unperturbed Hamiltonian H_0.

Again we consider the situation where the symmetry of H is smaller than the symmetry of $H_0 = \boldsymbol{p}^2/2m + V(\boldsymbol{r})$. We denote the symmetry group of H by G, the symmetry group of H_0 by G_0. As explained in Ch. 3, §3 there are two cases: for a strong crystal field, $W(\boldsymbol{r})$ is more important than V_s, for a weak crystal field the opposite holds. In the former case we denote $H_0 + W(\boldsymbol{r})$ by H' and its symmetry group by G'. In the latter case H'' denotes $H_0 + V_s$, with symmetry group G''. To determine the group G one considers a substitution operator P_g defined by its action on a function $\psi(\boldsymbol{r})$ as $P_g \psi(\boldsymbol{r}) = \psi(g^{-1}\boldsymbol{r})$. This operator commutes with $\boldsymbol{p}^2/2m$ if g is an element of the Euclidean group $E(3)$. For $g \in E(3)$ the operator commutes with $V(\boldsymbol{r})$ if and only if g is an orthogonal transformation, because $P_g V(|\boldsymbol{r}|) P_g^{-1} = V(|g^{-1}\boldsymbol{r}|)$. P_g commutes for the same reason with $W(\boldsymbol{r})$ if and only if g is an element of the space group of the crystal. Therefore, P_g commutes with $H_0 + W(\boldsymbol{r})$ if and only if $g = R$ is an element of the point group K, the symmetry of the atom site. Finally we have to look at $P_g V_s P_g^{-1}$ for $g \in K$. One has in this case $P_g V_s P_g^{-1} = v(|g^{-1}\boldsymbol{r}|)\boldsymbol{S} \cdot P_g \boldsymbol{L} P_g^{-1}$. Since $\boldsymbol{L}$ transforms as a pseudovector under orthogonal transformations, one has $P_g \boldsymbol{L} P_g^{-1} = (\det R) R^{-1} \boldsymbol{L}$ for $g = R \in K$. Consequently one has $P_R V_s P_R^{-1} = (\det R)(R\boldsymbol{S}) \cdot \boldsymbol{L}$. Here we see that P_R does not commute with V_s. However, we can combine P_R with a linear transformation $u(R)$ in the two-dimensional spin space. This operator $u(R)$ acts only on the spin components, not on the space variables. For $T_R = u(R)P_R$ one has $T_R V_s T_R^{-1} = (\det R) v(r) u(R)(R\boldsymbol{S}) u(R)^{-1} \cdot \boldsymbol{L}$, which is equal to V_s if $u(R)$ satisfies the relations

$$\sum_{j=1}^{3} (\det R) u(R) R_{ij} \sigma_j u(R)^{-1} = \sigma_i \qquad (i = 1, 2, 3) . \tag{5.6}$$

If R is a rotation, this is simply eq. (5.2). Then one can choose $u(R)$ as in eq. (5.3). When the elements R form a point group K of the first kind, the elements $\pm u(R)$ form the double group K^{d}. The operators T_R form a representation of K^{d}. When R has determinant -1, one can choose $u(R) = u(IR)$ according to eq. (5.6).

Choosing one element $u(R) \in SU(2)$ satisfying eq. (5.6) for each $R \in K$ the operators T_R commuting with H (5.5) satisfy

$$T_{R_1} T_{R_2} = u(R_1) u(R_2) P_{R_1} P_{R_2} = \omega_s(R_1, R_2)\, T_{R_1 R_2} \tag{5.7}$$

with $\omega_s(R_1, R_2) = \pm 1$ depending on the choice of $u(R)$. Thus the operators T_R form a projective representation of K with factor system ω_s. It may happen that this factor system is associated to the trivial one. In that case one can give the operators T_R phase factors such that they form an ordinary representation of K. This is the case for the rotation groups 1, 2, 3, 4, 6, and 32, the groups $m, \bar{4}, \bar{6}$, and $3m$ and the direct products $\bar{1}, 2/m, \bar{3}, 4/m, 6/m$ and $\bar{3}m$. The nonequivalent irreducible spin representations of the point groups with nontrivial factor system ω_s are given in the appendix.

Another approach is to extend the set of operators T_R in such a way that one obtains a group of operators. As we remarked already, if ω_s is associated with the trivial factor system it is sufficient to endow the operators with phase factors in order to obtain a group of operators commuting with H. When this is not possible, one can make a group by taking all operators $\pm T_R$. As we saw, for a point group of the first kind they form a representation of the double group K^{d}. When K is a point group of the second kind, to an element $R \in K \cap SO(3)$ is associated $\pm u(R) P_R$, to an element $IR \in K$ with

$R \in SO(3)$ correspond $\pm u(R)P_{IR}$. This means that for a point group of the second kind without I, the group of operators is a representation of K^{d} which is the double group of the isomorphic group of the 1st kind. For a group $K \times C_2$, where C_2 is the group of I, the group of operators is a representation of $K^{\mathrm{d}} \times C_2$. It contains as subgroups K^{d} (with elements $\pm u(R)P_R, R \in K$) and C_2 (with elements $\mathbb{1}$ and P_I). We call $K^{\mathrm{d}} \times C_2$ the double group of $K' = K \times C_2$ and denote it by K'^{d}.

Let us now consider the symmetry groups G_0, G', G'', and G. As we have taken the spin into account, and as any $u \in SU(2)$ commutes with H_0 and H', the symmetry group of H_0 is not merely $O(3)$, but the direct product $O(3) \times SU(2)$. Notice that an element of $SU(2)$ is not induced by a space transformation: the invariance group is $O(3)$. This is the case, because in H_0 the spin does not appear. For a s rong crystal field the group G' is, for the same reason, the direct product $K \times SU(2)$. For a weak crystal field the term V_s in H'' couples the transformations in ordinary space and in spin space. In this case the symmetry group G'' is the double group of $O(3)$, which is $SU(2) \times C_2$. The invariance group of H is K^{d}.

For a weak crystal field one has the sequence of Hamiltonians of increas ing accuracy: $H_0, H_0 + V_s, H$ with corresponding symmetry groups $O(3) \times SU(2)$, $SU(2)$, K^{d}. For H_0, both orbital angular momentum and spin are conserved quantities, which means that l and s determine an irreducible representation $D^{(l)} \times D^{(\frac{1}{2})}$ of G_0. For H'', the total angular momentum J remains a good quantum number. The $(l, \frac{1}{2})$ level splits up into two levels, with $J = l \pm \frac{1}{2}$. Finally, the splitting of these two levels under the influence of the crystal field is determined by the reduction of the representation of K^{d} subduced from $D^{(J)}$. As the character of $D^{(J)}$ has the property: $\chi_J(\alpha) = \chi_J(-\alpha)$ for integral J and $\chi_J(\alpha) = -\chi_J(-\alpha)$ for half-integral J, the character of the representation $D^{(J)}(K^{\mathrm{d}})$ is orthogonal to the character of the irreducible extra representations for integral J, and orthogonal to the other ones for half-integral J. This means that a level with half-integral J splits up into levels belonging to the extra representations.

As an example we consider an atom in a field of cubic symmetry. For half-integral J the characters of $D^{(J)}(O^{\mathrm{d}})$ are given by

Class:	$[\epsilon]$	$[-\epsilon]$	$[\beta]$	$[-\beta]$	$[\alpha^2]$	$[\alpha]$	$[-\alpha]$	$[\alpha\beta]$
J								
1/2	2	−2	1	−1	0	$\sqrt{2}$	$-\sqrt{2}$	0
3/2	4	−4	−1	1	0	0	0	0
5/2	6	−6	0	0	0	$-\sqrt{2}$	$\sqrt{2}$	0
7/2	8	−8	1	−1	0	0	0	0
9/2	10	−10	−1	0	0	$\sqrt{2}$	$-\sqrt{2}$	0
.....	....							

The reductions of the induced representations follow from the character table of O^{d} in the appendix.

$$D^{(1/2)} = \Gamma_6' \qquad\qquad D^{(7/2)} = \Gamma_6' \oplus \Gamma_7' \oplus \Gamma_8'$$

$$D^{(3/2)} = \Gamma_8' \qquad\qquad D^{(9/2)} = \Gamma_6' \oplus 2\Gamma_8'$$

$$D^{(5/2)} = \Gamma_7' \oplus \Gamma_8' \qquad\qquad \ldots\ldots\ldots\ldots$$

Thus levels with $J = 1/2$ or $3/2$ are not split in such a field.

For a strong crystal field, the sequence of Hamiltonians is $H_0, H_0 + V_K, H$ with symmetry groups $O(3) \times SU(2), K \times SU(2), K^{\mathrm{d}}$. By the crystal potential V_K the levels of H_0 characterized by $(l, \frac{1}{2})$ are split up into levels characterized by irreducible representations $\Gamma_i(K) \times D^{(1/2)}$ of $K \times SU(2)$. The reduction of $D^{(l)}(K)$ into irreducible components is as discussed in Ch. 3. The subsequent splitting by the spin-orbit coupling is determined from the reduction of $\Gamma_i \times D^{(1/2)}$ subduced to K^{d}.

As example we consider again an atom in a cubic field. The reduction of $D^{(l)}(O)$ is given in Ch. 3, §3.2. This also gives the reduction of $D^{(l)} \times D^{(1/2)}$. The subduced characters of O^{d} are obtained from the product $\Gamma_i(\alpha) \otimes D^{1/2}(\alpha)$. It is given by

Class:	$[\epsilon]$	$[-\epsilon]$	$[\beta]$	$[-\beta]$	$[\alpha^2]$	$[\alpha]$	$[-\alpha]$	$[\alpha\beta]$	
$\Gamma_1 \times D^{(1/2)}$	2	−2	1	−1	0	$\sqrt{2}$	$-\sqrt{2}$	0	$= \Gamma_6'$
$\Gamma_2 \times D^{(1/2)}$	2	−2	1	−1	0	$-\sqrt{2}$	$\sqrt{2}$	0	$= \Gamma_7'$
$\Gamma_3 \times D^{(1/2)}$	4	−4	−1	1	0	0	0	0	$= \Gamma_8'$
$\Gamma_4 \times D^{(1/2)}$	6	−6	0	0	0	$\sqrt{2}$	$-\sqrt{2}$	0	$= \Gamma_6' \oplus \Gamma_8'$
$\Gamma_5 \times D^{(1/2)}$	6	−6	0	0	0	$-\sqrt{2}$	$\sqrt{2}$	0	$= \Gamma_7' \oplus \Gamma_8'$

Thus a d-level is split by spin-orbit coupling into a $J = 3/2$ and a $J = 5/2$ level, which split in a weak cubic field into three levels: two with Γ_8', one with Γ_7'. In a strong cubic field the d-level splits up into a Γ_3 and a Γ_5 level, and is split further by spin-orbit coupling into the levels $\Gamma_8', \Gamma_7', \Gamma_8'$.

As in Ch. 3, one can construct symmetry adapted functions. For the double groups they can be found in Cracknell [1969a].

5.1.4. Double space groups

An electron in the electrostatic crystal potential $V(\boldsymbol{r})$ can be described by the Hamiltonian obtained in the nonrelativistic approximation of the Dirac Hamiltonian:

$$H = \frac{p^2}{2m} - \frac{p^4}{8m^3c^2} + V(\boldsymbol{r}) + \frac{\hbar e}{8m^2c^2} \nabla V \cdot \boldsymbol{p} - \frac{\hbar e}{4m^2c^2} \boldsymbol{\sigma} \cdot (\nabla V \times \boldsymbol{p}) . \qquad (5.8)$$

Here $p^2/2m$ is the nonrelativistic kinetic energy, the second term is a relativistic correction arising from the series expansion of $(m^2c^4 + c^2p^2)^{1/2} - m^2c^4$, $V(\boldsymbol{r})$ is the crystal potential, the 4th term is called the *Darwin term* and the last one is the *Thomas term* describing the spin-orbit interaction. In a centrally symmetric potential the Thomas term becomes $(dV/dr)\boldsymbol{S} \cdot \boldsymbol{L}/2m^2c^2r$. To determine the symmetry of the Hamiltonian we proceed in the same way as in the foregoing section. The substitution operator P_g commutes with the first two terms, if $g \in E(3)$, it commutes with the first three terms, if g is an element of the space group G. For $g \in G$ this P_g also commutes with the next term, because $P_g \nabla V(\boldsymbol{r}) \cdot \boldsymbol{p} P_g^{-1} = R^{-1} \nabla V(g^{-1}\boldsymbol{r}) \cdot R^{-1}\boldsymbol{p} = \nabla V(\boldsymbol{r}) \cdot \boldsymbol{p}$. Finally, P_g does not commute with the Thomas term: $P_g \boldsymbol{\sigma} \cdot (\nabla V \times \boldsymbol{p}) P_g^{-1} = \boldsymbol{\sigma} \cdot (R^{-1} \nabla V(g^{-1}\boldsymbol{r}) \times R^{-1}\boldsymbol{p}) = (\det R)\, \boldsymbol{\sigma} \cdot R^{-1}(\nabla V(\boldsymbol{r}) \times \boldsymbol{p}) = (\det R)(R\boldsymbol{\sigma}) \cdot (\nabla V \times \boldsymbol{p})$. However, combining P_g with the spinor operator $u(R)$ satisfying eq. (5.5) the operator $P_g u(R)$ commutes with the Thomas term. This operator $T_g = u(R)P_g$ also commutes with the other terms. To determine the symmetry group of H one can follow two different lines: one approach via the projective representations, the other via the double group.

Let us first consider the approach using projective representations. Then we choose for each $g = \{R|\boldsymbol{t}\} \in G$ one element $u(R)$ satisfying eq. (5.6). The operators $T_g = u(R)P_g$ form a projective representation with factor system $\omega_s(g,g') = \omega_s(R,R')$ given in eq. (5.7). To find the nonequivalent irreducible representations with this factor system we follow the same method as used in Ch. 4 for ordinary representations. Here we will only indicate the steps and omit rigorous proofs. These can be given by a generalization of the method of induction (Ch. 4, §2). Suppose an irreducible projective representation of G with factor system ω_s is realized by operators T_g on an n-dimensional space V. The representation can be chosen in such a way that the subduced representation of U is diagonal. This can be done since this subduced representation is similar to an ordinary one because $\omega_s(a,g) = \omega_s(g,a) = 1$ for any $a \in U$ and any $g \in G$. The irreducible components of $D(U)$ are characterized by vectors $\boldsymbol{k}$ in the Brillouin zone. When a basis function ψ belongs to the component with vector $\boldsymbol{k}$ the function $T_{\{R|\boldsymbol{t}\}}\psi$ belongs to the component with vector $R\boldsymbol{k}$ because $T_{\{\mathbb{1}|\boldsymbol{a}\}}T_{\{R|\boldsymbol{t}\}}\psi = \omega_s(\mathbb{1},R)\,T_{\{R|\boldsymbol{t}\}\{\mathbb{1}|R^{-1}\boldsymbol{a}\}}\psi = T_{\{R|\boldsymbol{t}\}}\exp(iR\boldsymbol{k}\cdot\boldsymbol{a})\psi$. So the irreducible components belong to the vectors of one star. The group of $\boldsymbol{k}$ is defined as for ordinary representations. The eigenspace belonging to the vector $\boldsymbol{k}$ (under U) is invariant under $G_{\boldsymbol{k}}$ and carries a d-dimensional projective representation $D_{\boldsymbol{k}}(G_{\boldsymbol{k}})$. Let $\psi_1, ..., \psi_d$ be basis functions for this representation. The sd functions $T_{g_i}\psi_\mu = \psi_{i\mu}$, where g_i are coset representatives of $G_{\boldsymbol{k}}$ in G, form a basis for V. The matrices for the (induced) representation $D(G)$ are given by

$$T_g\psi_{i\mu} = \frac{\omega_s(g,g_i)}{\omega_s(g_j,h)} \sum_{\nu=1}^{d} D_{\boldsymbol{k}}(h)_{\nu\mu}\psi_{j\nu}\,, \tag{5.9}$$

where j and h are determined by $gg_i = g_jh$ with $h \in G_{\boldsymbol{k}}$. Suppose that $D_{\boldsymbol{k}}(G_{\boldsymbol{k}})$ has factor system ω'. For $g, g' \in G$ one determines j, l, h, h' by $gg_i = g_jh$ and $g'g_j = g_lh'$. Then $g'gg_i = g'g_jh = g_lh'h$. One has the relations

$$T_{g'}T_g\psi_{i\mu} = \frac{\omega_s(g,g_i)\omega_s(g',g_j)}{\omega_s(g_j,h)\omega_s(g_l,h')} \sum_{\nu=1}^{d} [D_{\boldsymbol{k}}(h')D_{\boldsymbol{k}}(h)]_{\nu\mu}\psi_{l\nu}$$

$$= \omega_s(g',g)\, T_{g'g}\psi_{i\mu}$$

$$= \omega_s(g',g)\frac{\omega_s(g'g,g_i)}{\omega_s(g_l,h'h)} \sum_{\nu=1}^{d} D_{\boldsymbol{k}}(h'h)_{\nu\mu}\psi_{l\nu}\,.$$

Application of eq. (1.28) then leads with $D_{\boldsymbol{k}}(h')D_{\boldsymbol{k}}(h) = \omega'(h',h)D_{\boldsymbol{k}}(h'h)$ to

$$\omega'(h',h) = \omega_s(h',h) \qquad (\text{all } h,h' \in G_{\boldsymbol{k}})\,. \tag{5.10}$$

The irreducible allowable projective representations $D_{\boldsymbol{k}}(G_{\boldsymbol{k}})$ can be found from

$$D_{\boldsymbol{k}}(\{R\,|\,\boldsymbol{t}\}) = \mathrm{e}^{i\boldsymbol{k}\cdot\boldsymbol{t}}\Gamma(R)\,, \tag{5.11}$$

where now $\Gamma(K_{\boldsymbol{k}})$ is a projective representation of $K_{\boldsymbol{k}}$, the point group of $G_{\boldsymbol{k}}$. In order that $D_{\boldsymbol{k}}(G_{\boldsymbol{k}})$ has the factor system ω' (5.10) the representation $\Gamma(K_{\boldsymbol{k}})$ must have a factor system ω determined by

$$\omega(R,R') = \omega_s(R,R')\, \mathrm{e}^{i(R^{-1}\boldsymbol{k}-\boldsymbol{k})\cdot\boldsymbol{t}_{R'}}\,. \tag{5.12}$$

For the symmorphic space groups or for $\boldsymbol{k}$ inside the Brillouin zone the factor systems ω and ω_s are the same.

One can show (see Mackey [1958]) that all irreducible projective representations with ω_s (the spin representations) can be found in this way. Just as for ordinary representations one obtains all nonequivalent spin representations by taking all $\boldsymbol{k}$ in a fundamental region of the Brillouin zone, considering all irreducible representations of $K_{\boldsymbol{k}}$ with factor system (5.12), and then constructing the induced representations (5.9). Since ω or ω_s may be associated to the trivial factor system, the representations of G can be similar to ordinary representations. This may happen if the factor systems of the point group are all trivial, but also when ω_s is not trivial, but $\boldsymbol{k}$ is on the border of the zone such that $\mathrm{e}^{i(R^{-1}\boldsymbol{k}-\boldsymbol{k})\cdot\boldsymbol{t}_{R'}} \sim \omega_s(R,R')^{-1}$.

The second approach is by completion of the set of operators T_g to a group. This group consists of all operators $\pm T_g$. Then to each $g \in G$ correspond two operators. The elements form the *double space group* G^{d}. One can denote the elements by $\{\alpha\,|\,\boldsymbol{t}\}$ with product rule $\{\alpha\,|\,\boldsymbol{t}\}\{\alpha'\,|\,\boldsymbol{t}'\} = \{\alpha\alpha'\,|\,\boldsymbol{t}+\alpha\boldsymbol{t}'\}$, where $\alpha\boldsymbol{t} = R\boldsymbol{t}$ for the R corresponding to $\alpha \in K^{\mathrm{d}}$. To find the irreducible representations of the double space group one proceeds as for the ordinary space groups. Notice that G^{d} has an Abelian, invariant subgroup U^{d} which is the double group for U. U^{d} is Abelian and invariant, because for $\{\alpha|\boldsymbol{t}\} \in G^{\mathrm{d}}$ one has $\{\alpha|\boldsymbol{t}\}\{\pm\epsilon|\boldsymbol{a}\}\{\alpha^{-1}|-R^{-1}\boldsymbol{t}\} = \{\pm\epsilon|R\boldsymbol{t}\}$. A subgroup of U^{d} is formed by the elements $\{\epsilon|\boldsymbol{a}\}$. This group is isomorphic to U. When C_2 is the group of order two generated by $\{-\epsilon|\boldsymbol{O}\}$, U^{d} is isomorphic to the direct product $U \times C_2$. Therefore, the irreducible representations of U^{d} can be obtained in a trivial way from those of U. They are characterized by a vector $\boldsymbol{k}$ in the first Brillouin zone together with a + or – sign. The characters of the elements are

	$\{\epsilon\|\boldsymbol{a}\}$	$\{-\epsilon\|\boldsymbol{a}\}$
$D^{(\boldsymbol{k}+)}$	$e^{i\boldsymbol{k}\cdot\boldsymbol{a}}$	$e^{i\boldsymbol{k}\cdot\boldsymbol{a}}$
$D^{(\boldsymbol{k}-)}$	$e^{i\boldsymbol{k}\cdot\boldsymbol{a}}$	$-e^{i\boldsymbol{k}\cdot\boldsymbol{a}}$

The representations of G^{d} can be found from those of U^{d} by the method of induction. The orbit of the representation $(\boldsymbol{k}+)$ is characterized by the star of $\boldsymbol{k}$, all vectors of which have a + sign. The orbit of the representation $(\boldsymbol{k}-)$ is characterized by the same star, but with a – sign, because for $g = \{\alpha|\boldsymbol{t}\} \in G^{\mathrm{d}}$ one has the conjugate representations $D_g^{(\boldsymbol{k}\pm)}(\{-\epsilon|\boldsymbol{a}\}) = \pm D^{(R\boldsymbol{k}\pm)}(\{-c|\boldsymbol{a}\})$. The little group of the representation $D^{(\boldsymbol{k}+)}$ is the group $G_{\boldsymbol{k}}^{\mathrm{d}}$, the double group of the group of $\boldsymbol{k}$. The little group of $D^{(\boldsymbol{k}-)}$ is the same group $G_{\boldsymbol{k}}^{\mathrm{d}}$. The allowable representations of the little group are found in the same way as for $G_{\boldsymbol{k}}$. They are given by

$$D_{\boldsymbol{k}}(\{\alpha|\boldsymbol{t}\}) = e^{i\boldsymbol{k}\cdot\boldsymbol{t}}\Gamma(\alpha) \qquad \text{for} \quad \{\alpha|\boldsymbol{t}\} \in G_{\boldsymbol{k}}^{\mathrm{d}} ,$$

where $\Gamma(K_{\boldsymbol{k}}^{\mathrm{d}})$ is a representation of the double point group $K_{\boldsymbol{k}}^{\mathrm{d}}$, which is the double group of $K_{\boldsymbol{k}}$. For a symmorphic group $G_{\boldsymbol{k}}$ or for $\boldsymbol{k}$ inside the Brillouin zone the representation is an ordinary one. For $\boldsymbol{k}$ on the border of the Brillouin zone for a nonsymmorphic group $G_{\boldsymbol{k}}$ it is a projective representation. The projective representations of the double point groups were determined by Kitz [1965]. The representations of G^{d} with $D(\{-\epsilon|\boldsymbol{O}\}) = -D(\{+\epsilon|\boldsymbol{O}\})$ are called the *extra representations*.

5.2. Time reversal

5.2.1. Time reversal operators

The symmetries considered in the foregoing chapters affected only the spatial coordinates. However, often physical systems are also invariant under transformations involving the time. An example is the invariance under time translations, when the Hamiltonian is not time dependent. This invariance is the reason why in these cases the time independent Schrödinger equation can replace the time dependent one. We will not discuss these transformations. Another transformation is the *time reversal T*. The operator induced by this transformation is the time reversal operator, denoted here by θ.

Let us first consider a spinless particle in a time independent potential. The Hamiltonian for this particle is $H = p^2/2m + V(\boldsymbol{r})$. It is evident that this H commutes with the substitution operator P_T defined by $P_T\psi(\boldsymbol{r},t) = \psi(\boldsymbol{r},-t)$. In this case $\theta = P_T$. We can give another expression for θ. Consider the time evolution of a function $\psi(\boldsymbol{r},t)$. It is given by the Schrödinger equation

$$\hbar i \frac{\partial}{\partial t} \psi(\boldsymbol{r},t) = H\psi(\boldsymbol{r},t) . \tag{5.13}$$

Take the complex conjugate of this equation:

$$\hbar i \frac{\partial}{\partial(-t)} \psi^*(\boldsymbol{r},t) = H\psi^*(\boldsymbol{r},t) ,$$

or

$$\hbar i \frac{\partial}{\partial t} \psi^*(\boldsymbol{r},-t) = H\psi^*(\boldsymbol{r},-t) .$$

If $\psi(\boldsymbol{r},t)$ is a solution, $\psi^*(\boldsymbol{r},-t)$ is also a solution. The state $\psi^*(\boldsymbol{r},t)$ has the same evolution along the $(-t)$ axis as the state $\psi(\boldsymbol{r},t)$ along the t-axis. Therefore, in this case one can choose θ equal to the operator θ_0 which maps a function ψ on the function ψ^*. It is the operator of complex conjugation. One has

$$\theta = \theta_0 = P_T . \tag{5.14}$$

The commutation relations of θ_0 with the operators $\boldsymbol{p}$ and $\boldsymbol{r}$ are

$$\theta_0 \boldsymbol{r} \theta_0^{-1} = \boldsymbol{r} , \qquad \theta_0 \boldsymbol{p} \theta_0^{-1} = -\boldsymbol{p} . \tag{5.15}$$

Moreover it commutes with the substitution operators P_g, when g is a space transformation.

For a particle with spin, one has to require that apart from satisfying eq. (5.15) the time reversal operator θ has the same commutation relations with the spin operator $\boldsymbol{S}$ as it has with the operator of orbital angular momentum $\boldsymbol{L}$: $\theta \boldsymbol{L} \theta^{-1} = \theta \boldsymbol{r} \times \boldsymbol{p} \theta^{-1} = -(\boldsymbol{r} \times \boldsymbol{p}) = -\boldsymbol{L}$. Thus

$$\theta \boldsymbol{S} \theta^{-1} = -\boldsymbol{S} . \tag{5.16}$$

One has to consider a combination $\theta = u\theta_{\mathrm{o}}$ such that θ has the required commutation relations with $\boldsymbol{S}$. In the standard representation of Pauli matrices one has

$$\theta_{\mathrm{o}} S_1 \theta_{\mathrm{o}}^{-1} = \tfrac{1}{2}\hbar \theta_{\mathrm{o}} \sigma_1 \theta_{\mathrm{o}}^{-1} = S_1 ,$$

$$\theta_{\mathrm{o}} S_2 \theta_{\mathrm{o}}^{-1} = -S_2 , \qquad \theta_{\mathrm{o}} S_3 \theta_{\mathrm{o}}^{-1} = S_3 .$$

Hence for the two by two matrix u one has

$$u\sigma_1 u^{-1} = -\sigma_1 ; \qquad u\sigma_2 u^{-1} = \sigma_2 ; \qquad u\sigma_3 u^{-1} = -\sigma_3 .$$

It follows that u is a multiple of σ_2. So for a system with H (5.5), one has, choosing the phase factor equal to 1,

$$\theta = \sigma_2 \theta_{\mathrm{o}} . \tag{5.17}$$

We have now shown that the time reversal operator θ for a particle in a time independent potential is to be defined as

$$\theta = \begin{cases} \theta_{\mathrm{o}} & \text{for a system without spin,} \\ \sigma_2 \theta_0 & \text{for a system with spin } \tfrac{1}{2} . \end{cases}$$

The operator θ is antilinear and antiunitary, because
1) $\theta(\alpha\psi_1 + \beta\psi_2) = \alpha^* \theta\psi_1 + \beta^* \theta\psi_2$,
2) for a spinless particle $\langle \theta_{\mathrm{o}}\psi_1 | \theta_{\mathrm{o}}\psi_2 \rangle = \langle \psi_1^* | \psi_2^* \rangle = \langle \psi_2 | \psi_1 \rangle$, whereas for a particle with spin one has (ψ^+ and ψ^- denoting the spin-up and the spin-down component of ψ) $\langle \sigma_2\theta_{\mathrm{o}}\psi_1 | \sigma_2\theta_{\mathrm{o}}\psi_2 \rangle = \langle i\psi_1^{-*} | i\psi_2^{*-} \rangle + \langle i\psi_1^{*+} | i\psi_2^{+*} \rangle = \langle \psi_1^{+*} | \psi_2^{+*} \rangle + \langle \psi_1^{-*} | \psi_2^{-*} \rangle = \langle \psi_1 | \psi_2 \rangle^*$.

5.2.2. Time reversal symmetry

The Hamiltonian of an electron in a crystal is given by

$$H = H_o + \frac{\hbar}{4m^2c^2}(\nabla V \times \boldsymbol{p}) \cdot \boldsymbol{\sigma} = \frac{p^2}{2m} + V(\boldsymbol{r}) + V_s\,, \tag{5.18}$$

where $V(\boldsymbol{r})$ is the crystal potential which is invariant under the space group G and V_s is the spin term which is invariant under the double space group G^{d}. We want to determine the symmetry groups of H and H_o, including time reversal. The Hamiltonian H_o commutes with θ_o and with all elements P_g with $g \in G$. The Hamiltonian H commutes with θ and with all elements $\pm u(R)P_g$ of the double space group G^{d}. The operator θ_o commutes with all elements of the space group. Therefore, the invariance group of the Hamiltonian H_o is the direct product of the group G with the group generated by θ_o. As θ_o is of order two the symmetry group is $G \times C_2$.

For the Hamiltonian H we investigate θ. Since $\theta^2 = -\mathbb{I}$, it generates a cyclic group of order four. Moreover, θ commutes with all elements of $SU(2)$ and consequently with all elements of G^{d}. To see this we use eq. (5.1):

$$\sigma_2\theta_0 u = \begin{pmatrix} 0 & -i \\ i & 0 \end{pmatrix} \theta_o \begin{pmatrix} a & b \\ -b^* & a^* \end{pmatrix} = \begin{pmatrix} 0 & -i \\ i & 0 \end{pmatrix} \begin{pmatrix} a^* & b^* \\ -b & a \end{pmatrix} \theta_o$$

$$= \begin{pmatrix} a & b \\ -b^* & a^* \end{pmatrix} \begin{pmatrix} 0 & -i \\ i & 0 \end{pmatrix} \theta_o = u\sigma_2\theta_o\,.$$

However, the invariance group of H is not the direct product $G^{\mathrm{d}} \times C_4$, because $G^{\mathrm{d}} \cap C_4 = -\mathbb{I}$. It is an extension of C_4, generated by θ, by the space group G.

The group of underlying space-time transformations we have considered here is a subgroup of the direct product of the Euclidean group $E(3)$ and the group generated by the time reversal T. This direct product $E(3) \times C_2$ is called the *Shubnikov group* and is denoted by $S(4)$ †. It is a group of transformations of four-dimensional space-time. The subgroup of $S(4)$ which is the

† There seems to be no standard use for this term. Since this group $E(3) \times C_2 = \mathbb{R}^3 \boxtimes (O(3) \times C_2)$ occurs very often here, it is convenient to give it a name. However, sometimes the term (homogeneous) Shubnikov group is used for $O(3) \times C_2$, whereas sometimes by inhomogeneous Shubnikov group is denoted the semidirect product $\mathbb{R}^4 \boxtimes (O(3) \times C_2)$.

symmetry group of the system is the direct product of the space group G and C_2. For a spinless particle, the mapping which assigns P_g to $g \in G$ and $\theta_0 P_g$ to gT is a homomorphism on a group of operators commuting with the Hamiltonian. However, as some of the operators are antilinear, it is not a representation of $G \times C_2$. To see the properties of the homomorphism we consider an eigenspace of H_0 with a certain eigenvalue. Since the operators commute with H, any vector of this eigenspace is transformed into a vector of the same space by the operators. In particular for basis functions

$$P_s \psi_i = \sum_j C(s)_{ji} \psi_j \qquad (\text{any } s \in G \times C_2) .$$

The matrices $C(s)$ do not form a matrix representation because P_T is antilinear. For $g_1, g_2 \in G$ and with $P_T = \theta_0$, one has

$$\begin{aligned} P_{g_1} P_{g_2} \psi_i &= \sum_k \left[C(g_1)\, C(g_2)\right]_{ki} \psi_k , \\ P_{g_1} P_{g_2 T} \psi_i &= \sum_k \left[C(g_1)\, C(g_2 T)\right]_{ki} \psi_k , \\ P_{g_1 T} P_{g_2} \psi_i &= \sum_k \left[C(g_1 T)\, C^*(g_2)\right]_{ki} \psi_k . \end{aligned}$$

Then the matrices $C(s)$ form a set, related to the elements $s \in G \times C_2$ by

$$\begin{aligned} C(g_1 g_2) &= C(g_1)\, C(g_2) \\ C(g_1 g_2 T) &= C(g_1)\, C(g_2 T) \\ C(g_1 T g_2) &= C(g_1 T)\, C^*(g_2) \\ C(g_1 T g_2 T) &= C(g_1 T)\, C^*(g_2 T) . \end{aligned} \tag{5.19}$$

A set of matrices satisfying these relations is called a *co-representation* of the group. See Wigner [1959] or Jansen and Boon [1967] for details.

For a particle possessing spin one can choose one element $u(R)P_g$ of the double group G^{d} for any element $g \in G$. Then the mapping which assigns $T_g = u(R)P_g$ to $g \in G$ and $T_{gT} = \sigma_2 \theta_0 u(R) P_g$ to $gT \in G \times C_2$ is a mapping on a set of operators commuting with the Hamiltonian. Now the mapping of the product of two elements is the product of the mappings of these elements only up to a phase factor. Again, we do not obtain a (projective) representation, but rather a projective co-representation. We return to this subject in Ch. 6, §4.4.

5.2.3. Consequences of time reversal symmetry

When the symmetry group contains time reversal, an eigenspace belonging to an eigenvalue of the Hamiltonian is a co-representation space of this symmetry group. Consider first the spinless case. If ψ is a state of the eigenspace, the functions $P_g\psi$ $(g \in G)$ span an irreducible representation space for the space group G. Suppose that $\psi_1, ..., \psi_d$ is a basis of this space. Also the functions $\theta_0\psi_1, ..., \theta_0\psi_d$ form a basis for a representation of G. Now, either the two spaces spanned by $\psi_1, ..., \psi_d$ and $\theta_0\psi_1, ..., \theta_0\psi_d$ coincide, or they are different spaces with only the null vector in common. In the first case the functions $\theta_0\psi_i$ are linear combinations of the elements $\psi_1, ..., \psi_d$ and the level is d-fold degenerate. When the two spaces are different they span a $2d$-dimensional space belonging to one eigenvalue. Then the eigenvalue is $2d$-fold degenerate. With respect to the group G this is an accidental degeneracy. It is called *additional degeneracy*.

In the same way, for a particle with spin the elements $\pm u(R)P_g\psi$ span a representation space for the double group G^{d}. When $\psi_1, ..., \psi_d$ form a basis for this space, also $\theta\psi_1, ..., \theta\psi_d$ form the basis for a representation space. However, in this case the states ψ_i and $\theta\psi_i$ are always orthogonal, because the inner product of the spinors

$$\psi_i = \begin{pmatrix} \psi_i^+ \\ \psi_i^- \end{pmatrix}, \qquad \theta\psi_i = \begin{pmatrix} -i(\psi_i^-) \\ i(\psi_i^+) \end{pmatrix}$$

is zero. Hence for spin $\frac{1}{2}$ particles one has always at least two-fold degeneracy. This is called *Kramers degeneracy*. This does not mean that the sets $\psi_1, ..., \psi_d$ and $\theta\psi_1, ..., \theta\psi_d$ span different spaces. One has d-fold or $2d$-fold degeneracy.

The question of when additional degeneracy occurs is answered by *Herrings criterion*. When $\psi_1, ..., \psi_d$ form a basis for the representation $D(G^{(\mathrm{d})})$ of $G^{(\mathrm{d})}$, the functions $\theta\psi_1, ..., \theta\psi_d$ form a basis for the representation $D^*(G^{(\mathrm{d})})$. Either $D \sim D^*$, or $D \not\sim D^*$. When $D \sim D^*$, there is a matrix S such that $D^*(g) = SD(g)S^{-1}$ for any $g \in G^{(\mathrm{d})}$. It can be shown that $SS^* = \pm \mathbb{1}$. Then the dimension of the space spanned by $\psi_1, ..., \psi_d, \theta\psi_1, ..., \theta\psi_d$ is given by table 5.2. In practice it is often difficult to determine which of these possibilities occurs. Then one can make use of a *theorem by Frobenius and Schur*:

Table 5.2.
Herrings criterion.

Cases	Dimension irreducible space	
1) $D \sim D^*$ and $SS^* = +\mathbb{1}$	with spin: $2d$	without spin: d
2) $D \not\sim D^*$	with spin: $2d$	without spin: $2d$
3) $D \sim D$ and $SS^* = -\mathbb{1}$	with spin: d	without spin: $2d$

d is the dimension of an irreducible component of the subduced unitary representation.

PROPOSITION 5.2. When D is an irreducible representation of a group G of order N, one has

$$\sum_{g \in G} \chi(g^2) = \begin{cases} N, & \text{if } D \sim D^* \text{ and } SS^* = \mathbb{1}, \\ 0, & \text{if } D \not\sim D^*, \\ -N, & \text{if } D \sim D^* \text{ and } SS^* = -\mathbb{1}. \end{cases} \tag{5.20}$$

Proofs for Herrings criterion and for the Frobenius–Schur theorem can be found in Ch. 6, §4, where we will consider more general groups involving time reversal.

For groups of small order, such as the point groups, the Frobenius–Schur theorem is very convenient. However, for space groups the summation over all group elements requires some more study. Using eq. (4.33) one has

$$\sum_{g \in G} \chi(g^2) = \sum_{R \in K} \sum_{a \in U} \sum_{\substack{j \\ R^2 \boldsymbol{k}_j \equiv \boldsymbol{k}_j}} \operatorname{tr} D_{\boldsymbol{k}}(g_j^{-1} g^2 g_j)$$

$$= \sum_j \sum_{a \in U} \sum_{S \in K}' \operatorname{tr} D_{\boldsymbol{k}}(h^2) \quad \text{with} \quad h = g_j^{-1} g g_j = \{S | \boldsymbol{t}_S + \boldsymbol{a}\},$$

where the prime denotes summation over those elements S of the point group for which the square S^2 is in the point group of the group of $\boldsymbol{k}$: $S^2\boldsymbol{k} \equiv \boldsymbol{k}$, or equivalently $R^2\boldsymbol{k}_j = R^2R_j\boldsymbol{k}_1 = R_jS^2\boldsymbol{k}_1 \equiv \boldsymbol{k}_j$. The summation over the elements a of U can be performed:

$$\sum_{a\in U} \operatorname{tr} D_{\boldsymbol{k}}(h^2) = \sum_{a\in U} \operatorname{tr} D_{\boldsymbol{k}}(\{S^2|\boldsymbol{t}_S+\boldsymbol{a}+S\boldsymbol{t}_S+S\boldsymbol{a}\})$$

$$= \sum_{a\in U} e^{i\boldsymbol{k}\cdot(S\boldsymbol{a}+\boldsymbol{a})} \operatorname{tr} D_{\boldsymbol{k}}(\{S^2|\boldsymbol{t}_S+S\boldsymbol{t}_S\})$$

$$= \operatorname{tr} D_{\boldsymbol{k}}(\{S^2|\boldsymbol{t}_S+S\boldsymbol{t}_S\}) \sum_{a\in U} e^{i(S^{-1}\boldsymbol{k}+\boldsymbol{k})\cdot \boldsymbol{a}}$$

$$= \operatorname{tr} D_{\boldsymbol{k}}(\{S^2|\boldsymbol{t}_S+S\boldsymbol{t}_S\}) N^3 \sum_{\boldsymbol{K}\in\Lambda^*} \delta(S^{-1}\boldsymbol{k}+\boldsymbol{k}-\boldsymbol{K}) \,.$$

This gives

$$\sum_{g\in G} \chi(g^2) = sN^3 \sum_{\substack{S\in K \\ S^2\in K_{\boldsymbol{k}}}} \delta^*(S^{-1}\boldsymbol{k}+\boldsymbol{k}) \operatorname{tr} D_{\boldsymbol{k}}(\{S|\boldsymbol{t}_S\}^2) \,.$$

where $\delta^*(\boldsymbol{k}) = \Sigma\, \delta(\boldsymbol{k}-\boldsymbol{K})$, (summation over all reciprocal lattice vectors $\boldsymbol{K}$). We now introduce the set of all elements of the point group K which invert the vector $\boldsymbol{k}$:

$$M_{\boldsymbol{k}} = \{S\in K\,|\,S\boldsymbol{k}\equiv -\boldsymbol{k}\} \,.$$

This gives finally, since the contribution of S vanishes unless $S\boldsymbol{k}\equiv -\boldsymbol{k}$,

$$\sum_{g\in G} \chi(g^2) = sN^3 \sum_{S\in M_{\boldsymbol{k}}} \operatorname{tr} D_{\boldsymbol{k}}(\{S|\boldsymbol{t}_S\}^2) \,, \tag{5.21}$$

where for each $S\in M_{\boldsymbol{k}}$, one element $\{S|\boldsymbol{t}_S\}$ of the space group G is selected. This element is not uniquely determined, because $\boldsymbol{t}_S$ is determined only up to a primitive translation, but because of the definition of $M_{\boldsymbol{k}}$ another $\boldsymbol{t}_S$ gives the same result. In this way, we have an expression in which the summation is over point group elements only. Herrings criterion becomes with eq. (5.21)

$$\sum_{S\in M_{\boldsymbol{k}}} \operatorname{tr} D_{\boldsymbol{k}}(\{S|\boldsymbol{t}_S\}^2) = \begin{cases} n/s & \text{for case 1}\,, \\ 0 & \text{for case 2}\,, \\ -n/s & \text{for case 3}\,, \end{cases} \tag{5.22}$$

where n denotes the order of the point group and s the number of points of the star of $\boldsymbol{k}$. When $-\boldsymbol{k}$ does not occur in the star of $\boldsymbol{k}$, $M_{\boldsymbol{k}}$ is empty and we have case 2. This means additional degeneracy in spinless problems. When $-\boldsymbol{k}$ occurs in the star of $\boldsymbol{k}$, but $\boldsymbol{k}\neq -\boldsymbol{k}$ (i.e. $2\boldsymbol{k}$ is not a reciprocal lattice vector), then there is an element R of K with $-\boldsymbol{k}\equiv R\boldsymbol{k}$. Then $M_{\boldsymbol{k}} = RK_{\boldsymbol{k}}$. In particular, when the central inversion I occurs in the point group, one has

$$\sum_{S \in M_{\boldsymbol{k}}} \operatorname{tr} D_{\boldsymbol{k}}(\{S \mid \boldsymbol{t}_S\}^2) = \sum_{R \in K_{\boldsymbol{k}}} \operatorname{tr} D_{\boldsymbol{k}}(\{R \mid \boldsymbol{t}_R\}^2) .$$

CHAPTER VI

APPLICATIONS IN SOLID STATE PHYSICS

In the preceding chapters we have considered space-time transformation groups and their representations which play a role in crystal physics. In the present chapter we will discuss application of this theory to two important systems: we have taken as examples the cases of an electron in a crystal (electron band theory) and of lattice vibrations. To these examples we could add a large number of other systems, but the techniques will not be different. We assume some familiarity with the fundamentals of solid state theory and refer for a deeper understanding to other books.

In two further sections we shall give two generalizations of the notions and methods used so far. These generalized notions are important for other situations in solid state physics. In the third section we treat a charged particle in an electromagnetic field. In the equation of motion of such a particle, i.e. in the Schrödinger equation, occurs the potential of the field. It turns out that it is necessary to study the relation between the symmetries of field and potential. The theory is applied to a Bloch electron in a homogeneous magnetic field. In the fourth section we study crystals with localized magnetic moments. For such systems plain time reversal is not a symmetry transformation. However, one can define symmetry transformations which combine spatial transformations with time reversal. This leads to magnetic groups and their co-representations. In the Bibliography we shall give a list of papers on the use of group theory. In that list are given also papers on subjects not discussed in this book.

6.1. Electron bands

6.1.1. Electrons in a crystal

We can consider a crystal as a large system composed of nuclei and electrons. The Hamiltonian of the system is $H = H_n + H_e' + H_{ne}'$, where H_n describes the nuclei, H_e' the electrons and H_{ne}' their interaction. However, there is a great difference between the strongly bound electrons of the cores and the loosely bound outer electrons. Therefore, it is better to write

$$H = H_c + H_e + H_{ce} , \tag{6.1}$$

where H_c describes the cores of nuclei and strongly bound electrons, H_e the loosely bound electrons and H_{ce} their interaction. In the quasi-stationary approximation one can neglect H_c and consider the electrons as moving in a fixed crystal potential. Then

$$H = \sum_{i=1}^{N} \frac{p_i}{2m} + \sum_{i<j} V_{ij} + \sum_i W_i , \tag{6.2}$$

where V_{ij} is the Coulomb repulsion and W_i the potential energy. Finally in the Hartree–Fock approximation the N-electron wave function is given by a Slater determinant

$$\psi = \frac{1}{\sqrt{N!}} \det(\varphi_i(j)) ,$$

where $\varphi_i(j)$ is an eigenfunction for particle j of the Hamiltonian

$$H = \frac{p^2}{2m} + V(\boldsymbol{r}) + V_{\mathrm{so}} . \tag{6.3}$$

Here $V(\boldsymbol{r})$ is the sum of the potential due to the atom cores in the crystal and a smeared-out potential due to the repulsion of the outer electrons. The term V_{so} is the spin-orbit term

$$V_{\mathrm{so}} = \frac{\hbar}{4m^2c^2} (\nabla V \times \boldsymbol{p}) \cdot \boldsymbol{\sigma} , \tag{6.4}$$

which arises in the nonrelativistic approximation to the Dirac Hamiltonian. When we neglect the spin term, we obtain the Hamiltonian of eq. (4.34), because $V(\boldsymbol{r})$ is invariant under space group transformations (cf. Streitwolf [1967], p. 114). Taking the spin into account we obtain, apart from some relativistic corrections, eq. (5.5). Therefore the invariance group of H of (6.3) is the double space group G^{d}, or the space group G if we can neglect spin effects. Hence, the eigenfunctions of H belong to representations of the (double) space group. Barring accidental degeneracy these representations are irreducible. The basis functions for these representations are Bloch functions $\psi_{\boldsymbol{k}\nu}$, where $\boldsymbol{k}$ is in the first Brillouin zone and ν is the band index. It is possible to choose $\psi_{\boldsymbol{k}\nu}$ as basis functions of standard representations of G (or G^{d}) such that they form an orthonormal set

$$\langle\psi_{\boldsymbol{k}\nu}|\psi_{\boldsymbol{k}'\nu'}\rangle = \delta_{\boldsymbol{k}\boldsymbol{k}'}\delta_{\nu\nu'}\,. \tag{6.5}$$

As we discussed in Ch. 4 the function $\psi_{\boldsymbol{k}\nu}$ can be written as

$$\psi_{\boldsymbol{k}\nu} = \mathrm{e}^{-i\boldsymbol{k}\cdot\boldsymbol{r}}u_{\boldsymbol{k}\nu}(\boldsymbol{r})\,, \tag{6.6}$$

where $u_{\boldsymbol{k}\nu}(\boldsymbol{r})$ has the periodicity of the (direct) lattice.

Substitution of eq. (6.6) in the Schrödinger equation with H from eq. (6.3) yields

$$H'u_{\boldsymbol{k}\nu} = \{\frac{p^2}{2m} + V(\boldsymbol{r}) - \frac{\hbar}{m}\boldsymbol{k}\cdot\boldsymbol{p} - \frac{\hbar}{4m^2c^2}\boldsymbol{\sigma}\cdot[\nabla V(\boldsymbol{r})\times(\boldsymbol{k}-\boldsymbol{p})]\}u_{\boldsymbol{k}\nu} = E'_{\boldsymbol{k}\nu}u_{\boldsymbol{k}\nu} \tag{6.7}$$

with

$$E'_{\boldsymbol{k}\nu} = E_{\boldsymbol{k}\nu} - \frac{h^2k^2}{2m}\,.$$

The new Hamiltonian H' has a more involved structure than the old one, but the advantage is that one has only to solve for solutions inside one cell, because $u_{\boldsymbol{k}\nu}(\boldsymbol{r})$ is periodic. This situation is analogous to the Schrödinger problem for a particle in a spherically symmetric potential, where one derives from the original equation the more complicated radial equation, which has the advantage that it does not contain the angle variables. The first, second and fourth terms in eq. (6.7) have the symmetry of the (double) space group. This symmetry is lowered by the $\boldsymbol{k}$-dependent third and fifth terms. These terms have only the group of $\boldsymbol{k}$ as invariance group. This means that the energy levels are labelled with the irreducible representations of $G_{\boldsymbol{k}}$ or $G_{\boldsymbol{k}}^{\mathrm{d}}$. Here one sees immediately the dependence of the degeneracy on the $\boldsymbol{k}$-vector. Of course the description of wave functions and energy levels is equivalent to that given in Ch. 4, §3. In the present case we characterize the wave functions by irreducible representations of the subgroup $G_{\boldsymbol{k}}^{(\mathrm{d})}$ of the invariance group $G^{(\mathrm{d})}$ of H. This is called the "subgroup method". A wave function is labelled by a vector $\boldsymbol{k}$ from the first Brillouin zone and the band index ν which denotes, in a way described in Ch. 4, §3, both the irreducible representation $D_{\boldsymbol{k}}(G_{\boldsymbol{k}}^{(\mathrm{d})})$ and the row of the corresponding representation $\Gamma(K_{\boldsymbol{k}}^{(\mathrm{d})})$ to which $\psi_{\boldsymbol{k}\nu}$

belongs. Since the irreducible representations of $G^{(\mathrm{d})}$ may be obtained by induction from those of $G_{\boldsymbol{k}}^{(\mathrm{d})}$, the functions $\psi_{\boldsymbol{k}\nu}$, with $\boldsymbol{k}$ running through its star and ν denoting the rows of one representation of $K_{\boldsymbol{k}}$, form a basis for an irreducible representation of G. The eigenfunctions of H can then also be described by the star of $\boldsymbol{k}$ and the index ν, or more simply by the pair $\boldsymbol{k}\nu$. Now $\boldsymbol{k}$-star and ν denote an irreducible representation of the full group $G^{(\mathrm{d})}$. This is called the "full group method". As discussed in Ch. 4, §3 for selection rules the two methods show slight differences.

6.1.2. Compatibility relations

The eigenvalues of H in eq. (6.3) are denoted by $E_{\boldsymbol{k}\nu}$, where $\boldsymbol{k}$ is in the first Brillouin zone and ν the band index. The pair $\boldsymbol{k}\nu$ denotes a representation of the invariance group G. (We consider here the case where the spin may be neglected. The spin case is easily obtained by taking everywhere the double group instead of the ordinary symmetry group.) It is the representation induced from a representation of the group of $\boldsymbol{k}$. If the representation of $G_{\boldsymbol{k}}$ has a basis $\psi_{\boldsymbol{k}1}, ..., \psi_{\boldsymbol{k}d\nu}$, it induces a representation with basis $\psi_{\boldsymbol{k}_i 1}, ...$ $..., \psi_{\boldsymbol{k}_i d_\nu}$ $(i = 1, ..., s)$ where $\boldsymbol{k}_i$ runs through the set of vectors of the star of $\boldsymbol{k}$. This means that $E_{\boldsymbol{k}_i\nu}$ denotes the same energy level as $E_{\boldsymbol{k}\nu}$. Hence

$$E_{\boldsymbol{k}\nu} = E_{R\boldsymbol{k},\nu} \qquad (R \in K)$$

for any element R from the point group of G. Therefore, it is sufficient to know the function $E_{\boldsymbol{k}\nu}$ only in a part of the first Brillouin zone. We call a domain in the first Brillouin zone, containing exactly one vector from each star, a *fundamental region*. Like the unit cell of the reciprocal lattice it is not uniquely determined. Any domain of the zone which covers the complete zone exactly once under the action of the point group will do. However, in the literature the choice of the fundamental regions is more or less standardized, cf. Koster [1957].

The degeneracy of the level $E_{\boldsymbol{k}\nu}$ is the dimension of the corresponding representation of G. It is the product of the dimension of the representation of $G_{\boldsymbol{k}}$ from which the representation of G is induced, and the number of points of the star of $\boldsymbol{k}$. The dimension of the representation of the group of $\boldsymbol{k}$ is the degeneracy of the Hamiltonian H' (6.7). As this Hamiltonian depends on $\boldsymbol{k}$ via the third and fifth term, the degeneracy depends on $\boldsymbol{k}$. Moving over an energy surface $E_{\boldsymbol{k}\nu}$ for fixed ν, level splitting may occur along the path. However, the degeneracy and the transformation properties of the corresponding eigenfunctions, in short the associated representations of $G_{\boldsymbol{k}}$, have to obey certain rules, called *compatibility conditions*.

Consider a point $\boldsymbol{k}$ in the Brillouin zone and an irreducible representation $D_{\boldsymbol{k}}(G_{\boldsymbol{k}})$ of the group of $\boldsymbol{k}$. A basis of the representation is formed by $\psi_{\boldsymbol{k}1}$, ..., $\psi_{\boldsymbol{k}d}$. These may be chosen in the form of Bloch functions and in such a way that for another point $\boldsymbol{k}'$ the functions $\lim_{\boldsymbol{k}\to\boldsymbol{k}'}\psi_{\boldsymbol{k}i} = \psi_{\boldsymbol{k}'i}$ belong to a basis of a representation $D_{\boldsymbol{k}'}(G_{\boldsymbol{k}'})$. Suppose that, barring accidental degeneracy, the level $E_{\boldsymbol{k}\nu}$ corresponds to $D_{\boldsymbol{k}}(G_{\boldsymbol{k}})$ and $E_{\boldsymbol{k}'\nu}$ with the same ν to $D_{\boldsymbol{k}'}(G_{\boldsymbol{k}'})$. Furthermore, we assume that $G_{\boldsymbol{k}}$ is a subgroup of $G_{\boldsymbol{k}'}$. Then the subduced representation $D_{\boldsymbol{k}'}(G_{\boldsymbol{k}})$ contains in the limit for $\boldsymbol{k}\to\boldsymbol{k}'$ the representation $D_{\boldsymbol{k}}(G_{\boldsymbol{k}})$. This means that in the reduction of $D_{\boldsymbol{k}'}(G_{\boldsymbol{k}})$ the irreducible component $D_{\boldsymbol{k}}(G_{\boldsymbol{k}})$ must occur. If this happens, which can be determined from the character table, the representations are called compatible.

As an example we consider an electron in a crystal with space group $Pm3m$. Neglecting spin effects the invariance group is also $Pm3m$. This space group has a simple cubic lattice. The reciprocal lattice is also simple cubic. In fig. 6.1 a fundamental region (ΓXMR) is indicated. Its volume is 1/48-th of the volume of the complete Brillouin zone because the point group $m3m$ is of order 48. The $\boldsymbol{k}$-vectors with different groups $G_{\boldsymbol{k}}$ are also indicated. They are the points Γ, R, X and M, the points $\Lambda, S, \Delta, \Sigma, Z$ and T on symmetry lines, and the points in the symmetry planes ΓXR, ΓXM, ΓMR and RMX. Let us see what is happening if we move on a $E_{\boldsymbol{k}\nu}$ surface along the k_x-axis. The group of Γ, i.e. $\boldsymbol{k} = 0$, is the group $Pm3m$. Its point group is the direct product of the octahedral group O with the group of order two. Its ten irreducible representations are $\Gamma_1^\pm$, ..., $\Gamma_5^\pm$. Going to a point Δ, i.e. $\boldsymbol{k} = (\epsilon, 0, 0)$ for some number ϵ, the symmetry is lowered to $P4mm$ with point group $4mm$. This group is isomorphic to D_4 and has five irreducible representations, Δ_1, ..., Δ_5. Going from Γ to Δ the level corresponding to a representation $\Gamma_\beta^\pm$ splits into levels corresponding to representations of $4mm$. On the other hand in the limit $\epsilon\to 0$ the eigenfunctions of H corresponding to $\boldsymbol{k} = \Delta$ and given ν and Δ_α go continuously over into eigenfunctions at $\boldsymbol{k} = \Gamma$. These eigen-

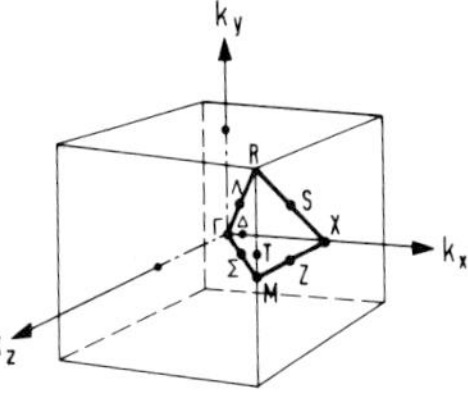

Fig. 6.1. Brillouin zone of the primitive cubic lattice.

Table 6.1
Compatibility conditions for points Δ and Γ in a cubic Brillouin zone.

Irreducible characters $4mm$						Characters of $4mm$ subduced from $m3m$						Compatibility
	$[\epsilon]$	$[\alpha^2]$	$[\alpha]$	$[\beta]$	$[\alpha\beta]$		$[\epsilon]$	$[\alpha^2]$	$[\alpha]$	$[\beta]$	$[\alpha\beta]$	
Δ_1	1	1	1	1	1	$\Gamma_1^{\pm}$	1	1	1	± 1	± 1	$= \Delta_1$ resp. Δ_2
Δ_2	1	1	1	-1	-1	$\Gamma_2^{\pm}$	1	1	-1	± 1	∓ 1	$= \Delta_3$ resp. Δ_4
Δ_3	1	1	-1	1	-1	$\Gamma_3^{\pm}$	2	2	0	± 2	0	$= \Delta_1 \oplus \Delta_3$ resp. $\Delta_2 \oplus \Delta_4$
Δ_4	1	1	-1	-1	1	$\Gamma_4^{\pm}$	3	-1	$+1$	∓ 1	∓ 1	$= \Delta_2 \oplus \Delta_5$ resp. $\Delta_1 \oplus \Delta_5$
Δ_5	2	-2	0	0	0	$\Gamma_5^{\pm}$	3	-1	-1	∓ 1	± 1	$= \Delta_4 \oplus \Delta_5$ resp. $\Delta_3 \oplus \Delta_5$

functions belong to a representation $\Gamma_\beta^{\pm}$ such that $\Gamma_\beta^{\pm}(4mm)$ contains Δ_α. The characters of the irreducible representations of $4mm$ and of the representations of $4mm$ subduced from a Γ_β are given in table 6.1. Compare the subduced characters with the character table of O and notice that $\epsilon, \alpha, \alpha^2, \alpha^3$ are rotations of order 1, 4, 2, 4 respectively, that β and $\alpha^2\beta$ are products of the central inversion I with rotations of order 2, and that $\alpha\beta$ and $\alpha^3\beta$ are products of I with the square of a rotation of order 4. From table 6.1 one sees that Δ_1 is compatible with Γ_1^+, Γ_3^+ and Γ_4^-, Δ_2 with Γ_1^-, Γ_3^- and Γ_4^+, Δ_3 with Γ_2^+, Γ_3^+ and Γ_5^-, Δ_4 with Γ_2^-, Γ_3^- and Γ_5^+, Δ_5 with $\Gamma_4^{\pm}$ and $\Gamma_5^{\pm}$. Moving along the k_x-axis over a $E_{k\nu}$-surface the representation Δ_α remains the same. Approaching the point Γ this representation goes over into a subrepresentation of $\Gamma_\beta^{\pm}$. At the point Γ several surfaces come together. On the other hand, starting from Γ the level splits up in general. One could obtain a behaviour like that sketched in fig. 6.2. The same can be done for other paths in the Brillouin zone. Leaving a point of high symmetry in the direction of lower symmetry the level splits up, in general. The splitting depends on the direction of the path. Inside the fundamental region, where the $\boldsymbol{k}$-vectors have only the identity as sym-

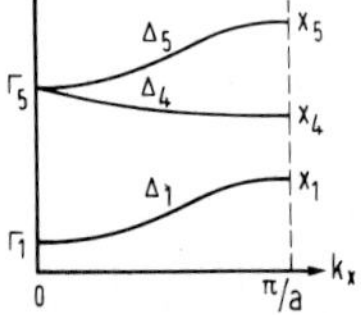

Fig. 6.2. Possible energy levels along the k_x-axis.

metry point group, the levels become nondegenerate.

6.1.3. Symmetrized plane waves

From translational symmetry it follows that the eigenfunctions of H in eq. (6.3) can be written as Bloch functions (6.6). The functions $u_{\boldsymbol{k}\nu}$ are solutions of a Schrödinger equation with Hamiltonian H' (6.7). To solve this equation is the goal of band structure calculations. As this cannot be done exactly one has to apply perturbation theory. The choice of the zeroth order approximations is then important. Here come in arguments which differ from case to case. For some solids one can apply the nearly-free electron model. Here the crystal potential is considered as a small perturbation on the kinetic energy. The zeroth order wave functions are in this case plane waves. As one has degeneracy in the unperturbed states, not every set of plane waves forms a good approximation. One has to construct symmetrized plane waves. The eigenvalue problem we will have to solve will become much easier using these symmetrized wave functions.

Which plane waves have to be considered to construct the functions $u_{\boldsymbol{k}\nu}(\boldsymbol{r})$? Since $u_{\boldsymbol{k}\nu}(\boldsymbol{r})$ has the periodicity of the lattice one has for a translation $a \in U$

$$P_a u_{\boldsymbol{k}\nu}(\boldsymbol{r}) = u_{\boldsymbol{k}\nu}(\boldsymbol{r}-\boldsymbol{a}) = u_{\boldsymbol{k}\nu}(\boldsymbol{r}) \,.$$

This means that the $u_{\boldsymbol{k}\nu}(\boldsymbol{r})$ form a basis for a representation of U which is equivalent to one with wave vector $\boldsymbol{k} = 0$. This means that in the Fourier decomposition of $u_{\boldsymbol{k}\nu}(\boldsymbol{r})$ into plane waves only those waves occur which have a wave vector from the reciprocal lattice. Then one has

$$\psi_{\boldsymbol{k}\nu} = \mathrm{e}^{-i\boldsymbol{k}\cdot\boldsymbol{r}} u_{\boldsymbol{k}\nu}(\boldsymbol{r}) = \sum_{\boldsymbol{K}\in\Lambda^*} a_{\boldsymbol{k}\nu}(\boldsymbol{K})\, \mathrm{e}^{-i(\boldsymbol{k}+\boldsymbol{K})\cdot\boldsymbol{r}} \,. \tag{6.8}$$

When the nearly-free electron model can be used the function $\psi_{\boldsymbol{k}\nu}$ can be approximated by a small number of plane waves. In this case it will not be necessary to calculate all coefficients $a_{\boldsymbol{k}\nu}(\boldsymbol{K})$. When the model is not valid, the convergence will be poor and we will have to consider other approximations.

Consider an element $g = \{R\,|\,\boldsymbol{a}\}$ from the group of $\boldsymbol{k}$. For such an element $P_g \exp[-i(\boldsymbol{k}+\boldsymbol{K})\cdot\boldsymbol{r}] = \exp[-i(\boldsymbol{k}+\boldsymbol{K})\cdot(R^{-1}\boldsymbol{r}-R^{-1}\boldsymbol{a})] =$ $\exp[-iR(\boldsymbol{k}+\boldsymbol{K})\cdot(\boldsymbol{r}-\boldsymbol{a})] = \exp[iR(\boldsymbol{k}+\boldsymbol{K})\cdot\boldsymbol{a}]\cdot\exp[-iR(\boldsymbol{k}+\boldsymbol{K})\cdot\boldsymbol{r}]$. Because $g \in G_{\boldsymbol{k}}$ one has $R\boldsymbol{k} + R\boldsymbol{K} = \boldsymbol{k} + \boldsymbol{K}' + \boldsymbol{K}'' = \boldsymbol{k} + \boldsymbol{K}'''$ for certain reciprocal lattice vectors $\boldsymbol{K}', \boldsymbol{K}'', \boldsymbol{K}'''$. Therefore, the functions $\exp[-iR(\boldsymbol{k}+\boldsymbol{K})\cdot\boldsymbol{r}]$ for $R \in K_{\boldsymbol{k}}$ form a basis of a representation of $G_{\boldsymbol{k}}$. In general, this representation is reducible. Moreover, it is an allowable representation since for a translation $a = \{\epsilon | \boldsymbol{a}\}$ one has $P_a \exp[-iR(\boldsymbol{k}+\boldsymbol{K})\cdot\boldsymbol{r}] = \exp[-iR(\boldsymbol{k}+\boldsymbol{K})\cdot(\boldsymbol{r}-\boldsymbol{a})] = \exp[iR(\boldsymbol{k}+\boldsymbol{K})\cdot\boldsymbol{a}]\cdot\exp[-iR(\boldsymbol{k}+\boldsymbol{K})\cdot\boldsymbol{r}] = \exp[iR\boldsymbol{k}\cdot\boldsymbol{a}]\exp[-iR(\boldsymbol{k}+\boldsymbol{K})\cdot\boldsymbol{r}]$ and $\exp[iR\boldsymbol{k}\cdot\boldsymbol{a}] = \exp[i(\boldsymbol{k}+\boldsymbol{K}')\cdot\boldsymbol{a}] = \exp(i\boldsymbol{k}\cdot\boldsymbol{a})$. The collection $R(\boldsymbol{k}+\boldsymbol{K})$ for $R \in K_{\boldsymbol{k}}$ and for given $\boldsymbol{K} \in \Lambda^*$ forms the $(\boldsymbol{k}+\boldsymbol{K})$-star. The plane waves corresponding to one $(\boldsymbol{k}+\boldsymbol{K})$-star transform according to a representation of $G_{\boldsymbol{k}}$. One can replace eq. (6.8) by

$$\psi_{\boldsymbol{k}} = \sum_{\text{stars}} \sum_{\substack{\boldsymbol{k}+\boldsymbol{K}\text{ in}\\ \text{one star}}} a_{\boldsymbol{k}\nu}(\boldsymbol{K})\,\mathrm{e}^{-i(\boldsymbol{k}+\boldsymbol{K})\cdot\boldsymbol{r}} . \tag{6.9}$$

As we will see, under the perturbation plane waves will combine to form perturbed states. Because the symmetry of the perturbation is $G_{\boldsymbol{k}}$, only plane waves belonging to equivalent irreducible representations of $G_{\boldsymbol{k}}$ will mix. Therefore, it is convenient to combine the plane waves in such a way that they form bases of irreducible representations of $G_{\boldsymbol{k}}$.

We consider eq. (6.7) and neglect here the spin terms. Generalization to the complete Hamiltonian H' is straightforward. Substituting the Fourier expansion (6.8) in eq. (6.7) one obtains

$$\left[\frac{p^2}{2m} + \frac{\hbar^2 k^2}{2m} + V(\boldsymbol{r}) - \frac{\hbar}{m}\boldsymbol{k}\cdot\boldsymbol{p} - E_{\boldsymbol{k}}\right] u_{\boldsymbol{k}\nu}$$

$$= \sum_{\boldsymbol{K}\in\Lambda^*} \left[\frac{\hbar^2}{2m}(\boldsymbol{k}+\boldsymbol{K})^2 + V(\boldsymbol{r}) - E_{\boldsymbol{k}}\right] a_{\boldsymbol{k}\nu}(\boldsymbol{K})\,\mathrm{e}^{-i\boldsymbol{K}\cdot\boldsymbol{r}} = 0 .$$

Multiplication from the left with $(1/\Omega)\,\mathrm{e}^{i\boldsymbol{K}'\cdot\boldsymbol{r}}$, where Ω is a normalization volume, and integration over $\boldsymbol{r}$ gives

$$\left[\frac{\hbar^2}{2m}(\boldsymbol{k}+\boldsymbol{K})^2 - E_{\boldsymbol{k}} + V_{\boldsymbol{K}\boldsymbol{K}}\right] a_{\boldsymbol{k}\nu}(\boldsymbol{K}) = -\sum_{\boldsymbol{K}'\neq\boldsymbol{K}} V_{\boldsymbol{K}\boldsymbol{K}'} a_{\boldsymbol{k}\nu}(\boldsymbol{K}') . \tag{6.10}$$

Here the matrix element $V_{\boldsymbol{K}\boldsymbol{K}'}$ is defined by

$$V_{\boldsymbol{K}\boldsymbol{K}'} = \frac{1}{\Omega}\int d_3\boldsymbol{r}\,\mathrm{e}^{i\boldsymbol{K}'\cdot\boldsymbol{r}}\,V(\boldsymbol{r})\,\mathrm{e}^{-i\boldsymbol{K}\cdot\boldsymbol{r}} = V_{\boldsymbol{K}-\boldsymbol{K}'} .$$

We assume that $V_{\boldsymbol{K}\boldsymbol{K}}$ is taken as part of the unperturbed energy $\hbar^2(\boldsymbol{k}+\boldsymbol{K})^2/2m$. Then one has the equation

$$\sum_{K'} \left[\frac{\hbar^2}{2m}(k+K')^2\delta_{KK'} - E_k\delta_{KK'} + V_{KK'}\right] a_{k\nu}(K') = 0 . \tag{6.11}$$

The solution of eq. (6.7) is obtained by diagonalization of eq. (6.11). This means diagonalization of an infinite-dimensional matrix. However, for the nearly-free electron one can restrict oneself to a limited number of stars.

In order to use the invariance group of the Hamiltonian H' one has to combine plane waves into basis sets of irreducible representations. We will show this symmetrization procedure on an example. Consider an electron in a crystal with space group $Pm3m$. If one wants to construct the function $u_{k\nu}$ for $k = 0$, one can consider the $(k+K)$-stars as in eq. (6.9). The first star contains only the point $\Gamma = (0, 0, 0)$. The second star is the one formed by the six reciprocal lattice vectors (fig. 6.3)

$$(2\pi/a, 0, 0), \quad (-2\pi/a, 0, 0), \quad (0, 2\pi/a, 0), \quad (0, -2\pi/a, 0),$$
$$(0, 0, 2\pi/a), \quad (0, 0, -2\pi/a) , \tag{6.12}$$

where a is the length of the side of the unit cell. We denote the corresponding six plane waves $e^{-iK\cdot r}$ by

$$(100), \quad (\bar{1}00), \quad (010), \quad (0\bar{1}0), \quad (001), \quad \text{and} \quad (00\bar{1}),$$

respectively. They span a six-dimensional representation of G_k which is $Pm3m$. The character of the representation is obtained from

$$P_{\{R|a\}} e^{-iK\cdot r} = e^{-iRK\cdot(r-a)} = e^{iRK\cdot a}\, e^{-iRK\cdot r} ,$$

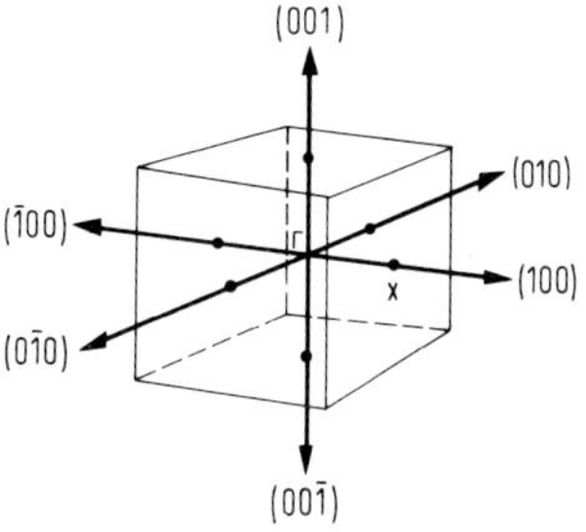

Fig. 6.3. The star of (100).

when $\{R|\boldsymbol{a}\} \in Pm3m$. As the space group is symmorphic one has

$$\chi(\{R|\boldsymbol{a}\}) = \chi(\{R|\boldsymbol{O}\}) = \text{number of } \boldsymbol{K}\text{-vectors with } R\boldsymbol{K} = \boldsymbol{K}\,.$$

The representation of $K_{\boldsymbol{k}} = m3m$ is then given by the character

$\{\epsilon\}$	$\{\beta\}$	$\{\alpha\}$	$\{\alpha^2\}$	$\{\alpha\beta\}$	$\{\gamma\}$	$\{\beta\gamma\}$	$\{\alpha\gamma\}$	$\{\alpha^2\gamma\}$	$\{\alpha\beta\gamma\}$
6	0	2	2	0	0	0	0	4	2

where $\{\beta\}$ denotes the class of 3-fold rotations, $\{\alpha\}$ that of 4-fold rotations, $\{\gamma\}$ that of the central inversion and so on. As one sees from the character table of $m3m$, which is isomorphic to $O \times C_2$, the representation Γ is reducible and

$$\Gamma = \Gamma_1^+ \oplus \Gamma_3^+ \oplus \Gamma_4^-\,.$$

The six basis vectors (6.12) do not form a basis of a reduced matrix representation. To find such a basis one can apply the projection operators from Ch. 2, §1. A basis for Γ_1^+ is given by

$$\{(100) + (\bar{1}00) + (010) + (0\bar{1}0) + (001) + (00\bar{1})\}/\sqrt{6}\,.$$

A basis for the irreducible representations Γ_3^+ is given by the two functions

$$\{(100) + (\bar{1}00) - (010) - (0\bar{1}0) - 2(001) - 2(00\bar{1})\}/\sqrt{8}\,,$$

$$\{2(100) + 2(\bar{1}00) - (010) - (0\bar{1}0) - (001) - (00\bar{1})\}/\sqrt{8}\,.$$

Finally a basis for Γ_4^- is given by the three functions

$$\{(100) - (\bar{1}00)\}/\sqrt{2}$$

$$\{(010) - (0\bar{1}0)\}/\sqrt{2}$$

$$\{(001) - (00\bar{1})\}/\sqrt{2}\,.$$

One can do this for any star. For the lowest stars one finds:

star of (000) gives Γ_1^+ ,

star of (100) gives $\Gamma_1^+ + \Gamma_3^+ + \Gamma_4^-$,

star of (110) gives $\Gamma_1^+ + \Gamma_3^+ + \Gamma_4^- + \Gamma_5^+ + \Gamma_5^-$, etc. .

If we want to approximate $u_{\boldsymbol{k}\nu}$ by a linear combination of these 3 stars we have to diagonalize the 19×19 matrix $V_{\boldsymbol{KK}'}$ ($\boldsymbol{K}, \boldsymbol{K}'$ in one of the three stars) in eq. (6.11). However, by taking the symmetrized plane waves as basis functions instead of the plane waves one has also to diagonalize a 19×19 matrix, but as V has only nonvanishing matrix elements between functions belonging to the same row of the same irreducible representation, the matrix of V between symmetrized plane waves falls apart in a 3×3 block (connecting the 3 Γ_1^+ functions), two 2×2 blocks (connecting the rows of Γ_3^+), three 2×2 blocks (between Γ_4^- functions), and six diagonal elements between the functions of Γ_5^- and Γ_5^+. Thus the diagonalization is simply a diagonalization of at most a 3×3 matrix. This spectacular reduction of the V matrix is caused by the high symmetry of the point Γ. For a less symmetric point the reduction is not as great. For a $\boldsymbol{k}$-vector in general position the representation is the trivial one and the diagonalization is not at all simplified by symmetry considerations. Hence the calculation is much easier at a point in the Brillouin zone with large point symmetry.

6.1.4. Other band structure models

Qualitatively, an electron in a crystal will behave as quasi-free only far away from the ions. In the neighbourhood of an atom its wave function will vary more like an atomic wave function. This means that the quasi-free electron model will not be very good. Although it can give in certain circumstances a fairly good overall impression, the number of plane waves needed in the expansion of $u_{\boldsymbol{k}\nu}$ for an accurate description will be to great. To describe the behaviour near the ions, plane waves with big $\boldsymbol{k}$-vectors would be required. To get an expansion which converges more rapidly, Herring (Herring [1940]) proposed another set of expansion functions which have a better behaviour near the ions. The lowest energy bands have their origin in the core electron states. Usually there is a clear distinction between these core states and the outer electron states. Suppose $\psi_{\boldsymbol{k}\mu}$ are the bands corresponding to the core states. They can be approximated by Bloch sums.

$$\psi_{\boldsymbol{k}\mu}(\boldsymbol{r}) \approx \sum_{a \in U} \mathrm{e}^{i\boldsymbol{k}\cdot\boldsymbol{a}} \phi_\mu(\boldsymbol{r}-\boldsymbol{a}) ,$$

where the summation is over all lattice vectors and $\phi_\mu(\boldsymbol{r}-\boldsymbol{a})$ is the μ-th orbital of an electron centered around the atom at site $\boldsymbol{a}$. Anyhow, we suppose the $\psi_{\boldsymbol{k}\mu}(\boldsymbol{r})$ core band functions to be known. Herring proposed as basis functions

$$\chi_{\boldsymbol{k},\boldsymbol{K}}(\boldsymbol{r}) = \mathrm{e}^{-i(\boldsymbol{k}+\boldsymbol{K})\cdot\boldsymbol{r}} - \sum_\mu \langle \boldsymbol{k}\mu | \boldsymbol{k}+\boldsymbol{K}\rangle \psi_{\boldsymbol{k}\mu}(\boldsymbol{r}) , \tag{6.13}$$

where the summation is over all core states, and $\langle \boldsymbol{k}\mu | \boldsymbol{k}+\boldsymbol{K}\rangle$ is the inner product of $\psi_{\boldsymbol{k}\mu}(\boldsymbol{r})$ and $\exp\{-i(\boldsymbol{k}+\boldsymbol{K})\cdot\boldsymbol{r}\}$. This function is orthogonal to the core states and behaves in the neighbourhood of the core as an orbital function. We can expect that an expansion of the conduction electron wave functions

$$\psi_{\boldsymbol{k}\nu}(\boldsymbol{r}) = \sum_{\boldsymbol{K}\in\Lambda^*} a_{\boldsymbol{k}\nu}(\boldsymbol{K})\chi_{\boldsymbol{k},\boldsymbol{K}}(\boldsymbol{r}) \tag{6.14}$$

converges more rapidly than an expansion in plane waves. One can substitute eq. (6.14) in the Schrödinger equation. Again one has to diagonalize an infinite-dimensional matrix which can be approximated by a finite one. To simplify the interaction matrix one must consider the transformation properties of the *orthogonalized plane waves* (OPW) $\chi_{\boldsymbol{k},\boldsymbol{K}}(\boldsymbol{r})$. For an element $g = \{R|\boldsymbol{a}\}$ from the group $G_{\boldsymbol{k}}$ of $\boldsymbol{k}$ one has

$$P_g\chi_{\boldsymbol{k},\boldsymbol{K}}(\boldsymbol{r}) = \mathrm{e}^{-i(\boldsymbol{k}+\boldsymbol{K})\cdot(R^{-1}\boldsymbol{r}-R^{-1}\boldsymbol{a})} - \sum_\mu \langle \boldsymbol{k}\mu | \boldsymbol{k}+\boldsymbol{K}\rangle P_g\psi_{\boldsymbol{k}\mu}(\boldsymbol{r}) .$$

Now we can choose the functions $\psi_{\boldsymbol{k}\mu}(\boldsymbol{r})$ as basis functions for a (unitary) representation $D_{\boldsymbol{k}}(G_{\boldsymbol{k}})$. This means

$$P_g\psi_{\boldsymbol{k}\mu}(\boldsymbol{r}) = \sum_\rho D_{\boldsymbol{k}}(g)_{\rho\mu}\psi_{\boldsymbol{k}\rho}(\boldsymbol{r}) .$$

Using the definition of $\langle \boldsymbol{k}\mu | \boldsymbol{k}+\boldsymbol{K}\rangle$ and the transformation property of the plane waves, there is a $\boldsymbol{k} + \boldsymbol{K}' = R(\boldsymbol{k}+\boldsymbol{K})$ in the $\boldsymbol{k}+\boldsymbol{K}$-star such that

$$\begin{aligned} P_g\chi_{\boldsymbol{k},\boldsymbol{K}}(\boldsymbol{r}) &= \mathrm{e}^{-i(\boldsymbol{k}+\boldsymbol{K}')\cdot(\boldsymbol{r}-\boldsymbol{a})} - \sum_{\mu\rho} D_{\boldsymbol{k}}(g)_{\rho\mu}\langle P_g\psi_{\boldsymbol{k}\mu}|P_g \exp\{-i(\boldsymbol{k}+\boldsymbol{K})\cdot\boldsymbol{r}\}\rangle \psi_{\boldsymbol{k}\rho} \\ &= \mathrm{e}^{i(\boldsymbol{k}+\boldsymbol{K}')\cdot\boldsymbol{a}}\{\mathrm{e}^{-i(\boldsymbol{k}+\boldsymbol{K}')\cdot\boldsymbol{r}} - \sum_{\mu\rho\sigma} D_{\boldsymbol{k}}(g)_{\rho\mu}D_{\boldsymbol{k}}(g)^*_{\sigma\mu}\langle \boldsymbol{k}\sigma|\boldsymbol{k}+\boldsymbol{K}'\rangle\psi_{\boldsymbol{k}\rho}(\boldsymbol{r})\} \\ &= \mathrm{e}^{i(\boldsymbol{k}+\boldsymbol{K}')\cdot\boldsymbol{a}}\{\mathrm{e}^{-i(\boldsymbol{k}+\boldsymbol{K}')\cdot\boldsymbol{r}} - \sum_\rho \langle \boldsymbol{k}\rho|\boldsymbol{k}+\boldsymbol{K}'\rangle\psi_{\boldsymbol{k}\rho}(\boldsymbol{r})\} \\ &= \mathrm{e}^{i(\boldsymbol{k}+\boldsymbol{K}')\cdot\boldsymbol{a}}\chi_{\boldsymbol{k},\boldsymbol{K}'}(\boldsymbol{r}) . \end{aligned} \tag{6.15}$$

Here we see that the functions $\chi_{\boldsymbol{k},\boldsymbol{K}}(\boldsymbol{r})$ with $\boldsymbol{k}+\boldsymbol{K}$ in one $\boldsymbol{k}+\boldsymbol{K}$-star form a basis for a representation of $G_{\boldsymbol{k}}$, just as the plane waves $\exp\{-i(\boldsymbol{k}+\boldsymbol{K})\cdot\boldsymbol{r}\}$ did. As the interaction V is invariant under $G_{\boldsymbol{k}}$, its matrix is simplified if we use symmetrized combinations of the $\chi_{\boldsymbol{k},\boldsymbol{K}}(\boldsymbol{r})$ which form bases of irreducible representations of $G_{\boldsymbol{k}}$. Moreover, as $\chi_{\boldsymbol{k},\boldsymbol{K}}(\boldsymbol{r})$ transform in exactly the same way as the plane waves, the linear combinations one has to take are the same as those for plane waves. This is the reason why the study of symmetrized plane waves in the preceding section is so useful, although the nearly-free electron model is such a crude approximation.

Apart from NFE and OPW models there exist a great number of other methods for band structure calculations. In these methods as well use can be made of the existing symmetry of the problem. A further discussion of the role of symmetry considerations in band structure calculations can be found in Nussbaum [1966], Luehrmann [1968] and Cornwell [1969].

6.2. Lattice vibrations

6.2.1. A crystal in the harmonic approximation

Contrary to the case of electron states discussed in the preceding section, where we neglected the contribution of the cores H_c, in this section we will be concerned with the dynamics of the system of cores. Here the electron contribution is taken as a uniform charge background. The cores are considered as rigid particles with an interaction which can be described by a potential. As in Ch. 4, § 1.7 the positions of the cores (which can also form complete atoms or ions) are denoted by $\boldsymbol{x}\binom{\boldsymbol{n}}{i}$ with $\boldsymbol{n}\in\Lambda$ and $i=1,\ldots,s$. The Hamiltonian of the system of sN^3 particles is

$$H=\sum_{\boldsymbol{n}i}\frac{p_{\boldsymbol{n}i}^2}{2m_i}+\Phi(\boldsymbol{x}\binom{\boldsymbol{n}}{i})\,. \tag{6.16}$$

Here $\boldsymbol{p}_{\boldsymbol{n}i}$ is the momentum of the particle with position $\boldsymbol{x}\binom{\boldsymbol{n}}{i}$ and Φ is the potential energy. As the system vibrates, the positions are time-dependent. One can write $\boldsymbol{x}\binom{\boldsymbol{n}}{i}_t=\boldsymbol{x}\binom{\boldsymbol{n}}{i}_0+\boldsymbol{u}\binom{\boldsymbol{n}}{i}_t$, where $\boldsymbol{u}\binom{\boldsymbol{n}}{i}$ is the displacement from the equilibrium position. For small displacements one can expand Φ in a power series. The zeroth order gives only a shift in the energy, the first order vanishes because we expand in displacements from a stable equilibrium position. When the third and higher order terms are neglected, one considers the crystal in the so-called harmonic approximation. When we denote the second derivative by

$$\Phi_{\alpha_1\alpha_2}\binom{n_1\ n_2}{i_1\ i_2} = \frac{\partial^2\Phi}{\partial u_{\alpha_1}\binom{n_1}{i_1}\ \partial u_{\alpha_2}\binom{n_2}{i_2}},$$

the Hamiltonian in the harmonic approximation is

$$H = \frac{1}{2}\sum_{ni\alpha}\frac{(p_{ni})_\alpha^2}{m_i} + \frac{1}{2}\sum_{\substack{n_1n_2i_1\\ i_2\alpha_1\alpha_2}} \Phi_{\alpha_1\alpha_2}\binom{n_1\ n_2}{i_1\ i_2} u_{\alpha_1}\binom{n_1}{i_1} u_{\alpha_2}\binom{n_2}{i_2},$$

or in vector notation

$$H = \frac{1}{2}\sum_{ni}\frac{p_{ni}^2}{m_i} + \frac{1}{2}\sum_{n_1i_1n_2i_2} \tilde{u}\binom{n_1}{i_1}\Phi\binom{n_1\ n_2}{i_1\ i_2}u\binom{n_2}{i_2}. \tag{6.17}$$

For the theory of lattice vibrations we refer to Born and Huang [1954] or Maradudin et al. [1963]. Here we give only a brief sketch in order to see the role of symmetry in this problem.

When we consider the system in the frame-work of classical mechanics, the equation of motion is

$$m_i\ddot{u}_\alpha\binom{n}{i} = -\sum_{n'i'\alpha'}\Phi_{\alpha\alpha'}\binom{n\ n'}{i\ i'}u_{\alpha'}\binom{n'}{i'}. \tag{6.18}$$

We notice that the tensor $\Phi\binom{n\ n'}{i\ i'}$ has a number of symmetry properties. Among them

$$\Phi_{\alpha\alpha'}\binom{n\ n'}{i\ i'} = \Phi_{\alpha'\alpha}\binom{n'\ n}{i'\ i}, \tag{6.19}$$

which follows from the definition, and

$$\Phi_{\alpha\alpha'}\binom{n\ n'}{i\ i'} = \Phi_{\alpha\alpha'}\binom{n-n'}{i\ \ i'} \tag{6.20}$$

because the potential energy depends only on the mutual distance of the particles. We use the translational invariance by making the Ansatz

$$u_\alpha^{q}\binom{n}{j} = (m_j)^{-1/2}u_\alpha^{q}\binom{0}{j}\,e^{i(q\cdot n-\omega(q)t)} \tag{6.21}$$

for the solutions of eq. (6.18). Here $\boldsymbol{q}$ is a vector in the reciprocal space. We will give an argument for the choice (6.21) in the next section. Substitution of (6.21) in eq. (6.18) gives

$$-\omega(\boldsymbol{q})^2 (m_j)^{+1/2} u_\alpha^{\boldsymbol{q}}\binom{\boldsymbol{O}}{j} \mathrm{e}^{i\boldsymbol{q}\cdot\boldsymbol{n}} = -\sum_{\boldsymbol{n}'\alpha' j'} \Phi_{\alpha\alpha'}\binom{\boldsymbol{n}-\boldsymbol{n}'}{j \quad j'} (m_{j'})^{-1/2} u_{\alpha'}^{\boldsymbol{q}}\binom{\boldsymbol{O}}{j'} \mathrm{e}^{i\boldsymbol{q}\cdot\boldsymbol{n}'} .$$

We now introduce a matrix $D(\boldsymbol{q})$, called the dynamical matrix, by

$$D_{\alpha\alpha'}(jj'|\boldsymbol{q}) = \sum_{\boldsymbol{n}} \Phi_{\alpha\alpha'}\binom{\boldsymbol{n}}{j\ j'} (m_j m_{j'})^{-1/2} \mathrm{e}^{-i\boldsymbol{q}\cdot\boldsymbol{n}} .$$

Then the equation of motion becomes

$$\omega(\boldsymbol{q})^2 u_\alpha^{\boldsymbol{q}}\binom{\boldsymbol{O}}{j} = \sum_{\alpha' j'} D_{\alpha\alpha'}(j\,j'|\boldsymbol{q}) u_{\alpha'}^{\boldsymbol{q}}\binom{\boldsymbol{O}}{j'} . \tag{6.22}$$

Forming a $3s$-dimensional vector $u(\boldsymbol{q})$ from the components $u_\alpha^{\boldsymbol{q}}\binom{\boldsymbol{O}}{j}$, one has

$$\omega(\boldsymbol{q})^2 u(\boldsymbol{q}) = D(\boldsymbol{q}) u(\boldsymbol{q}) .$$

The frequencies $\omega^2(\boldsymbol{q})$ are the $3s$ eigenvalues of the matrix $D(\boldsymbol{q})$. In this way use of the translational invariance leads to a $3s$-dimensional problem instead of the original $3sN^3$-dimensional problem. For each of the N^3 vectors $\boldsymbol{q}$ in the first Brillouin zone there are $3s$ eigenvalues denoted by $\omega_j(\boldsymbol{q})^2$. The corresponding eigenvectors are $e(\boldsymbol{q}|j)$. The matrix $D(\boldsymbol{q})$ has the following properties.

1) It is Hermitian:

$$\begin{aligned} D_{\alpha\alpha'}(j\,j'|\boldsymbol{q}) &= \sum_{\boldsymbol{n}} (m_j m_{j'})^{-1/2} \Phi_{\alpha\alpha'}\binom{\boldsymbol{n}}{j\ j'} \mathrm{e}^{-i\boldsymbol{q}\cdot\boldsymbol{n}} \\ &= \sum_{\boldsymbol{n}} (m_j m_{j'})^{-1/2} \Phi_{\alpha'\alpha}\binom{-\boldsymbol{n}}{j'\ j} \mathrm{e}^{-i\boldsymbol{q}\cdot\boldsymbol{n}} \qquad \text{(cf. eq. (6.19))} \\ &= \sum_{\boldsymbol{n}} (m_j m_{j'})^{-1/2} \Phi_{\alpha'\alpha}\binom{\boldsymbol{n}}{j'\ j} \mathrm{e}^{i\boldsymbol{q}\cdot\boldsymbol{n}} = D_{\alpha'\alpha}(j'j|\boldsymbol{q})^* . \end{aligned} \tag{6.23}$$

2) From the definition of $D(\boldsymbol{q})$ follow

$$\begin{aligned} &D(\boldsymbol{q}) = D(-\boldsymbol{q})^* , \\ &D(\boldsymbol{q}+\boldsymbol{K}) = D(\boldsymbol{q}) \qquad (\text{any } \boldsymbol{K} \in \Lambda^*) . \end{aligned} \tag{6.24}$$

The Hermiticity implies that the $3s$ eigenvectors can be chosen in such a way that

$$\sum_{\alpha i} e_{\alpha i}(\boldsymbol{q}\,|\,j)^* \, e_{\alpha i}(\boldsymbol{q}\,|\,j') = \delta_{jj'} \,. \tag{6.25}$$

When we form from the $3s$ eigenvectors a matrix $e(\boldsymbol{q})$ with columns $e(\boldsymbol{q}\,|\,j)$, the eigenvalue equation is

$$D(\boldsymbol{q})\, e(\boldsymbol{q}) = e(\boldsymbol{q})\, \Omega(\boldsymbol{q}) \,, \tag{6.26}$$

where $\Omega(\boldsymbol{q})$ is a diagonal matrix with elements $\omega_j(\boldsymbol{q})^2$. Solution of the eigenvalue equation (6.26) is equivalent to a solution of the equation of motion (6.18).

6.2.2. *Symmetry of the dynamical matrix*

After this brief introduction to the theory of lattice vibrations in the harmonic approximation, we want to discuss the role of the symmetry. Because the particles vibrate, only for some special modes the symmetry of the real crystal will exactly be a space group. However, the configuration of the equilibrium positions (which are the time averages of the positions) is transformed into itself by an element of the space group of the crystal. As always, we assume an infinite perfect crystal. Suppose that $\boldsymbol{x}\binom{\boldsymbol{m}}{j}_0$ is the equilibrium position of an atom of a certain kind. Then for an element $g = \{R\,|\,\boldsymbol{t}\}$ from the space group G the position

$$\{R\,|\,\boldsymbol{t}\}\, \boldsymbol{x}\binom{\boldsymbol{m}}{j}_0 = R\boldsymbol{m} + R\boldsymbol{\tau}_j + \boldsymbol{t}$$

is also the equilibrium position of an atom of the same kind. Then there is a pair $(\boldsymbol{m}', j')$ such that

$$\boldsymbol{x}\binom{\boldsymbol{m}'}{j'}_0 = \{R\,|\,\boldsymbol{t}\}\, \boldsymbol{x}\binom{\boldsymbol{m}}{j}_0 \,. \tag{6.27}$$

The pair $(\boldsymbol{m}', j')$ is uniquely determined by the pair $(\boldsymbol{m}, j)$ and the element g. We suppress the information about g to have a not too cumbersome notation.

When $\boldsymbol{\tau}_j$ is the position of an atom in a unit cell, $R\boldsymbol{\tau}_j + \boldsymbol{t}$ is also an atom position:

$$R\boldsymbol{\tau}_j + \boldsymbol{t} = \boldsymbol{\tau}_{j'} + \boldsymbol{u}(g, j)$$

for some primitive translation $\boldsymbol{u}(g, j)$. Then $\boldsymbol{\tau}_{j'}$ is again in the unit cell. From eq. (6.27) it follows that

$$R\boldsymbol{m} + \boldsymbol{u}(g, j) = \boldsymbol{m}' . \tag{6.28}$$

As the lattice is invariant under elements of the point group, $R\boldsymbol{m}$ is again a lattice point.

Under a point group element R, the displacements $\boldsymbol{u}\binom{\boldsymbol{m}}{j}$ transform according to

$$\boldsymbol{u}'\binom{\boldsymbol{m}'}{j'} = R\boldsymbol{u}\binom{\boldsymbol{m}}{j} ,$$

because $\boldsymbol{u}\binom{\boldsymbol{m}}{j}$ is a vector. The potential energy Φ is a scalar function and is transformed into itself by elements of the space group:

$$\Phi(\ldots \boldsymbol{u}'\binom{\boldsymbol{m}'}{j'} \ldots) = \Phi(\ldots \boldsymbol{u}\binom{\boldsymbol{m}}{j} \ldots) .$$

For the second order term one has consequently

$$\Phi_{\alpha_1\alpha_2}\binom{\boldsymbol{m}_1\ \boldsymbol{m}_2}{j_1\ \ j_2} = \sum_{\beta_1\beta_2} R_{\beta_1\alpha_1} R_{\beta_2\alpha_2} \Phi_{\beta_1\beta_2}\binom{\boldsymbol{m}_1'\ \boldsymbol{m}_2'}{j_1'\ \ j_2'} . \tag{6.29}$$

In particular, one has for a primitive translation $a = \{\mathbb{1} | \boldsymbol{a}\}$

$$\Phi_{\alpha_1\alpha_2}\binom{\boldsymbol{m}_1\ \boldsymbol{m}_2}{j_1\ \ j_2} = \Phi_{\alpha_1\alpha_2}\binom{\boldsymbol{m}_1+\boldsymbol{a}\ \ \boldsymbol{m}_2+\boldsymbol{a}}{j_1\qquad j_2} ,$$

as we have seen in eq. (6.20). As the equation of motion is a linear one, the solutions form a linear vector space which carries a representation of the invariance group. Among other things this means that one can choose a basis for the space with transformation property

$$P_{\{\mathbb{1}|\boldsymbol{a}\}} \boldsymbol{u}\binom{\boldsymbol{m}}{j} = \mathrm{e}^{i\boldsymbol{q}\cdot\boldsymbol{a}} \boldsymbol{u}\binom{\boldsymbol{m}}{j} .$$

This implies that there is a basis consisting of solutions of the form (6.21). Here we see that the values of $\boldsymbol{q}$ can be restricted to vectors in the first Brillouin zone. Choosing periodic boundary conditions means restriction to the lattice Λ^*/N inside the Brillouin zone.

The transformation law (6.29) can be used to simplify the tensor Φ, because it gives relations between its components. As an example we consider a crystal with diamond structure. We assume an interaction between the atoms for which the range is restricted such that only nearest neighbours have interaction. For an atom at position [000] there are four nearest neighbours:

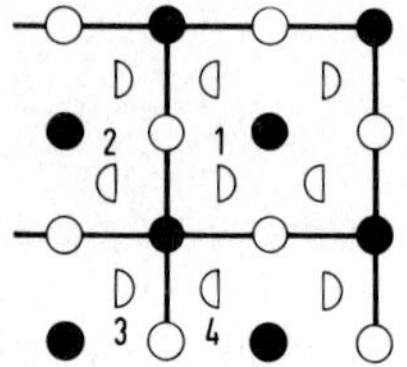

Fig. 6.4. The four nearest neighbours in the diamond crystal.

$[\frac{1}{4}\frac{1}{4}\frac{1}{4}]$, $[\frac{1}{4}-\frac{3}{4}\frac{1}{4}]$, $[-\frac{3}{4}\frac{1}{4}\frac{1}{4}]$, and $[\frac{1}{4}\frac{1}{4}-\frac{3}{4}]$ (fig. 6.4). Hence only the 36 elements of $\Phi^{(1)} = \Phi\binom{000}{01}$, $\Phi^{(2)} = \Phi\binom{010}{01}$, $\Phi^{(3)} = \Phi\binom{100}{01}$, $\Phi^{(4)} = \Phi\binom{001}{01}$ can be different from zero. Using eq. (6.29) where we take for R three generators of the point group $m3m$, respectively, we get:

1) $$R = \begin{pmatrix} 0 & 1 & 0 \\ -1 & 0 & 0 \\ 0 & 0 & -1 \end{pmatrix} \text{ gives } \Phi^{(1)} = R\Phi^{(2)}\tilde{R} = R^2\Phi^{(3)}\tilde{R}^2 = R^3\Phi^{(4)}\tilde{R}^3 \;;$$

2) $$R = \begin{pmatrix} 0 & 0 & 1 \\ -1 & 0 & 0 \\ 0 & -1 & 0 \end{pmatrix} \text{ gives } \Phi^{(1)} = R\Phi^{(4)}\tilde{R} = R^2\Phi^{(3)}\tilde{R}^2,\ \ \Phi^{(2)} = R\Phi^{(2)}\tilde{R} \;;$$

3) the central inversion I interchanges the two sublattices, which means that for $g = \{I|\boldsymbol{\tau}\}$ with $\boldsymbol{\tau} = [\frac{1}{4}\frac{1}{4}\frac{1}{4}]$ one has

$$\boldsymbol{x}\binom{0\,0\,0}{0} \to \boldsymbol{x}\binom{0\,0\,0}{1}$$

$$\boldsymbol{x}\binom{0\,0\,0}{1} \to \boldsymbol{x}\binom{0\,0\,0}{0}$$

which leads (using eq. (6.19)) to $\Phi_{\alpha\beta}\binom{\boldsymbol{n}}{0\;1} = \Phi_{\alpha\beta}\binom{-\boldsymbol{n}}{1\;0} = \Phi_{\beta\alpha}\binom{\boldsymbol{n}}{0\;1}$. Because the three elements used above are generators of the point group, all other relations follow from the ones obtained. Use of these relations gives the four tensors $\Phi^{(i)}$:

$$\Phi^{(1)} = \begin{pmatrix} \alpha & \beta & -\beta \\ \beta & \alpha & -\beta \\ -\beta & -\beta & \alpha \end{pmatrix} \qquad \Phi^{(2)} = \begin{pmatrix} \alpha & -\beta & -\beta \\ -\beta & \alpha & \beta \\ -\beta & \beta & \alpha \end{pmatrix}$$

$$\Phi^{(3)} = \begin{pmatrix} \alpha & \beta & \beta \\ \beta & \alpha & \beta \\ \beta & \beta & \alpha \end{pmatrix} \qquad \Phi^{(4)} = \begin{pmatrix} \alpha & -\beta & -\beta \\ -\beta & \alpha & \beta \\ -\beta & \beta & \alpha \end{pmatrix}$$

for arbitrary real numbers α and β. From the 36 elements of the 4 tensors only two are independent. The same procedure can of course be used to simplify the form of the other matrices. Herman (Herman [1959]) has done this for the five nearest neighbours. In that case there are 15 independent elements.

The transformation (6.29) of Φ leads to the transformation of the dynamical matrix. For the space group element g one finds with eq. (6.29)

$$D_{\alpha\beta}(j_1 j_2 | \boldsymbol{q}) = \sum_{\boldsymbol{n}} \Phi_{\alpha\beta}\binom{\boldsymbol{n}}{j_1 j_2} (m_{j_1} m_{j_2})^{-1/2}\, \mathrm{e}^{-i\boldsymbol{q}\cdot\boldsymbol{n}}$$

$$= \sum_{\boldsymbol{n}_1} \sum_{\mu\nu} R_{\mu\alpha} R_{\nu\beta} \Phi_{\mu\nu}\binom{\boldsymbol{n}'_1 - \boldsymbol{n}'_2}{j'_1 \quad j'_2} (m_{j_1} m_{j_2})^{-1/2}\, \mathrm{e}^{-i\boldsymbol{q}\cdot(\boldsymbol{n}_1 - \boldsymbol{n}_2)} .$$

Eq. (6.28) gives $\boldsymbol{n}'_1 - \boldsymbol{n}'_2 = R(\boldsymbol{n}_1 - \boldsymbol{n}_2) + \boldsymbol{u}(g, j_1) - \boldsymbol{u}(g, j_2)$. Hence

$$D_{\alpha\beta}(j_1 j_2 | \boldsymbol{q}) = \sum_{\boldsymbol{n}'_1} \sum_{\mu\nu} R_{\mu\alpha} R_{\nu\beta} \Phi_{\mu\nu}\binom{\boldsymbol{n}'_1 - \boldsymbol{n}'_2}{j'_1 \quad j'_2} (m_{j'_1 j'_2})^{-1/2}$$

$$\times \exp\{-iR\boldsymbol{q}\cdot(\boldsymbol{n}'_1 - \boldsymbol{n}'_2 - \boldsymbol{u}(g, j_1) + \boldsymbol{u}(g, j_2)\}$$

$$= \sum_{\mu\nu} R_{\mu\alpha} R_{\nu\beta} D_{\mu\nu}(j'_1 j'_2 | R\boldsymbol{q}) \exp\{iR\boldsymbol{q}\cdot(\boldsymbol{u}(g, j_1) - \boldsymbol{u}(g, j_2))\}$$

$$= \sum_{\mu\nu} \sum_{l_1 l_2} \delta_{l_1 j'_1} R_{\mu\alpha} \mathrm{e}^{iR\boldsymbol{q}\cdot\boldsymbol{u}(g, j_1)} D_{\mu\nu}(l_1 l_2 | R\boldsymbol{q}) R_{\nu\beta} \mathrm{e}^{-iR\boldsymbol{q}\cdot\boldsymbol{u}(g, j_2)} \delta_{l_2 j'_2} ,$$

where δ_{lj} is the Kronecker symbol. The last line is only a trivial remodelling of the preceding one. Then we define a $3s \times 3s$ matrix $\Gamma(\boldsymbol{q}, g)$

$$\Gamma(\boldsymbol{q}, g)^{\beta\alpha}_{lj} = R_{\beta\alpha}\, \mathrm{e}^{-iR\boldsymbol{q}\cdot\boldsymbol{u}(g, j)} \delta_{lj'} \tag{6.30}$$

which is composed of s^2 blocks of dimension 3. In the l-th row of blocks only the j'-th column is different from zero, where j' is determined by j and $g \in G$. The block lj' is the matrix $R \exp(-iR\boldsymbol{q}\cdot\boldsymbol{u}(g, j))$. For a primitive translation one has $j = j'$, which means that $\Gamma(\boldsymbol{q}, g)$ is the direct sum of s three-dimensional matrices. One can write

$$D_{\alpha\beta}(j_1 j_2 | \boldsymbol{q}) = \sum_{\mu\nu} (\Gamma(\boldsymbol{q}, g)^{\mu\alpha}_{l_1 j_1})^* D_{\mu\nu}(l_1 l_2 | R\boldsymbol{q}) \Gamma(\boldsymbol{q}, g)^{\nu\beta}_{l_2 j_2} ,$$

or in matrix form

$$D(\boldsymbol{q}) = \Gamma(\boldsymbol{q}, g)^\dagger\, D(R\boldsymbol{q})\, \Gamma(\boldsymbol{q}, g) . \tag{6.31}$$

Because Γ is unitary, as follows from the definition (6.30), one also has the relation $D(\boldsymbol{q}) = \Gamma(\boldsymbol{q},g)^{-1}D(R\boldsymbol{q})\,\Gamma(\boldsymbol{q},g)$. This means that $D(\boldsymbol{q})$ and $D(R\boldsymbol{q})$ have the same eigenvalues. We can choose the order of the $3s$ eigenvalues such that $\omega_j(\boldsymbol{q}) = \omega_j(R\boldsymbol{q})$. Therefore, the functions $\omega_j(\boldsymbol{q})$ have the symmetry of the point group. Moreover, since $D(\boldsymbol{q}) = D(-\boldsymbol{q})^*$, one can order the eigenvalues $\omega_j(-\boldsymbol{q})$ in such a way that $\omega_j(\boldsymbol{q}) = \omega_j(-\boldsymbol{q})$. Then we have the relations

$$\omega_j(\boldsymbol{q}) = \omega_j(\boldsymbol{q}+\boldsymbol{K}) \qquad \text{for any } \boldsymbol{K} \in \Lambda^* ,$$

$$\omega_j(\boldsymbol{q}) = \omega_j(R\boldsymbol{q}) \qquad \text{for any } R \text{ in the point group,}$$

$$\omega_j(\boldsymbol{q}) = \omega_j(-\boldsymbol{q}) .$$

6.2.3. Transformation properties of the normal modes

Using the matrices $\Gamma(\boldsymbol{q},g)$ for $g \in G$ one can construct eigenvectors of $D(R\boldsymbol{q})$ from those of $D(\boldsymbol{q})$. For the polarisation matrix $e(\boldsymbol{q})$, for which the $3s$ columns are the eigenvectors of $D(\boldsymbol{q})$, one has $D(R\boldsymbol{q})\,\Gamma(\boldsymbol{q},g)\,e(\boldsymbol{q}) = \Gamma(\boldsymbol{q},g)\,D(\boldsymbol{q})\,e(\boldsymbol{q})$ by eq. (6.31). From eq. (6.26) it follows that $e(R\boldsymbol{q}) = \Gamma(\boldsymbol{q},g)\,e(\boldsymbol{q})$ is a polarisation matrix for the wave vector $R\boldsymbol{q}$. In the special case that g is an element of $G_{\boldsymbol{q}}$, the group of $\boldsymbol{q}$, the matrices $\Gamma(\boldsymbol{q},g)$ and $D(\boldsymbol{q})$ commute. Then

$$D(\boldsymbol{q})\,\Gamma(\boldsymbol{q},g)\,e(\boldsymbol{q}) = \Gamma(\boldsymbol{q},g)\,e(\boldsymbol{q})\,\Omega(\boldsymbol{q}) ,$$

which means that both $e(\boldsymbol{q})$ and $\Gamma(\boldsymbol{q},g)\,e(\boldsymbol{q})$ are polarisation matrices for $\boldsymbol{q}$. It also means that the columns of $\Gamma(\boldsymbol{q},g)\,e(\boldsymbol{q})$, which are eigenvectors of $D(\boldsymbol{q})$, are linear combinations of the eigenvectors $e(\boldsymbol{q}\,|\,j)$. One can write

$$\Gamma(\boldsymbol{q},g)\,e(\boldsymbol{q}) = e(\boldsymbol{q})\,\Delta(\boldsymbol{q},g) \qquad \text{for } g \in G_{\boldsymbol{q}} . \tag{6.32}$$

Putting the eigenvectors with the same eigenvalues next to each other, the matrices $\Delta(\boldsymbol{q},g)$ become direct sums of a number of matrices. This number is the number of different eigenvalues of $D(\boldsymbol{q})$, because a column of $\Gamma(\boldsymbol{q},g)\,e(\boldsymbol{q})$ is a linear combination of those columns of $e(\boldsymbol{q})$ which correspond to the same eigenvalue. The dimension of a component of $\Delta(\boldsymbol{q},g)$ is the degeneracy of the eigenvalue to which it belongs.

The matrices $\Gamma(\boldsymbol{q},g)$ and consequently also $\Delta(\boldsymbol{q},g)$ form a representation of the group $G_{\boldsymbol{q}}$. To prove this, suppose that $g_1 = \{R_1|\boldsymbol{t}_1\}$ and $g_2 = \{R_2|\boldsymbol{t}_2\}$ are elements of $G_{\boldsymbol{q}}$. One has $\boldsymbol{\tau}_{j'} = R_1\boldsymbol{\tau}_j + \boldsymbol{t}_1 - \boldsymbol{u}(g_1,j)$ and $\boldsymbol{\tau}_{j''} = R_2\boldsymbol{\tau}_{j'} + \boldsymbol{t}_2 - \boldsymbol{u}(g_2,j')$. On the other hand $\boldsymbol{\tau}_{j''} = R_2R_1\boldsymbol{\tau}_j + \boldsymbol{t}_2 + R_2\boldsymbol{t}_1 - \boldsymbol{u}(g_2g_1,j)$. Therefore,

$$\boldsymbol{u}(g_2g_1,j) = \boldsymbol{u}(g_2,j') + R_2\boldsymbol{u}(g_1,j)\,. \tag{6.33}$$

We now use this relation to calculate

$$\begin{aligned}
&[\Gamma(\boldsymbol{q},g_1)\Gamma(\boldsymbol{q},g_2)]_{ij}^{\alpha\beta} \\
&= \sum_{\gamma k} (R_1)_{\alpha\gamma}(R_2)_{\gamma\beta}\delta_{ik'}\delta_{kj'} \exp\{-i(R_1\boldsymbol{q}\cdot\boldsymbol{u}(g_1,k) + R_2\boldsymbol{q}\cdot\boldsymbol{u}(g_2,j))\} \\
&= (R_1R_2)_{\alpha\beta}\delta_{ij''} \exp\{-i\boldsymbol{q}\cdot(\boldsymbol{u}(g_1,j') + \boldsymbol{u}(g_2,j))\} \\
&= (R_1R_2)_{\alpha\beta}\delta_{ij''} \exp\{-i\boldsymbol{q}\cdot\boldsymbol{u}(g_1g_2,j)\} = \Gamma(\boldsymbol{q},g_1g_2)_{ij}^{\alpha\beta}\,.
\end{aligned} \tag{6.34}$$

This shows that $\Gamma(\boldsymbol{q},G_{\boldsymbol{q}})$ forms a unitary representation of $G_{\boldsymbol{q}}$, equivalent to the representation $\Delta(\boldsymbol{q},G_{\boldsymbol{q}})$. The representation $\Delta(\boldsymbol{q},G_{\boldsymbol{q}})$ is already in block form and each block corresponds to a number of coincident eigenvalues $\omega_j(\boldsymbol{q})^2$. The reduction of the representation $\Gamma(\boldsymbol{q},G_{\boldsymbol{q}})$ would also give a representation in block form, and there also the eigenvalues corresponding to a given block are the same. If one has natural degeneracy, blocks belonging to different components correspond to different eigenvalues. If two blocks belong to the same eigenvalue, one has accidental degeneracy. For each $\boldsymbol{q}$ in the first Brillouin zone, $\omega(\boldsymbol{q})$ takes $3s$ values. Degeneracy is determined by the irreducible components of $\Gamma(\boldsymbol{q},G_{\boldsymbol{q}})$. The functions $\omega_j(\boldsymbol{q})$ can be seen as a number of hypersurfaces in the four-dimensional $(\boldsymbol{q},\omega)$-space. The various sheets correspond to various branches of phonons. This picture shows many analogies with electron bands. In both cases one has a multi-valued function ($E_{\boldsymbol{k}}$ or $\omega(\boldsymbol{q})$) with point group symmetry. However, in the electron case the number of sheets is, in principle, without limit, whereas it is at most $3s$ in the phonon case.

Here we have described the normal modes (basis solutions) with the symmetry group $G_{\boldsymbol{q}}$. This is called the subgroup method, as we did not consider the full space group. If we take the whole group G into consideration, one has the full-group method. To see the relation between the two, notice that for a primitive translation a the matrix $\Gamma(\boldsymbol{q},a)$ has the form

$$\Gamma(\boldsymbol{q},a)_{lj}^{\alpha\beta} = \delta_{\alpha\beta}\delta_{lj}\,e^{-i\boldsymbol{q}\cdot\boldsymbol{a}}\,.$$

Therefore, $\Gamma(\boldsymbol{q}, G_{\boldsymbol{q}})$ is an allowable representation of $G_{\boldsymbol{q}}$, as are the irreducible components $\Delta^{(i)}(\boldsymbol{q}, G_{\boldsymbol{q}})$ of $\Delta(\boldsymbol{q}, G_{\boldsymbol{q}})$. These representations $\Delta^{(i)}(\boldsymbol{q}, G_{\boldsymbol{q}})$ induce irreducible representations of G. When the eigenvectors $e(\boldsymbol{q}\,|\,j)$ with $j = i_1, ..., i_{d_i}$ carry the representation $\Delta^{(i)}(\boldsymbol{q}, G_{\boldsymbol{q}})$, and when the points of the star of $\boldsymbol{q}$ are denoted by $\boldsymbol{q}, R_2\boldsymbol{q}, ..., R_s\boldsymbol{q}$, the induced representation is carried by $\Gamma(\boldsymbol{q}, R_l)\, e(\boldsymbol{q}\,|\,j)$ with $l = 1, ..., s$ and $j = i_1, ..., i_{d_i}$. These eigenvectors of respectively $D(\boldsymbol{q}), ..., D(R_s\boldsymbol{q})$ have all the same eigenvalue $\omega_{i_1}(\boldsymbol{q})$. The dimension of the induced representation is sd_i, where s is the number of points of the star and d_i is the dimension of $\Delta^{(i)}(\boldsymbol{q}, G_{\boldsymbol{q}})$.

For a path on a sheet, to every point $\boldsymbol{q}$ of the sheet corresponds a representation of $G_{\boldsymbol{q}}$. For two adjacent points with the same $G_{\boldsymbol{q}}$ the representations of $G_{\boldsymbol{q}}$ are the same. On the other hand, if for any point of a path the $G_{\boldsymbol{q}}$ is the same except in some limit point $\boldsymbol{q}'$, where the group is $G_{\boldsymbol{q}'}$ and if $G_{\boldsymbol{q}'}$ contains $G_{\boldsymbol{q}}$, the limit representation of $G_{\boldsymbol{q}}$ for $\boldsymbol{q} \to \boldsymbol{q}'$ must be a component of the restriction of the representation of $G_{\boldsymbol{q}'}$ to $G_{\boldsymbol{q}}$. The representations belonging to different points are therefore not independent. They have to satisfy *compatibility relations* like those discussed in § 1 for electron bands. The compatibility relations are determined in the same way as in the case of electron bands. An example of an assignment of representations to sheets of $\omega(\boldsymbol{q})$ is given in fig. 6.5 for a crystal with diamond structure.

We will now treat an example of the use of the representation $\Gamma(\boldsymbol{q}, G_{\boldsymbol{q}})$. First we determine the matrices $\Gamma(\boldsymbol{q}, g)$ for a crystal with diamond structure. One can distinguish two cases.

1) For an element $g = \{R\,|\,\boldsymbol{t}_R + \boldsymbol{a}\}$ in the symmorphic subgroup $F\bar{4}3m$ one has $j' = j$ and $\boldsymbol{t}_R = 0$. Therefore, $\boldsymbol{u}(g, j) = (R - \mathbb{1})\boldsymbol{\tau}_j + \boldsymbol{a}$ and consequently $\exp(-iR\boldsymbol{q}\cdot\boldsymbol{u}(g,j)) = \exp(-iR\boldsymbol{q}\cdot\boldsymbol{a})\exp(-iR\boldsymbol{q}\cdot R\boldsymbol{\tau}_j + iR\boldsymbol{q}\cdot\boldsymbol{\tau}_j) =$ $\exp(-iR\boldsymbol{q}\cdot\boldsymbol{a})\exp(iR\boldsymbol{q}\cdot\boldsymbol{\tau}_j - i\boldsymbol{q}\cdot\boldsymbol{\tau}_j) = \exp(-i\boldsymbol{q}\cdot\boldsymbol{a})\exp(i\boldsymbol{K}_R\cdot\boldsymbol{\tau}_j)$ with

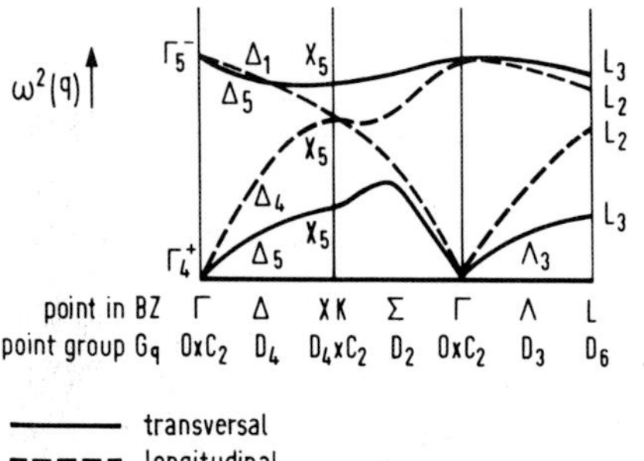

Fig. 6.5. Phonon branches and the representations of $G_{\boldsymbol{q}}$ for silicium.

$Rq = q + K_R$ and $K_R \in \Lambda^*$. Then the matrix $\Gamma(q,g)$ becomes

$$\Gamma(q,g) = \begin{pmatrix} R & 0 \\ 0 & R\mathrm{e}^{iK_R\cdot\tau} \end{pmatrix} \mathrm{e}^{-iq\cdot a} \qquad \text{with } \tau = [\tfrac{1}{4}\tfrac{1}{4}\tfrac{1}{4}]\,, \tag{6.35}$$

2) When g is not in the symmorphic subgroup, the two sublattices are interchanged: $t_R = \tau$ and $j' = 1$ for $j = 0, j' = 0$ for $j = 1$. In this case $\exp(-iRq\cdot u(g,j)) = \exp(-iq\cdot a)\exp(-iRq\cdot R\tau_j + iRq\cdot\tau_{j\pm 1} - iRq\cdot\tau)$, i.e. $\exp(-iRq\cdot u(g,0) = \exp(-iq\cdot a)$ and $\exp(-iRq\cdot u(g,1)) = \exp(-iq\cdot a)\exp(-iRq\cdot[R\tau+\tau]) = \exp(-iq\cdot a)\exp(-i(2q+K_R)\cdot\tau)$. Then one finds for the matrix $\Gamma(q,g)$

$$\Gamma(q,g) = \begin{pmatrix} 0 & R \\ R\mathrm{e}^{-i(2q+K_R)\cdot\tau} & 0 \end{pmatrix} \mathrm{e}^{-iq\cdot a}\,. \tag{6.36}$$

The representation $\Gamma(q, G_q)$ is then specified by the character

$$\chi(g) = \begin{cases} \chi(R)(1+\exp\{iK_R\cdot\tau\})\exp(-iq\cdot a) & \text{for } g \in F\bar{4}3m \\ 0 & \text{for } g \notin F\bar{4}3m\,. \end{cases}$$

For q inside the first Brillouin zone one has $K_R = 0$ or

$$\chi(g) = \begin{cases} 2\chi(R)\exp(-iq\cdot a) & \text{for } g \in F\bar{4}3m \\ 0 & \text{for } g \notin F\bar{4}3m\,. \end{cases}$$

As an example we take $q = 0$ (the point Γ in the Brillouin zone). The group G_q is the full space group. The allowable irreducible representations of G_q have character

$$\chi_\alpha(g) = \mathrm{e}^{-iq\cdot a}\chi_\alpha(R) = \chi_\alpha(R)\,,$$

whereas for $\Gamma(q, G_q)$ one has the character

$$\chi(g) = \begin{cases} 2\chi(R) = 2\,\mathrm{tr}\,R & \text{for } g \in F\bar{4}3m \\ 0 & \text{for } g \notin F\bar{4}3m\,. \end{cases}$$

The matrices R form a three-dimensional representation Γ_4^- of the point group $O \times C_2$. Hence, from the character table it follows that $\Gamma = \Gamma_5^+ \oplus \Gamma_4^-$. Then one can say that for $q = 0$ there are two three-fold degenerate phonon branches. To find the polarisation matrix $e(q)$ we have to look for a matrix which diagonalizes $D(\boldsymbol{q})$. This is also a matrix which reduces the representation $\Gamma(\boldsymbol{q}, G_{\boldsymbol{q}})$ as we see from eq. (6.32). However, although $e(\boldsymbol{q})$ is such a matrix, not every diagonalizing matrix is a polarisation matrix. First we look for a matrix S which reduces $\Gamma(\boldsymbol{0},g)$ by

$$S\Gamma(0,g)S^{-1} = \Gamma_4^-(g) \oplus \Gamma_5^+(g) .$$

Taking for g two generators of the point group 432 one obtains for the matrices $\Gamma(\boldsymbol{0},g)$:

$$\Gamma(\boldsymbol{0},g) = \begin{pmatrix} 0 & \Gamma_4(R_1) \\ \Gamma_4(R_1) & 0 \end{pmatrix} \quad \text{for } R_1 = \Gamma_4(R_1) = \begin{pmatrix} 0 & 1 & 0 \\ -1 & 0 & 0 \\ 0 & 0 & 1 \end{pmatrix},$$

$$\Gamma(\boldsymbol{0},g) = \begin{pmatrix} \Gamma_4(R_2) & 0 \\ 0 & \Gamma_4(R_2) \end{pmatrix} \quad \text{for } R_2 = \Gamma_4(R_2) = \begin{pmatrix} 0 & 1 & 0 \\ 0 & 0 & 1 \\ 1 & 0 & 0 \end{pmatrix}.$$

Putting

$$S = \begin{pmatrix} A & B \\ C & D \end{pmatrix}$$

one obtains the equations

$$\begin{pmatrix} B\Gamma_4(R_1) & A\Gamma_4(R_1) \\ D\Gamma_4(R_1) & C\Gamma_4(R_1) \end{pmatrix} = \begin{pmatrix} \Gamma_4(R_1)A & \Gamma_4(R_1)B \\ \Gamma_5(R_1)C & \Gamma_5(R_1)D \end{pmatrix}$$

and

$$\begin{pmatrix} A\Gamma_4(R_2) & B\Gamma_4(R_2) \\ C\Gamma_4(R_2) & D\Gamma_4(R_2) \end{pmatrix} = \begin{pmatrix} \Gamma_4(R_2)A & \Gamma_4(R_2)B \\ \Gamma_5(R_2)C & \Gamma_5(R_2)D \end{pmatrix}.$$

Then $\Gamma_4(R_1^2)A = A\Gamma_4(R_1^2)$ and $\Gamma_4(R_2)A = A\Gamma_4(R_2)$. As now A commutes with all the matrices of the irreducible group generated by $\Gamma_4(R_1^2)$ and $\Gamma_4(R_2)$, according to Schur's lemma it is $A = \lambda \mathbb{1}$ and consequently $B = \Gamma_4(R_1)A\Gamma_4(R_1)^{-1} = \lambda \mathbb{1}$. Moreover, $\Gamma_5(R_1^2)C = C\Gamma_5(R_1^2)$ and $\Gamma_5(R_2)C = C\Gamma_5(R_2)$, because $\Gamma_4(R_1^2) = \Gamma_5(R_1^2)$ and $\Gamma_4(R_2) = \Gamma_5(R_2)$. Then $C = \mu \mathbb{1}$ and $D = -\mu \mathbb{1}$. This means that the matrix S must be of the form

$$S = \begin{pmatrix} \lambda \mathbb{1} & \lambda \mathbb{1} \\ \mu \mathbb{1} & -\mu \mathbb{1} \end{pmatrix}.$$

However, in general $e(\boldsymbol{q})$ does not effect a reduction into the form $\Gamma_4 \oplus \Gamma_5$ but in an equivalent form $P\Gamma_4 P^{-1} \oplus Q\Gamma_5 Q^{-1}$, for certain nonsingular matrices P and Q. This means that $e(\boldsymbol{q})$ is of the form

$$e(\boldsymbol{q}) = \begin{pmatrix} P & Q \\ P & -Q \end{pmatrix}.$$

We see that in the first 3 columns the eigenvectors for one eigenvalue describe vibrations where both atoms in a unit cell move in the same way (acoustical phonons), whereas the 3 last columns describe vibrations in which the two atoms in a unit cell move with opposite phase (optical phonons). The acoustical phonons for $\boldsymbol{q} = 0$ describe a translation of the whole crystal. For these modes the frequency is zero. In this case we could determine the degeneracy and the possible forms of the polarisation matrix using the symmetry of the problem. Of course one needs more in order to evaluate explicitly the modes.

6.2.4. *The phonon wave functions*

In the frame of quantum mechanics the problem is described by the Hamiltonian

$$H = \sum_{\boldsymbol{n}\alpha l} \frac{p_\alpha\binom{\boldsymbol{n}}{l}^\dagger p_\alpha\binom{\boldsymbol{n}}{l}}{2m_l} + \frac{1}{2} \sum_{\substack{\boldsymbol{n}_1 l_1 \alpha_1 \\ \boldsymbol{n}_2 l_2 \alpha_2}} \Phi\binom{\boldsymbol{n}_1 - \boldsymbol{n}_2}{l_1 \quad l_2} u_{\alpha_1}\binom{\boldsymbol{n}_1}{l_1}^\dagger u_{\alpha_2}\binom{\boldsymbol{n}_2}{l_2} \tag{6.37}$$

where now $p_\alpha\binom{\boldsymbol{n}}{l}$ and $u_\alpha\binom{\boldsymbol{n}}{l}$ are operators with commutation relations

$$[p_\alpha\binom{\boldsymbol{n}}{l}, u_\beta\binom{\boldsymbol{m}}{s}] = \hbar i\, \delta_{\alpha\beta}\delta_{\boldsymbol{mn}}\delta_{ls}\,.$$

A state of the system is given by a wave function $\psi(\ldots \boldsymbol{x}\binom{\boldsymbol{n}}{l} \ldots)$, or with the variables $\boldsymbol{u}\binom{\boldsymbol{n}}{l}$ by $\phi(\ldots \boldsymbol{u}\binom{\boldsymbol{n}}{l} \ldots) = \psi(\ldots \boldsymbol{x}\binom{\boldsymbol{n}}{l}_0 + \boldsymbol{u}\binom{\boldsymbol{n}}{l} \ldots)$. How does such a wave function transform under an element g of the space group G? Consider the action of the substitution operator P_g. One has $P_g\phi(\ldots \boldsymbol{u}\binom{\boldsymbol{n}}{l} \ldots) = \psi(\ldots g^{-1}[\boldsymbol{x}\binom{\boldsymbol{n}}{l}_0 + \boldsymbol{u}\binom{\boldsymbol{n}}{l}] \ldots)$. Notice that g is not a linear transformation, but $g^{-1}(\boldsymbol{x}_0 + \boldsymbol{u}) = R^{-1}\boldsymbol{x}_0 + R^{-1}\boldsymbol{u} - R^{-1}\boldsymbol{t} = g^{-1}\boldsymbol{x}_0 + R^{-1}\boldsymbol{u}$, when $g = \{R\,|\,\boldsymbol{t}\}$. Then $P_g\phi(\ldots \boldsymbol{u}\binom{\boldsymbol{n}}{l} \ldots) = \phi(\ldots R^{-1}\boldsymbol{u}\binom{\boldsymbol{n}}{l} + g^{-1}\boldsymbol{x}\binom{\boldsymbol{n}}{l}_0 - \boldsymbol{x}\binom{\boldsymbol{n}}{l}_0 \ldots)$. Furthermore, $P_g\boldsymbol{u}\binom{\boldsymbol{n}}{l}P_g^{-1} = R^{-1}\boldsymbol{u}\binom{\boldsymbol{n}}{l} + g^{-1}\boldsymbol{x}\binom{\boldsymbol{n}}{l}_0 - \boldsymbol{x}\binom{\boldsymbol{n}}{l}_0$. Consequently the operator P_g does not commute with H. However, if one defines a permutation operator S_g defined by $S_g\boldsymbol{x}\binom{\boldsymbol{n}}{l}S_g^{-1} = \boldsymbol{x}\binom{\boldsymbol{n}'}{l'}$ which transforms the particle with equilibrium position $\boldsymbol{x}\binom{\boldsymbol{n}}{l}_0$ into the one at $\boldsymbol{x}\binom{\boldsymbol{n}'}{l'}_0$ (cf. eq. (6.27)), one has for the combined operator $T_g = S_g P_g$ the action

$$T_g u\binom{n}{l} T_g^{-1} = R^{-1} u\binom{n}{l} + g^{-1} x\binom{n'}{l'}_0 - x\binom{n}{l}_0 = R^{-1} u\binom{n}{l} . \tag{6.38}$$

The action of the operator T_g on the wave function $\phi(\ldots u\binom{n}{l} \ldots)$ is

$$T_g \phi(\ldots u\binom{n}{l} \ldots) = \phi(\ldots R^{-1} u\binom{n}{l} \ldots) .$$

The operators commute with the Hamiltonian and form a group homomorphic to G, since for $g = \{R|\boldsymbol{t}\}$ and $h = \{S|\boldsymbol{v}\}$ one has

$$T_g T_h \phi(\ldots u\binom{n}{l} \ldots) = T_h \phi(\ldots R^{-1} u\binom{n}{l} \ldots) = T_{gh} \phi(\ldots u\binom{n}{l} \ldots) .$$

They form the invariance operator group of H.

The eigenvalue problem for H can be solved by a diagonalization procedure. We give here the steps of this procedure, but we refer for details e.g. to Maradudin et al. [1963]. We combine the $3s$ operators $p_\alpha\binom{n}{l}$ into a column $\boldsymbol{p}(\boldsymbol{n})$, the operators $u_\alpha\binom{n}{l}$ into $\boldsymbol{u}(\boldsymbol{n})$. Then

$$H = \tfrac{1}{2} \sum_{\boldsymbol{n}} \boldsymbol{p}(\boldsymbol{n})^\dagger M^{-1} \boldsymbol{p}(\boldsymbol{n}) + \tfrac{1}{2} \sum_{\boldsymbol{n}_1 \boldsymbol{n}_2} \boldsymbol{u}(\boldsymbol{n}_1)^\dagger \Phi(\boldsymbol{n}_1 - \boldsymbol{n}_2) \boldsymbol{u}(\boldsymbol{n}_2) , \tag{6.39}$$

where M is a $3s$ dimensional diagonal matrix with elements $m_1, m_1, m_1, m_2, m_2, m_2, \ldots, m_s, m_s, m_s$. With the substitution

$$\boldsymbol{u}(\boldsymbol{n}) = N^{-3/2} \sum_{\boldsymbol{q}} e^{i\boldsymbol{q}\cdot\boldsymbol{n}} \boldsymbol{U}(\boldsymbol{q})$$

$$\boldsymbol{p}(\boldsymbol{n}) = N^{-3/2} \sum_{\boldsymbol{q}} e^{-i\boldsymbol{q}\cdot\boldsymbol{n}} \boldsymbol{P}(\boldsymbol{q})$$

and the definition of the dynamical matrix $D(\boldsymbol{q})$ one obtains

$$H = \frac{1}{2N^3} \sum_{\boldsymbol{q}} \boldsymbol{P}(\boldsymbol{q})^\dagger M^{-1} \boldsymbol{P}(\boldsymbol{q}) + \frac{1}{2N^3} \sum_{\boldsymbol{q}} \boldsymbol{U}(\boldsymbol{q})^\dagger M^{1/2} D(\boldsymbol{q}) M^{1/2} \boldsymbol{U}(\boldsymbol{q}) .$$

One can use eq. (6.26) to diagonalize this Hamiltonian by

$$\boldsymbol{P}(\boldsymbol{q}) = M^{1/2} e(\boldsymbol{q}) \boldsymbol{\Pi}(\boldsymbol{q}) N^{3/2}$$

$$\boldsymbol{U}(\boldsymbol{q}) = M^{-1/2} e(\boldsymbol{q}) \boldsymbol{V}(\boldsymbol{q}) N^{3/2} .$$

Then the Hamiltonian becomes

$$H = \tfrac{1}{2} \sum_{\boldsymbol{q}} \{\boldsymbol{\Pi}(\boldsymbol{q})^\dagger \boldsymbol{\Pi}(\boldsymbol{q}) + V(\boldsymbol{q})^\dagger \Omega(\boldsymbol{q}) V(\boldsymbol{q})\} .$$

Introducing

$$a(\boldsymbol{q}) = i(\tfrac{1}{2}\hbar)^{-\frac{1}{2}} [\Omega(\boldsymbol{q})^{-\frac{1}{4}} \boldsymbol{\Pi}(-\boldsymbol{q}) - i\Omega(\boldsymbol{q})^{\frac{1}{4}} V(\boldsymbol{q})]$$

$$a(\boldsymbol{q})^\dagger = -i(\tfrac{1}{2}\hbar)^{-\frac{1}{2}} [\Omega(\boldsymbol{q})^{-\frac{1}{4}} \boldsymbol{\Pi}(\boldsymbol{q}) + i\Omega(\boldsymbol{q})^{\frac{1}{4}} V(-\boldsymbol{q})]$$

one obtains the well known form of the Hamiltonian of a system of uncoupled harmonic oscillators

$$H = \tfrac{1}{2}\hbar \sum_{\boldsymbol{q}} \{a(\boldsymbol{q})^\dagger \Omega(\boldsymbol{q})^{\frac{1}{2}} a(\boldsymbol{q}) + \operatorname{tr} \Omega(\boldsymbol{q})^{\frac{1}{2}}\}$$

$$= \tfrac{1}{2} \sum_{\boldsymbol{q}j} \hbar\omega_j(\boldsymbol{q}) \{a^\dagger_{\boldsymbol{q}j} a_{\boldsymbol{q}j} + \tfrac{1}{2}\} . \tag{6.40}$$

The operators $a_{\boldsymbol{q}j}$ and $a^\dagger_{\boldsymbol{q}j}$ are the creation and absorption operators with commutation relations

$$[a_{\boldsymbol{q}j}, a_{\boldsymbol{q}'j'}] = [a^\dagger_{\boldsymbol{q}j}, a^\dagger_{\boldsymbol{q}'j'}] = 0 , \qquad [a_{\boldsymbol{q}j}, a^\dagger_{\boldsymbol{q}'j'}] = \delta_{\boldsymbol{q}\boldsymbol{q}'} \delta_{jj'} .$$

As the operators $N_{\boldsymbol{q}j} = a^\dagger_{\boldsymbol{q}j} a_{\boldsymbol{q}j}$ commute with H, the eigenfunctions of H can be chosen to be simultaneous eigenfunctions of $N_{\boldsymbol{q}j}$. They are denoted by $|\ldots n_{\boldsymbol{q}j} \ldots\rangle$, where $n_{\boldsymbol{q}j}$ is the eigenvalue of the operator $N_{\boldsymbol{q}j}$. The normalized eigenfunctions are

$$|n_1 n_2 \ldots\rangle = (n_1! n_2! \ldots)^{-\frac{1}{2}} (a^\dagger_{\boldsymbol{q}_1 j_1})^{n_1} (a^\dagger_{\boldsymbol{q}_2 j_2})^{n_2} \ldots |0\rangle , \tag{6.41}$$

where the vacuum state $|0\rangle$ is defined as the state with $a_{\boldsymbol{q}j}|0\rangle = 0$ for any $\boldsymbol{q}$ and j. The function (6.41) is an n-phonon state, when $n_1 + n_2 + \ldots = n$. As they are eigenfunctions of H, they are stationary states. This means that the quantum numbers $n_{\boldsymbol{q}j}$ are constants of motion. This is no longer true if anharmonic terms are taken into account.

The relation between the operators $u_\alpha(^{\boldsymbol{n}}_{l})$, $p_\alpha(^{\boldsymbol{n}}_{l})$ and the operators $a_{\boldsymbol{q}j}$ and $a^\dagger_{\boldsymbol{q}j}$ is given by

$$u_\alpha(^{\boldsymbol{n}}_{l}) = (\hbar/2N^3 m_l)^{\frac{1}{2}} \sum_{\boldsymbol{q}j} e^{i\boldsymbol{q}\cdot\boldsymbol{n}} e_{\alpha l}(\boldsymbol{q}|j) \omega_j(\boldsymbol{q})^{-\frac{1}{2}} (a_{\boldsymbol{q}j} + a^\dagger_{-\boldsymbol{q}j})$$

$$p_\alpha(^{\boldsymbol{n}}_{l}) = (\hbar m_l/2N^3)^{\frac{1}{2}} \sum_{\boldsymbol{q}j} e^{-i\boldsymbol{q}\cdot\boldsymbol{n}} e_{\alpha l}(\boldsymbol{q}|j) \omega_j(\boldsymbol{q})^{\frac{1}{2}} i(a^\dagger_{\boldsymbol{q}j} - a_{-\boldsymbol{q}j})$$

$$a_{\boldsymbol{q}j} = i(2/\hbar N^3)^{1/2} \sum_{\boldsymbol{n}\alpha l} \mathrm{e}^{-i\boldsymbol{q}\cdot\boldsymbol{n}} \{\omega_j(\boldsymbol{q})^{-1/2} e_{\alpha l}(\boldsymbol{q}\,|\,j) m_l^{-1/2} p_\alpha\binom{\boldsymbol{n}}{l}$$
$$- i\,\omega_j(\boldsymbol{q})^{1/2} e_{\alpha l}(-\boldsymbol{q}\,|\,j) m_l^{1/2} u_\alpha\binom{\boldsymbol{n}}{l}\} \,. \tag{6.42}$$

From the transformation properties of $u_\alpha\binom{\boldsymbol{n}}{l}$ and $p_\alpha\binom{\boldsymbol{n}}{l}$ (cf. eq. (6.38))

$$T_g u_\alpha\binom{\boldsymbol{n}}{l} T_g^{-1} = R_{\beta\alpha} u_\beta\binom{\boldsymbol{n}'}{l'} \qquad \text{and} \qquad T_g p_\alpha\binom{\boldsymbol{n}}{l} T_g^{-1} = R_{\beta\alpha} p_\beta\binom{\boldsymbol{n}'}{l'}$$

the transformation properties of the creation and absorption operators follow. We define $B_{\boldsymbol{q}j} = a_{\boldsymbol{q}j} + a^\dagger_{-\boldsymbol{q}j}$ and $A_{\boldsymbol{q}j} = i(a^\dagger_{\boldsymbol{q}j} - a_{-\boldsymbol{q}j})$. They transform according to

$$T_g B_{\boldsymbol{q}j} T_g^{-1} = T_g\{(2/\hbar N^3)^{1/2} \sum_{\boldsymbol{n}\alpha l} \mathrm{e}^{-i\boldsymbol{q}\cdot\boldsymbol{n}} \omega_j(\boldsymbol{q})^{1/2} e_{\alpha l}(-\boldsymbol{q}\,|\,j) m_l^{1/2} u_\alpha\binom{\boldsymbol{n}}{l}\} T_g^{-1}$$
$$= (2/\hbar N^3)^{1/2} \sum_{\boldsymbol{n}\alpha\beta l} R_{\beta\alpha} \mathrm{e}^{-i\boldsymbol{q}\cdot\boldsymbol{n}} \omega_j(\boldsymbol{q})^{1/2} e_{\alpha l}(-\boldsymbol{q}\,|\,j) m_l^{1/2} u_\beta\binom{\boldsymbol{n}'}{l'} \,.$$

Using $\boldsymbol{n}' = R\boldsymbol{n} + \boldsymbol{u}(g,l)$ and the expression for $u_\beta\binom{\boldsymbol{n}'}{l'}$ this becomes

$$T_g B_{\boldsymbol{q}j} T_g^{-1} = \sum_{\alpha\beta l j'} \Gamma(\boldsymbol{q},g)^{\dagger\alpha\beta}_{ll'} \omega_j(\boldsymbol{q})^{1/2} e_{\alpha l}(-\boldsymbol{q}\,|\,j) e_{\beta l'}(R\boldsymbol{q}\,|\,j') \omega_{j'}(R\boldsymbol{q})^{-1/2} B_{R\boldsymbol{q}j'}$$

or, in matrix notation,

$$T_g B(\boldsymbol{q}) T_g^{-1} = \Omega^{1/4}(\boldsymbol{q})\, e(\boldsymbol{q})^{-1} \Gamma(\boldsymbol{q},g)^{-1} e(R\boldsymbol{q}) \Omega(\boldsymbol{q})^{-1/4} B(R\boldsymbol{q}) \,. \tag{6.43}$$

To simplify this expression we recall that $\Gamma(\boldsymbol{q},g)\, e(\boldsymbol{q}) = e(R\boldsymbol{q}) \Delta(\boldsymbol{q},g)$ and that the matrix $\Delta(\boldsymbol{q},g)$ commutes with $\Omega(\boldsymbol{q})$. Then eq. (6.43) becomes

$$T_g B(\boldsymbol{q}) T_g^{-1} = \Delta(\boldsymbol{q},g)^{-1} B(R\boldsymbol{q}) = e(\boldsymbol{q})^{-1} \Gamma(\boldsymbol{q},g)^{-1} e(R\boldsymbol{q})\, B(R\boldsymbol{q}) \,. \tag{6.44}$$

Analogously we find for the matrix $A(\boldsymbol{q})$ with elements $A_{\boldsymbol{q}j}$

$$T_g A(\boldsymbol{q}) T_g^{-1} = e(\boldsymbol{q})^{-1} \Gamma(-\boldsymbol{q},g)^{-1} e(R\boldsymbol{q})\, A(R\boldsymbol{q}) \,. \tag{6.45}$$

We recall that the elements $\Delta(\boldsymbol{q},g)$ for $g \in G_{\boldsymbol{q}}$ form a representation of $G_{\boldsymbol{q}}$. As the one-phonon state $|\boldsymbol{q}j\rangle = a^\dagger_{\boldsymbol{q}j}|0\rangle = B_{-\boldsymbol{q}j}|0\rangle$ transforms according to

$$T_g|\boldsymbol{q}j\rangle = T_g B_{-\boldsymbol{q}j} T_g^{-1} T_g |0\rangle = \Delta(\boldsymbol{q},g)^{-1}_{jj'} B_{-R\boldsymbol{q}j'}|0\rangle = \Delta(\boldsymbol{q},g)^{-1}_{jj'}|R\boldsymbol{q}j'\rangle$$

the one phonon states with $\boldsymbol{q}$ in one star and j labelling modes with the same frequency span an irreducible representation of $G_{\boldsymbol{q}}$. The states with more phonons transform according to product representations.

We have discussed here some only of the symmetry properties of phonon states. Other properties are treated in Streitwolf [1964] and Waeber [1969]. A slightly different approach to the symmetry properties of lattice vibrations is given in Raghavacharyulu [1961], Maradudin and Vosko [1968] and Warren [1968]. They are mainly concerned with the properties of the dynamical matrix and of the polarisation vectors. Because of eqs. (6.40) and (6.42) the quantum mechanical eigenvalues of H and the operators $u_\alpha\binom{\boldsymbol{n}}{l}$ are determined by $e(\boldsymbol{q})$ and $\omega_j(\boldsymbol{q})$; which gives the reason for our interest in these quantities.

6.3. Electrons in electromagnetic fields

6.3.1. Symmetries of electromagnetic fields and potentials

The symmetry of a physical system in an electromagnetic field is determined by the symmetry of the force fields $\boldsymbol{E}(\boldsymbol{r},t)$ and $\boldsymbol{H}(\boldsymbol{r},t)$. However, in the equations of motion like the Schrödinger equation it is the potentials $\boldsymbol{A}(\boldsymbol{r},t)$ and $\phi(\boldsymbol{r},t)$ that appear rather than $\boldsymbol{E}(\boldsymbol{r},t)$ and $\boldsymbol{H}(\boldsymbol{r},t)$. As the symmetries of fields and potentials are in general different, we want to discuss the relation between these symmetries. As is well known, the fields can be obtained from the potentials by

$$\begin{aligned}\boldsymbol{E}(\boldsymbol{r},t) &= -\nabla\phi(\boldsymbol{r},t) - \frac{1}{c}\frac{\partial}{\partial t}\boldsymbol{A}(\boldsymbol{r},t)\\ \boldsymbol{H}(\boldsymbol{r},t) &= \nabla\times\boldsymbol{A}(\boldsymbol{r},t)\,.\end{aligned} \tag{6.46}$$

The potentials are not uniquely determined by the fields. The same fields are obtained from potentials

$$\begin{aligned}\boldsymbol{A}'(\boldsymbol{r},t) &= \boldsymbol{A}(\boldsymbol{r},t) + \nabla\chi(\boldsymbol{r},t)\\ \phi'(\boldsymbol{r},t) &= \phi(\boldsymbol{r},t) - \frac{\partial}{c\,\partial t}\chi(\boldsymbol{r},t)\,,\end{aligned} \tag{6.47}$$

where $\chi(\boldsymbol{r},t)$ is an arbitrary three times differentiable real function of space and time. The transformation of eq. (6.47) is called a *gauge transformation*. On the other hand, two potentials giving the same fields are related by such a gauge transformation. We will often denote the two potentials $\boldsymbol{A}(\boldsymbol{r},t)$ and $\phi(\boldsymbol{r},t)$ by one four-potential $A = (\phi, \boldsymbol{A})$.

Let us consider the transformation properties of potentials and fields under elements of the *Shubnikov group* $S(4)$, i.e. under elements $gT^{\frac{1}{2}(1-\epsilon)}$ with g an element of $E(3)$, T the time reversal and $\epsilon = \pm 1$. The potentials transform according to

$$\begin{aligned}(g,\epsilon)\boldsymbol{A}(\boldsymbol{r},t) &= \epsilon R\boldsymbol{A}(g^{-1}\boldsymbol{r},\epsilon t)\\ (g,\epsilon)\phi(\boldsymbol{r},t) &= \phi(g^{-1}\boldsymbol{r},\epsilon t)\,,\end{aligned} \tag{6.48}$$

where $g = \{R\,|\,\boldsymbol{u}\} \in E(3)$. The fields transform according to

$$\begin{aligned}(g,\epsilon)\,\boldsymbol{E}(\boldsymbol{r},t) &= R\boldsymbol{E}(g^{-1}\boldsymbol{r},\epsilon t)\\ (g,\epsilon)\boldsymbol{H}(\boldsymbol{r},t) &= \epsilon(\det R)R\,\boldsymbol{H}(g^{-1}\boldsymbol{r},\epsilon t)\,.\end{aligned} \tag{6.49}$$

When (g,ϵ) leaves the potentials invariant, it also leaves the fields invariant, but the converse is, in general, not true. Consider e.g. a uniform electromagnetic field. It is invariant under arbitrary translations, but a potential which gives this field can not be invariant under all translations, unless the fields vanish, because of eq. (6.46). From the physical point of view the fields are the important quantities. Hence symmetry transformations of the field must have a significance, even when they are not symmetry transformations of the potentials. In fact, when (g,ϵ) leaves $(\boldsymbol{E},\boldsymbol{H})$ invariant, it transforms the four-potential A into A' which gives the same fields. Therefore, A and A' are related by a gauge transformation. We define the action of a pair (χ, s) with χ a gauge function and $s = (g,\epsilon) \in S(4)$ on potentials by

$$\begin{aligned}(\chi,s)\boldsymbol{A}(\boldsymbol{r},t) &= (s\boldsymbol{A})(\boldsymbol{r},t) - \nabla\chi(\boldsymbol{r},t)\\ (\chi,s)\,\phi(\boldsymbol{r},t) &= (s\phi)(\boldsymbol{r},t) + \frac{\partial}{c\partial t}\chi(\boldsymbol{r},t)\,.\end{aligned} \tag{6.50}$$

When $(\chi,s)A = A$, the function χ is called a *compensating gauge function* for $s \in S(4)$. This implies that s leaves both $\boldsymbol{E}$ and $\boldsymbol{H}$ invariant.

The subgroup of $S(4)$ which leaves the fields $\boldsymbol{E}$ and $\boldsymbol{H}$ invariant is called the *symmetry group of the field* $(\boldsymbol{E},\boldsymbol{H})$. For each element $g \in G$, the symmetry group of $(\boldsymbol{E},\boldsymbol{H})$, there exists a gauge function which is compensating. However, this function is not uniquely determined.

The functions χ form a group in a natural way by the definition $(\chi_1+\chi_2)(\boldsymbol{r},t)=\chi_1(\boldsymbol{r},t)+\chi_2(\boldsymbol{r},t)$. Then one can also give the set of pairs (χ,s) with $s\in S(4)$ the structure of a group by

$$(\chi,s)(\chi',s')=(\chi+s\chi',ss')\,, \tag{6.51}$$

where the action of $s=(g,\epsilon)$ on $\chi(\boldsymbol{r},t)$ is given by

$$s\chi(\boldsymbol{r},t)=\epsilon\chi(g^{-1}\boldsymbol{r},\epsilon t)\,.$$

It is easily verified that eq. (6.51) gives the set of pairs (χ,s) the structure of a semi-direct product of the Abelian group of functions and the group $S(4)$. Moreover, one has

$$(\chi,s)\{(\chi',s')A\}=\{(\chi,s)(\chi',s')\}A\,, \tag{6.52}$$

as should be the case. However, (6.51) is not the only possibility of obtaining a group with the property (6.52). From (6.52) one obtains

$$(\chi,s)(\chi',s')=(\chi+s\chi'+n(s,s'),ss')\,, \tag{6.53}$$

where $n(s,s')$ is an arbitrary real constant. As constants are also constant gauge functions, they transform under s according to

$$sn(s',s'')=\epsilon n(s',s''),\quad (s,s',s''\in S(4))\,.$$

Eq. (6.53) determines a group if and only if (cf. Ch. 4, § 1.8)

$$n(s,s')+n(ss',s'')=sn(s',s'')+n(s,s's'')\,, \tag{6.54}$$

for any $s,s',s''\in S(4)$. The group determined in this way is denoted by J_n. It is the group of pairs (χ,s), when χ is a gauge function and $s\in S(4)$ with product as defined by eq. (6.53). It is evident that J_n depends on the choice of the factor system n. In particular, the choice $n=0$ gives the semi-direct product J_0 of eq. (6.51). The group J_n is an extension of the group of Abelian functions by $S(4)$ with factor system n. We now define the *symmetry group of the four-dimensional potential* A as the subgroup Q_n of J_n which leaves A invariant. It consists of all pairs (χ,s) such that $(\chi,s)A=A$. The symmetry group clearly depends on the choice of the factor system n. A subgroup $R\subseteq Q_n$ is formed by all elements (φ,e) with φ a real constant function and e the identity of $S(4)$. Moreover, the subgroup R is an invariant one, because for any $(\chi, s) \in Q_n$ one has, because $n(s,e) = n(e,s) = 0$ for all s,

$$(\chi, s)(\varphi, e)(\chi, s)^{-1} = (\chi + \epsilon\varphi, s)(-s^{-1}\chi - \epsilon n(s, s^{-1}), s^{-1}) = (\epsilon\varphi, e)$$

and

$$(\varphi, e)(\varphi', e) = (\varphi + \varphi', e) = (\varphi', e)(\varphi, e) .$$

The factor group Q_n/R is isomorphic to the symmetry group of the field G, because (χ, s) and (χ', s') are in the same coset if and only if $s = s'$ and $\chi' = \chi + \varphi$. Thus the group Q_n is an extension of the Abelian group R by the group G. Its structure has some analogy with that of a space group. The group R corresponds to the group of translations, the group G to the point group. For every element $g \in G$ there is a compensating gauge transformation χ_g as there is for every element R of the point group a nonprimitive transtaion $\boldsymbol{t}_R$. The compensating gauge is only determined up to a constant: χ_g and $\chi_g + \varphi$ are both compensating for g, just like the nonprimitive translations are determined up to a primitive translation. One can choose for any $g \in G$ a compensating function χ_g with $(\chi_g, g) \in Q_n$. Such a set of functions $\{\chi_g\} = \chi_G$ is called a *system of compensating gauges,* just as one can obtain a system of nonprimitive translations for a space group.

Evidently the group Q_n depends on the choice of the gauge. When the same field is derived from a potential $A' = (\eta, e)A$, the symmetry group Q_n' is different from Q_n. If $(\chi, g) \in Q_n$, the element g is a symmetry for the fields. Then a symmetry element for the potential A' is $(\chi + (1-g)\eta, g)$, as

$$(\chi + (1-g)\eta, g)A' = g(A - \partial\eta) - \partial(\chi + \eta - g\eta)$$

$$= gA - \partial\chi - \partial\eta = A - \partial\eta = A' ,$$

where $A + \partial\chi$ means $(\phi - c^{-1}\partial\chi/\partial t, \boldsymbol{A} + \nabla\chi)$. Again χ' in $(\chi', g) \in Q_n'$ is determined up to a constant. Hence $\chi' = \chi + (1-g)\eta + \varphi$. For a system of compensating gauges χ_G one has

$$(\chi_g, g)(\chi_{g'}, g') = (\chi_g + g\chi_{g'} - \chi_{gg'} + n(g, g'), e)(\chi_{gg'}, gg') .$$

Denoting $\chi_g + g\chi_{g'} - \chi_{gg'}$ by $f(g, g')$ one obtains the relations

$$(\chi_{g_1}, g_1)(\chi_{g_2}, g_2) = (f(g_1, g_2) + n(g_1, g_2), e)(\chi_{g_1 g_2}, g_1 g_2) . \tag{6.55}$$

It is readily verified that $f(g_1, g_2)$ satisfies relation (4.10), which means that f is a factor system. Therefore, the group Q_n is an extension of R by the group G with factor system $f+n$. For another gauge $A' = A + \partial\eta$ one gets a system χ'_G of compensating gauges related to χ_G by

$$\chi'_G = \chi_g + (1-g)\eta + \varphi_g \,, \qquad (\text{all } g \in G) \tag{6.56}$$

for some constant φ_g. Choosing $\varphi_g = 0$ for any $g \in G$ one obtains a system of compensating gauges which determines a product as in eq. (6.55), with the same factor system f. This means that Q_n and Q'_n are equivalent extensions of R by G. *A fortiori*, they are isomorphic. In this way the isomorphism class of the symmetry group of the potential is determined by the field and by the choice of the factor system n. We will make use of the freedom we still have for n to obtain a simple relation between the symmetry group G and the invariance operator group of the Hamiltonian describing a system in such a field.

6.3.2. *Invariance group of the Hamiltonian*

In the following we restrict ourselves to *time-independent electromagnetic fields*. We consider an electron in such a field. For a particle for which we can forget about the spin the Hamiltonian is

$$H = \frac{1}{2m}\left(p - \frac{e}{c}A\right)^2 + e\phi \,. \tag{6.57}$$

The spin is taken into account in the nonrelativistic approximation of the Dirac equation, i.e. in the *Pauli equation*. This is a Schrödinger equation with Hamiltonian

$$H = \frac{1}{2m}\left(p - \frac{e}{c}A\right)^2 + e\phi - \frac{\hbar e}{2mc}\,\boldsymbol{\sigma}\cdot H + \frac{\hbar e}{4m^2c^2}\,\boldsymbol{\sigma}\cdot\left(\nabla\phi \times \left(p - \frac{e}{c}A\right)\right). \tag{6.58}$$

Notice that in the Hamiltonian the potentials A and ϕ occur.

We denote the Hamiltonians by $H(\boldsymbol{p}, \boldsymbol{r}, \boldsymbol{A}, \phi)$. Then for any Shubnikov group element s which leaves the potential A invariant, the substitution operator P_s commutes with H (6.57) since

$$P_s H P_s^{-1} = \frac{1}{2m}\left(\epsilon R^{-1}\boldsymbol{p} - \frac{e}{c}A(g^{-1}\boldsymbol{r})\right)^2 + e\phi(g^{-1}\boldsymbol{r})$$
$$= \frac{1}{2m}\left(\boldsymbol{p} - \frac{e}{c}\epsilon R A(g^{-1}\boldsymbol{r})\right)^2 + e\phi(\boldsymbol{r}) = \frac{1}{2m}\left(\boldsymbol{p} - \frac{e}{c}A(\boldsymbol{r})\right)^2 + e\phi(\boldsymbol{r}) = H\,, \tag{6.59}$$

where $s = (g, \epsilon) \in S(4)$. However, the symmetry of the system is the symmetry of the field and not only that of the potential. For a symmetry element of the field $g \in G$ one has $gA = A + \partial\chi$. Now we consider the operator $\exp(f(\boldsymbol{r}))$, where $f(\boldsymbol{r})$ is an arbitrary function of $\boldsymbol{r}$. Then one has the following commutation relation

$$\left(\boldsymbol{p} - \frac{e}{c}A\right)^2 \exp(f(\boldsymbol{r})) = \left(\boldsymbol{p} - \frac{e}{c}A\right)\exp(f(\boldsymbol{r}))\left(\boldsymbol{p} - \hbar i\nabla f - \frac{e}{c}A\right)$$
$$= \exp(f(\boldsymbol{r}))\left(\boldsymbol{p} - \frac{e}{c}A - \hbar i\nabla f\right)^2 .$$

Therefore, the operator $U_\chi = \exp\{(-ie/\hbar c)\chi\}$ satisfies the relation

$$U_\chi H U_\chi^{-1} = \frac{1}{2m}\left(\boldsymbol{p} + \frac{e}{c}\nabla\chi - \frac{e}{c}A\right)^2 + e\phi\,. \tag{6.60}$$

So, if χ_g is a compensating gauge for $g \in G$ and $T_g = U_{\chi_g} P_g$, one has the relation

$$T_g H T_g^{-1} = \frac{1}{2m}\left(\boldsymbol{p} - \frac{e}{c}(gA) + \frac{e}{c}\nabla\chi_g\right)^2 + e(g\phi) = H - \frac{e}{c}\frac{\partial}{\partial t}\chi_g\,. \tag{6.61}$$

When $g\phi = \phi - (1/c)(\partial/\partial t)\chi_g = \phi$, we have found operators commuting with the Hamiltonian. However, when $g\phi \neq \phi$, the operator T_g does not commute with H but with the operator $H + \hbar i(\partial/\partial t)$:

$$T_g\left(H - \hbar i\frac{\partial}{\partial t}\right)T_g^{-1} = H - \frac{e}{c}\frac{\partial}{\partial t}\chi_g + \hbar i U_{\chi_g}\left(\frac{\partial}{\partial t}U_{\chi_g}^{-1}\right) + \hbar i\frac{\partial}{\partial t}$$
$$= H - \hbar i\frac{\partial}{\partial t}\,.$$

This means that the solutions of the time-dependent Schrödinger equation transform according to irreducible representations of G. We will not discuss this problem here (see Janner and Janssen [1971]), but we assume that the scalar potential ϕ is left invariant by G. Then the product of two operators T_g and $T_{g'}$ commuting with the Hamiltonian H is given by

$$T_g T_{g'} = U_{\chi_g} P_g U_{\chi_{g'}} P_{g'} = \exp\left(\frac{ie}{\hbar c}(\chi_g + g\chi_{g'} - \chi_{gg'})\right) T_{gg'} .$$

Recalling the definition of the factor system f one has

$$T_g T_{g'} = \exp\left(\frac{-ie}{\hbar c} f(g,g')\right) T_{gg'} = \omega_f(g,g')\, T_{gg'} . \qquad (6.62)$$

Consequently, the operators T_G form a projective representation of G with factor system $\omega_f(g,g')$. However, they form an ordinary representation of the symmetry group Q_0 of the potential, when one assigns to the element $(\chi, g) \in Q_0$ the operator $T(\chi,g) = U_\chi P_g$. Then

$$T(\chi',g')T(\chi,g) = U_{\chi'+g'\chi} P_{g'g} = T[(\chi',g')(\chi,g)] , \qquad (6.63)$$

which shows that these operators form an ordinary representation of Q_0. Thus for an electromagnetic field $(\boldsymbol{E},\boldsymbol{H})$ with potential A the symmetry group G of the field and the symmetry group Q_0 of the potential form an invariance group for the Hamiltonian H (6.57) (always assuming that ϕ is G-invariant).

For the Hamiltonian H (6.58), the substitution operator P_g for an element g, leaving the potential invariant, is nevertheless not a symmetry operator for H. One has $P_g H(\boldsymbol{p},\boldsymbol{r},\boldsymbol{A},\phi)P_g^{-1} = H(\epsilon R^{-1}\boldsymbol{p}, h^{-1}\boldsymbol{r}, \boldsymbol{A}, \phi)$, when $g = (h,\epsilon)$ and $h = \{R\,|\,\boldsymbol{u}\}$. This implies

$$P_g H P_g^{-1} = \frac{1}{2m}\left(\boldsymbol{p} - \epsilon\frac{e}{c} R\boldsymbol{A}(h^{-1}\boldsymbol{r})\right)^2 + e\phi(h^{-1}\boldsymbol{r}) - \frac{\hbar e}{2mc}\boldsymbol{\sigma}\cdot \boldsymbol{H}(h^{-1}\boldsymbol{r})$$

$$+ \frac{\hbar e}{4m^2c^2}(\det R)(R\boldsymbol{\sigma})\cdot(\nabla\phi \times \boldsymbol{p}) . \qquad (6.64)$$

This means that P_g does not commute with H. However, according to Ch. 5, §1.4 the combination $u(R)P_g$ does commute if g leaves the potential invariant. When $g \in G$, the symmetry group of the field, one has

$$P_g u(R) H u(R)^{-1} P_g^{-1} = \frac{1}{2m}\left(\boldsymbol{p} - \frac{e}{c}(gA)\right)^2 + e(g\phi) - \frac{\hbar e}{2mc}\,\boldsymbol{\sigma}\cdot H$$

$$+ \frac{\hbar e}{4m^2c^2}\,\boldsymbol{\sigma}\cdot(\nabla\phi \times \boldsymbol{p})\,.$$

Considering again a compensating gauge χ for g one sees that the operator $U_\chi P_g u(R)$ commutes with H (6.58). For a system of compensating gauges χ_G the product of two operators $T_g = U_{\chi_g} P_g u(R)$ is given by

$$T_g T_{g'} = \exp\left[\frac{-ie}{\hbar c}\,(f(g,g') + s(g,g'))\right] T_{gg'}\,, \tag{6.65}$$

where $\omega_s(g,g') = \exp\{(-ie/\hbar c)\, s(g,g')\}$ is the factor system determined by $u(R)u(R') = \omega_s(g,g')u(RR')$. So the operators T_g $(g \in G)$ commute with H and form a projective representation of G.

We consider next the group Q_s which is the set of elements (χ, g) leaving the potential A invariant with group structure given by eq. (6.53) taking $n = s$. Define the operator $T(\chi,g) = U_\chi P_g u(R)$. Then

$$T(\chi,g)\,T(\chi',g') = U_{\chi+g\chi'} P_{gg'}\, \omega_s(g,g') u(RR') = T[(\chi,g)(\chi',g')]\,.$$

This means that the operators $T(\chi,g)$ form an ordinary representation of the group Q_s. So Q_s is an invariance group of the Hamiltonian H (6.58). Since the isomorphism class of Q_s is determined by the field, the invariance is gauge-independent.

6.3.3. A Bloch electron in a homogeneous magnetic field

The properties of solid state materials in an external magnetic field are very important for several effects. In this section we will study such a system from a group theoretical point of view. The fundamentals of this analysis were developed in the preceding section. When one considers the electromagnetic field in which the electron moves as a superposition of a crystal potential $V(\boldsymbol{r})$ and a homogeneous field, the four-potential appearing in the Hamiltonian is $(V(\boldsymbol{r}), \boldsymbol{A}(\boldsymbol{r}))$. As we have seen, the gauge can easily be dealt with. We choose the so-called *symmetric gauge*

$$A(\boldsymbol{r}) = \tfrac{1}{2} H \times \boldsymbol{r}\,. \tag{6.66}$$

The symmetry group of the magnetic field, i.e. the group of Shubnikov transformations leaving the field invariant, contains as a subgroup the full three-dimensional translation group $T(3)$. Besides this it contains all rotations around the direction of the field, the central inversion and combinations of 180° rotations perpendicular to the field direction and time reversal. Here we will consider the translation symmetry only. The crystal potential $V(\boldsymbol{r})$ has the periodicity of the lattice and can be written as

$$V(\boldsymbol{r}) = \sum_{\boldsymbol{K}\in\Lambda^*} a_{\boldsymbol{K}}\, \mathrm{e}^{-i\boldsymbol{K}\cdot\boldsymbol{r}} \,.$$

The electric field derived from $V(\boldsymbol{r})$ using eq. (6.46) has also the periodicity of the lattice. So the symmetry group of the electromagnetic field contains the lattice group U, the translation subgroup of the space group of the crystal. As the translational symmetry of $V(\boldsymbol{r})$ and $\boldsymbol{E}(\boldsymbol{r})$ is the same, one only needs compensating gauge transformations for $\boldsymbol{A}(\boldsymbol{r})$. To determine these compensating gauge transformations for $\boldsymbol{A}(\boldsymbol{r})$ consider the transformation of $\boldsymbol{A}(\boldsymbol{r})$ under a translation $\boldsymbol{t}$: $\boldsymbol{A}'(\boldsymbol{r}) = \boldsymbol{A}(\boldsymbol{r}-\boldsymbol{t}) = \boldsymbol{A}(\boldsymbol{r}) - \frac{1}{2}(\boldsymbol{H}\times\boldsymbol{t})$. The pair (χ, g) leaves $\boldsymbol{A}$ invariant according to eq. (6.48) if $\boldsymbol{\nabla}\chi = \frac{1}{2}(\boldsymbol{H}\times\boldsymbol{t})$. One can choose

$$\chi_t = \tfrac{1}{2}(\boldsymbol{H}\times\boldsymbol{t})\cdot\boldsymbol{r} = -\tfrac{1}{2}(\boldsymbol{H}\times\boldsymbol{r})\cdot\boldsymbol{t} \,. \tag{6.67}$$

The factor systems f and ω are given by eqs. (6.55) and (6.62).

$$\begin{aligned} f(\boldsymbol{t}_1, \boldsymbol{t}_2) &= \tfrac{1}{2}(\boldsymbol{t}_1\times\boldsymbol{t}_2)\cdot\boldsymbol{H} \\ \omega(\boldsymbol{t}_1, \boldsymbol{t}_2) &= \exp\left[\frac{ie}{2\hbar c}(\boldsymbol{t}_1\times\boldsymbol{t}_2)\cdot\boldsymbol{H}\right]. \end{aligned} \tag{6.68}$$

The operators $T_{\boldsymbol{t}} = U_{\chi_{\boldsymbol{t}}} P_{\boldsymbol{t}}$ (with $t\in U$) commute with the Hamiltonian (6.57) and form a projective representation of U with factor system ω (6.68). One can proceed in two ways. Either one studies this projective representation or one tries to construct a group of operators commuting with H. We choose here the second method. We have found already such a group of operators. It is the group Q_0 with elements $(\chi_{\boldsymbol{t}} + \varphi, \boldsymbol{t})$. It is an infinite group. However, imposing periodic boundary conditions one can get a finite invariance group.

Imposing periodic boundary conditions means that we consider only those representations of U which give the identity for all elements a^N for some integer N and for all $a\in U$. This means that for any $a\in U$ one has $T_{N\boldsymbol{a}}\psi = \psi$. In particular, one has

$$T_{Na}T_{a'}\psi = \omega(Na,a')\, T_{Na+a'}\psi = \frac{\omega(Na,a')}{\omega(a',Na)}\, T_{a'}T_{Na}\psi$$

$$= \frac{\omega(Na,a')}{\omega(a',Na)}\, T_{a'}\psi\ ,$$

which implies

$$\omega(Na,a') = \omega(a',Na) \qquad \text{for all} \quad a,a' \in U\ .$$

The application of eq. (6.68) leads to

$$\frac{ie}{\hbar c}(a \times a') \cdot H = \frac{2\pi\, im}{N} \qquad \text{for some integer } m\ . \tag{6.69}$$

This holds in particular for $a = a_i$ and $a' = a_j$, where a_i, a_j are basis elements of U. When we write the magnetic field in components with respect to a basis a_1, a_2, a_3 one has $H = h_1 a_1 + h_2 a_2 + h_3 a_3$. Then eq. (6.69) gives

$$\frac{eN}{\hbar c}(a_i \times a_j) \cdot H = \frac{eN}{\hbar c}\Omega_0 h_k = 2\pi m_k\ ,$$

where $i,j,k = 1,2,3$ or a cyclic permutation, Ω_0 is the volume of the unit cell spanned by a_1, a_2, a_3: $\Omega_0 = (a_i \times a_j) \cdot a_k$ and m_k is some integer. So the numbers $e\Omega_0 h_k/2\pi\hbar c$ are rational numbers m_k/N. One can choose a basis with $h_1 = h_2 = 0$. Then

$$H = \frac{2\pi}{\Omega_0}\frac{\hbar c}{e}\frac{p}{q}\, a_3 \tag{6.70}$$

for some rational number p/q, where p and q are relatively prime. So the periodic boundary conditions put restrictions on the magnetic field. In order to have these conditions one has to require the rationality relation (6.70). This means that in general it is not possible to put periodic boundary conditions on this system. However, it is always possible, also for infinite crystals, to consider rational fields (6.70).

The field (6.70) gives with $a = n_1 a_1 + n_2 a_2 + n_3 a_3$ for the factor system (6.68) the value

$$\omega(a,a') = \exp(\pi i(n_1 n_2' - n_2 n_1')\, p/q)\ .$$

An associated factor system consisting of q-th roots is obtained by eq. (1.29) for $u(a) = \exp(-i\pi n_1 n_2 p/q)$. Then one has

$$\omega'(\boldsymbol{a}, \boldsymbol{a}') = \exp(2\pi i n_1 n_2' p/q) . \tag{6.71}$$

To find the nonequivalent irreducible representations with this factor system, one notices that the elements $\omega(\boldsymbol{a}, \boldsymbol{a}')$ are q-th roots of one. In general one can show that there is a group F which is an extension of the cyclic group C_q by U such that the projective representations of U with factor system ω can be obtained from ordinary representations of the group F, in the way discussed in Ch. 1, §3.5. The group F can be obtained by adding phase factors to the operators $T_{\boldsymbol{a}}$ such that the set of operators with phases form a group. Here we put phase factors $\exp(2\pi im/q)$ with $m = 0, 1, ..., q-1$ in front of each operator $T_{\boldsymbol{a}}$. The operator $\exp(2\pi mi/q)T_{\boldsymbol{a}}$ is denoted here by $(m, \boldsymbol{a})$. Then the couples $(m, \boldsymbol{a})$ with $m = 0, ..., q-1$ and $a \in U$ form a group F with product rule

$$(m, \boldsymbol{a})(m', \boldsymbol{a}') = (m + m' - n_1 n_2' p, \boldsymbol{a} + \boldsymbol{a}') , \tag{6.72}$$

where $\boldsymbol{a} = \Sigma_i n_i \boldsymbol{a}_i$ and $\boldsymbol{a}' = \Sigma_i n_i' \boldsymbol{a}_i$. The elements $(m, \boldsymbol{O})$ form a subgroup isomorphic to C_q, and they commute with all elements of F. Notice that

$$\begin{aligned} &(m, \boldsymbol{a})^{-1} = (-m - n_1 n_2 p, -\boldsymbol{a}) \\ &(m, \boldsymbol{a})(m', \boldsymbol{a}')(m, \boldsymbol{a})^{-1} = (m' - n_1 n_2' p, \boldsymbol{a}') . \end{aligned} \tag{6.73}$$

For any irreducible representation $D(F)$ one obtains a projective representation $P(U)$ by $P(a) = D((O, \boldsymbol{a}))$ with factor system determined by $\omega(\boldsymbol{a}, \boldsymbol{a}') = D((-n_1 n_2' p, \boldsymbol{O}))$. So we have to look for the irreducible representations of F with $D((m, \boldsymbol{O})) = \exp(2\pi mi/q)\mathbb{1}$.

Representations of F can be constructed by the method of induction. First we look for an invariant, Abelian subgroup of F. From eqs. (6.72) and (6.73) it follows that such a group F^s is formed by the elements $(m, \boldsymbol{A})$ with $m = 0, ..., q-1$ and $\boldsymbol{A} = n_1 q \boldsymbol{a}_1 + n_2 \boldsymbol{a}_2 + n_3 \boldsymbol{a}_3$. This means that F^s is a subgroup of index q in F. The elements A form a subgroup $U^s \subseteq U$, also of index q. The group F^s is an Abelian, invariant subgroup since $(m, \boldsymbol{A})(m', \boldsymbol{A}') = (m + m' - n_1 q n_2' p, \boldsymbol{A} + \boldsymbol{A}') = (m + m', \boldsymbol{A} + \boldsymbol{A}') = (m', \boldsymbol{A}')(m, \boldsymbol{A})$ and for any $(m, \boldsymbol{a}) \in F$ and any $(m', \boldsymbol{A}) \in F^s$

$$(m, \boldsymbol{a})(m', \boldsymbol{A}')(m, \boldsymbol{a})^{-1} = (m' - n_1 n_2' p, \boldsymbol{A}') \in F^s . \tag{6.74}$$

As F^s contains $(m, \boldsymbol{O})$ and $(0, \boldsymbol{A})$ for any m and any $A \in U^s$,

$$F^s = C_q \times C_{N/q} \times C_N \times C_N \,.$$

The irreducible representations of F^s are given by

$$D_{\boldsymbol{q},r}(m, \boldsymbol{A}) = \mathrm{e}^{2\pi m i r/q}\, \mathrm{e}^{i\boldsymbol{q}\cdot\boldsymbol{A}} \,, \tag{6.75}$$

where $r = 0, 1, ..., q-1$ and $\boldsymbol{q}$ a vector inside the Brillouin zone of the reciprocal lattice $(\Lambda^s)^*$ for U^s. The vectors $\boldsymbol{b}_1, \boldsymbol{b}_2, \boldsymbol{b}_3$ span the reciprocal lattice Λ^* belonging to U and $\boldsymbol{b}_1/q, \boldsymbol{b}_2, \boldsymbol{b}_3$ the lattice $(\Lambda^s)^*$. The little group of the representation $D_{\boldsymbol{q},r}$ is obtained from eq. (6.74):

$$D_{\boldsymbol{q},r}\,[(m, \boldsymbol{a})(m', \boldsymbol{A}')(m, \boldsymbol{a})^{-1}] = \mathrm{e}^{2\pi m' i r/q - 2\pi i r n_1 n_2' p/q}\, \mathrm{e}^{i\boldsymbol{q}\cdot\boldsymbol{A}'} \,.$$

Because the element $(m, \boldsymbol{O})$ must be represented by $\exp(2\pi m i/q)\,\mathbb{1}$ one has to take $r = 1$. The expression may be simplified using $2\pi n_2' = \boldsymbol{b}_2 \cdot \boldsymbol{A}'$. Then

$$\begin{aligned} D_{\boldsymbol{q},1}[(m, \boldsymbol{a})(m', \boldsymbol{A}')(m, \boldsymbol{a})^{-1}] &= \mathrm{e}^{2\pi m' i/q}\, \mathrm{e}^{i(\boldsymbol{q} - n_1 p \boldsymbol{b}_2/q)\cdot\boldsymbol{A}'} \\ &= D_{\boldsymbol{q} - n_1 p \boldsymbol{b}_2/q, 1}(m', \boldsymbol{A}') \,. \end{aligned}$$

Therefore, $(m, \boldsymbol{a})$ is an element of the little group if $-n_1 p \boldsymbol{b}_2/q \in (\Lambda^s)^*$. As p and q are relatively prime, this is only the case when n_1 is a multiple of q, i.e. for $(m, \boldsymbol{a}) \in F^s$. This means that F^s is the little group of $D_{\boldsymbol{q},1}(F^s)$. The orbits of the representations of F^s are characterized by q points in a unit cell of $(\Lambda^s)^*$. A fundamental region is the parallelepiped spanned by $\boldsymbol{b}_1/q, \boldsymbol{b}_2/q, \boldsymbol{b}_3$. This is called a *magnetic Brillouin zone* or *magnetic cell* (fig. 6.6). Each point in this region characterizes an irreducible representation of F.

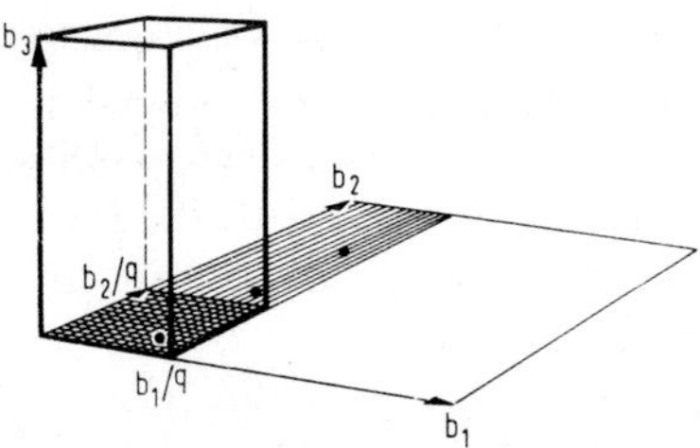

Fig. 6.6. The magnetic Brillouin zone.

The decomposition of F in cosets of F^s is given by

$$F = F^s + (0, \boldsymbol{a}_1)F^s + (0, 2\boldsymbol{a}_1)F^s + \ldots + (0, (q-1)\boldsymbol{a}_1)F^s . \tag{6.76}$$

Then from $(m, \boldsymbol{a})(0, l\boldsymbol{a}_1) \in (0, j\boldsymbol{a}_1)F^s$ it follows that $j = n_1 + l$ and $(m, \boldsymbol{a})(0, l\boldsymbol{a}_1) = (0, (n_1 + l)\boldsymbol{a}_1)(m - n_1 lp, \boldsymbol{a} - n_1\boldsymbol{a}_1)$. Thus the induced representation $D_{\boldsymbol{q},1} \uparrow F$ is given by

$$[D_{\boldsymbol{q}}(m, \boldsymbol{a})]_{lj} = D_{\boldsymbol{q},1}(m + jn_2 p, \boldsymbol{a} + (l-j)\boldsymbol{a}_1)\,\delta_{j,n_1+l} \,.$$

Notice that j is determined only up to a multiple of q. One can put $D_{\boldsymbol{q}}(m, \boldsymbol{a})$ into the form

$$[D_{\boldsymbol{k}}(m, \boldsymbol{a})]_{lj} = \mathrm{e}^{2\pi m i/q}\, \mathrm{e}^{i(\boldsymbol{k} + jp\boldsymbol{b}_2/q)\cdot(\boldsymbol{a} + l\boldsymbol{a}_1 - j\boldsymbol{a}_1)}\, \delta_{j,n_1+l} \,. \tag{6.77}$$

The induced representation $D_{\boldsymbol{k}}(F)$ is a q-dimensional irreducible representation. The nonequivalent representations are characterized by vectors $\boldsymbol{k}$ in the magnetic Brillouin zone.

It is a straightforward calculation to show that the representation $D_{\boldsymbol{k}}(F)$ gives a projective representation of U with factor system (6.71). As there are N^3/q^2 vectors in the magnetic Brillouin zone and each representation is q-dimensional, the sum of the squares of the dimensions of the nonequivalent irreducible representations is N^3, the order of U. This should be so according to exercise 1.9. This proves once more that we have obtained all irreducible projective representations with factor system (6.71). These representations are called *physical representations*. This name is also given to the representations of F from which the projective representations are derived. The other representations of F give representations with nonassociated factor systems. Since the equivalence class of the factor system ω is determined by the electromagnetic field, these other representations are called nonphysical.

Knowing the invariance group of a Bloch electron in a homogeneous magnetic field and the irreducible physical representations of this group, one can proceed to apply the usual group-theoretical machinery, to construct basis functions and to derive selection rules. We will not go further here, but refer to the reviews of the problem of Bloch electrons in a homogeneous magnetic field given by Brown [1968] and Fischbeck [1970]. The group theoretical problem goes back to Harper [1955], Fischbeck [1963], Zak [1964] and Brown [1964]. In these papers the non-Abelian group F called the magnetic translation group was introduced, although in various formulations. The physical irreducible representations of the group were discussed there for $\boldsymbol{A}$

in the symmetric gauge. As we have seen in §3.2 other gauges are easily dealt with. The problem for arbitrary gauge and arbitrary (i.e. also nonrational) fields was discussed in Opechowski and Tam [1969].

Here we have considered the Hamiltonian (6.57), it is clear that the invariance group of H (6.58) is the same, because $s(\boldsymbol{t}, \boldsymbol{t}') = 0$ in eq. (6.65). The spin does not affect the translational symmetry. The situation is different when one considers symmetry transformations of the field which are not translations. When one considers the group G which is the intersection of the space group of the crystal and the symmetry group of the magnetic field, the operators T_g commuting with H are $T_g = U_{\chi_g} P_g$ $(g \in G)$ for H (6.57). G is the subgroup of all elements $\{R\,|\,\boldsymbol{t}\}$ in the space group such that R leaves H invariant. Thus G is a space group itself. The operators T_g form a projective representation of G with factor system ω. One has to determine the irreducible representations of G with this factor system as was done for the translation group U. The group G and its corresponding operator group were treated by Tam [1969] and Overhof and Roessler [1968].

6.4. Magnetic groups

6.4.1. Magnetic symmetry

In the chapters 3 and 4 we discussed transformations of the three-dimensional space in the rôle of symmetry transformations of physical systems. In chapter 5 we also took time reversal into consideration. The group of space-time transformations considered is the *Shubnikov group* $S(4) = E(3) \times C_2$. Any element $(g, \epsilon) \in S(4)$ with $g = \{R\,|\,\boldsymbol{v}\} \in E(3)$ and $\epsilon = \pm 1$ acts on a point of the four-dimensional space-time $(\boldsymbol{r}, t)$ according to

$$g(\boldsymbol{r}, t) = (R\boldsymbol{r} + \boldsymbol{v}, \epsilon t) \,.$$

In particular, $(\{\mathbb{1}\,|\,0\}, -1)$ is the time reversal T. In §3.1 we discussed the transformation properties of electromagnetic fields under elements of $S(4)$. According to eq. (6.49) a time-independent electric field $\boldsymbol{E}(\boldsymbol{r})$ is left invariant by its spatial symmetry group (i.e. the subgroup of $E(3)$ leaving $\boldsymbol{E}(\boldsymbol{r})$ invariant) and by time reversal. When $\boldsymbol{E}(\boldsymbol{r})$ is the electric field inside a crystal, its symmetry group is the direct product of the space group G of the crystal and the group generated by T. A magnetic field $\boldsymbol{H}(\boldsymbol{r})$, however, is not invariant under time reversal. Nevertheless, it is often possible to combine Euclidean motions which do not leave $\boldsymbol{H}(\boldsymbol{r})$ invariant with T, such that the combinations are symmetry transformations for $\boldsymbol{H}(\boldsymbol{r})$. In general we define the *magnetic symmetry group* of an electromagnetic field as the subgroup of $S(4)$ which leaves the field invariant. Analogously the magnetic symmetry group of an electromagnetic potential is defined.

A subgroup of $E(3)$ can also be considered as a subgroup of $S(4)$. Mathematically speaking there is a natural monomorphism which maps the subgroup of $E(3)$ onto a subgroup of $S(4)$ of elements with $\epsilon = +1$. Such a group is called a *nonmagnetic group*. Examples are the space groups and point groups considered in Chs. 3 and 4. A subgroup of $S(4)$ which contains T is the direct product of a nonmagnetic group and the group C_2 generated by T. Such a group is called a *trivial magnetic group*. A subgroup of $S(4)$ which contains elements with $\epsilon = -1$, but which does not contain T itself, is called a *nontrivial magnetic group*. Nontrivial magnetic groups occur as symmetry groups of magnetic fields. As an example, one can consider a crystal in which the constituting particles carry a magnetic moment. Since the magnetic moment is reversed by T, this is not a symmetry element for such a crystal. An element $g = \{R \mid \boldsymbol{t}\}$ of the ordinary space group of the crystal carries each particle position into the position of a particle of the same kind. However, to the particles are attached magnetic moment vectors transforming according to $\boldsymbol{M}'(\boldsymbol{r}) = R\boldsymbol{M}(g^{-1}\boldsymbol{r})$. When the orientation of the magnetic moment vectors is arbitrary $\boldsymbol{M}'(\boldsymbol{r}) \neq \boldsymbol{M}(\boldsymbol{r})$. When $\boldsymbol{M}'(\boldsymbol{r}) = \boldsymbol{M}(\boldsymbol{r})$ the element g is a symmetry element of the crystal with moments. Now, it can also happen that $\boldsymbol{M}'(\boldsymbol{r}) = -\boldsymbol{M}(\boldsymbol{r})$ for any particle position $\boldsymbol{r}$. Since T reverses the direction of $\boldsymbol{M}$, the combination gT leaves the vector field invariant in that case. This situation has much in common with the problem of symmetry of a pattern consisting of black and white particles. If a space group element carries each white particle into the position of a black one and vice versa, its combination with an operator which changes the colour of the particles, is a symmetry element of the coloured pattern. Actually magnetic groups were studied for the first time by Heesch [1929] and Hermann [1928] as such groups. This is the reason why nontrivial magnetic groups are also called *black-and-white groups*. The trivial magnetic groups are called *gray groups*. Both kinds are also denoted by the name *Shubnikov groups*.

The elements of a magnetic group with $\epsilon = -1$ are denoted by a prime: $g' = gT$, the product of an Euclidean motion g and time reversal T. As T commutes with all elements of $E(3)$, the product of two primed elements is an unprimed element. This means that a magnetic group has a subgroup H of index two formed by its unprimed elements. This yields a method of constructing the magnetic groups. Take a subgroup G of $E(3)$. A trivial magnetic group $G \times C_2$ is generated by G and T. If H is a subgroup of index two in G, a nontrivial magnetic group is $G' = H + (gT)H$, when $G = H + gH$. All magnetic groups which give the same nonmagnetic group G form a set called the family of G.

When G' is a magnetic symmetry group which transforms a crystal into itself, we obtain a space group G by omitting all primes on the elements, because G leaves invariant the crystal obtained from the first one by neglecting the magnetic moments. The group G' is called a *magnetic space group*. It is a nontrivial magnetic group, if it describes the symmetry of a crystal with magnetic moments. It has a subgroup of index two which is a nonmagnetic space group. The magnetic space group G' belongs to the family of the nonmagnetic space group G. In the same way, a nontrivial magnetic space group which belongs to the family of a (crystallographic) point group is called a *(crystallographic) magnetic point group*. The properties of magnetic space and point groups are treated in Belov et al. [1957] and in Opechowski and Guccione [1965]. Some properties are discussed in the next section.

6.4.2. *Crystallographic magnetic groups*

a. *Magnetic point groups*. These groups can be constructed from the nonmagnetic point groups by the general method explained in the foregoing section. Take a point group K and consider all its subgroups of index two. If H is such a subgroup, and gH its coset, one obtains a nontrivial magnetic point group by giving all elements of gH a prime, i.e. by multiplying all these elements with T. The elements of such a group are of the form

$$\begin{pmatrix} R & 0 \\ 0 & \epsilon \end{pmatrix} = \begin{cases} R & \text{if} \quad \epsilon = +1 \\ R' & \text{if} \quad \epsilon = -1 \end{cases} \qquad (R \in K) . \tag{6.78}$$

The orthogonal transformations R form the point group K.

Two magnetic point groups are *geometrically equivalent* if they are conjugate subgroups of $S(4)$, i.e. if the nonmagnetic point groups to which they correspond are geometrically equivalent in such a way that corresponding elements of the two groups are either both primed or both unprimed. The equivalence classes of the crystallographic magnetic point groups are called the *magnetic crystal classes*. There are 58 nontrivial magnetic crystal classes, 32 trivial magnetic and 32 nonmagnetic crystal classes. They are given in table 6.2 in the international notation. Elements in the group with $\epsilon = -1$ are denoted by a prime. If T is an element of the group, it is denoted by $1'$. One can visualize magnetic point groups in the same way as is done for nonmag-

Table 6.2
The 122 nonmagnetic and magnetic geometric crystal classes.

Nonmagnetic	Trivial magnetic	Nontrivial magnetic (in parentheses subgroup of elements with $\epsilon = +1$)
1	$1'$	
$\bar{1}$	$\bar{1}1'$	$\bar{1}'(1)$
2	$21'$	$2'(1)$
m	$m1'$	$m'(1)$
$2/m$	$2/m1'$	$2'/m(m)$, $2/m'(2)$, $2'/m'(\bar{1})$
222	$2221'$	$22'2'(2)$
$2mm$	$2mm1'$	$2'mm'(m)$, $2m'm'(2)$
mmm	$mmm1'$	$m'mm(2mm)$, $m'm'm(2/m)$, $m'm'm'(222)$
4	$41'$	$4'(2)$
422	$4221'$	$4'22'(222)$, $42'2'(4)$
$\bar{4}$	$\bar{4}1'$	$\bar{4}'(2)$
$4/m$	$4/m1'$	$4'/m(2/m)$, $4/m'(4)$, $4'/m'(\bar{4})$
$4mm$	$4mm1'$	$4'mm'(2mm)$, $4m'm'(4)$
$\bar{4}2m$	$\bar{4}2m1'$	$\bar{4}'2'm(2mm)$, $\bar{4}'2m'(222)$, $\bar{4}2'm'(\bar{4})$
$4/mmm$	$4/mmm1'$	$4'/mmm'(mmm)$, $4'/m'm'm(\bar{4}2m)$, $4/m'mm(4mm)$, $4/mm'm'(4/m)$, $4/m'm'm'(422)$
3	$31'$	
32	$321'$	$32'(3)$
$\bar{3}$	$\bar{3}1'$	$\bar{3}'(3)$
$3m$	$3m1'$	$3m'(3)$
$\bar{3}m$	$\bar{3}m1'$	$\bar{3}'m(3m)$, $\bar{3}m'(\bar{3})$, $\bar{3}'m'(32)$

Table 6.2 (continued)

Nonmagnetic	Trivial magnetic	Nontrivial magnetic
6	$61'$	$6'(3)$
622	$6221'$	$6'22'(32)$, $62'2'(6)$
$\bar{6}$	$\bar{6}1'$	$\bar{6}'(3)$
$6/m$	$6/m1'$	$6'/m(\bar{6})$, $6/m'(6)$, $6'/m'(\bar{3})$
$6mm$	$6mm1'$	$6'mm'(3m)$, $6m'm'(6)$
$\bar{6}m2$	$\bar{6}m21'$	$\bar{6}'m2'(3m)$, $\bar{6}'m'2(32)$, $\bar{6}m'2'(\bar{6})$
$6/mmm$	$6/mmm1'$	$6'/mmm'(\bar{6}m2)$, $6'/m'm'(\bar{3}m)$, $6/m'mm(6mm)$, $6/mm'm'(6/m)$, $6/m'm'm'(622)$
23	$231'$	
432	$4321'$	$4'32'(23)$
$m3$	$m31'$	$m'3(23)$
$\bar{4}3m$	$\bar{4}3m1'$	$\bar{4}'3m'(23)$
$m3m$	$m3m1'$	$m'3m(\bar{4}3m)$, $m3m'(m3)$, $m'3m'(432)$

netic point groups, when one makes a distinction between points obtained from the first point by a primed transformation and those obtained by an unprimed one. Diagrams can be found in Koptzik [1966].

b. *Magnetic lattices*. These are obtained from the nonmagnetic lattices by considering sublattices of index two and by application of the general construction. On giving primes to elements which do not belong to this sublattice one obtains a magnetic lattice. For each lattice there are 7 sublattices of index two. The corresponding 7 magnetic lattices are generated by $\boldsymbol{a}_1'\boldsymbol{a}_2\boldsymbol{a}_3$, $\boldsymbol{a}_1\boldsymbol{a}_2'\boldsymbol{a}_3$, $\boldsymbol{a}_1\boldsymbol{a}_2\boldsymbol{a}_3'$, $\boldsymbol{a}_1'\boldsymbol{a}_2'\boldsymbol{a}_3$, $\boldsymbol{a}_1'\boldsymbol{a}_2\boldsymbol{a}_3'$, $\boldsymbol{a}_1\boldsymbol{a}_2'\boldsymbol{a}_3'$, $\boldsymbol{a}_1'\boldsymbol{a}_2'\boldsymbol{a}_3'$, when $\boldsymbol{a}_1\boldsymbol{a}_2\boldsymbol{a}_3$ generates the nonmagnetic lattice. The *holohedry of a magnetic lattice* is the subgroup of $O(3)$ which transforms the lattice into itself taking account of the primes. The holohedry is a nonmagnetic point group. As an example, consider a cubic lattice generated by three mutually orthogonal vectors of equal length. The holohedries of the 7 magnetic lattices belonging to its family are

1) generators $\boldsymbol{a}_1'\boldsymbol{a}_2\boldsymbol{a}_3$, $\boldsymbol{a}_1\boldsymbol{a}_2'\boldsymbol{a}_3$ or $\boldsymbol{a}_1\boldsymbol{a}_2\boldsymbol{a}_3'$: holohedry $4/mmm$;
2) generators $\boldsymbol{a}_1'\boldsymbol{a}_2'\boldsymbol{a}_3$, $\boldsymbol{a}_1'\boldsymbol{a}_2\boldsymbol{a}_3'$ or $\boldsymbol{a}_1\boldsymbol{a}_2'\boldsymbol{a}_3'$: holohedry $4/mmm$;
3) generators $\boldsymbol{a}_1'\boldsymbol{a}_2'\boldsymbol{a}_3'$: holohedry $m3m$.

We will not give a discussion of the notions of system and Bravais class. These concepts are treated in Opechowski and Guccione [1965] and Janner [1966].

c. *Magnetic space groups.* A complete list of magnetic space groups was given by Belov et al. [1955]. In principle, the procedure for the derivation of the magnetic space groups from the ordinary space groups by considering subgroups of index two is the same as for all magnetic groups. However, a precise derivation is rather involved. A subgroup of index two has either the translation subgroup or the point group of the space group. Two nontrivial magnetic space groups are called *equivalent* if there is an isomorphism between them that maps primed elements onto primed elements and unprimed ones onto unprimed ones. Just as for ordinary space groups, this equivalence relation is similar to the one which puts into correspondence space groups which are conjugated subgroups of $A(3)$ such that by the conjugation primed elements are mapped onto primed elements. A finer equivalence relation is: two nontrivial magnetic space groups are *equivalent* if they are conjugated subgroups of $A(3) \times C_2$ such that the conjugating element is (g, ϵ) with $g = \{S \mid \boldsymbol{t}\} \in A(3)$ and $\det S > 0$. With this definition there are 1191 nontrivial magnetic space groups and 230 trivial magnetic space groups. A discussion of their derivation can be found in Opechowski and Guccione [1965] and Janner [1966]. The properties of the 1421 magnetic space groups are tabulated in Koptzik [1966].

6.4.3. *Corepresentations of magnetic groups*

We now consider once more the problem of an electron in an electromagnetic field, in particular in a crystal with charges and magnetic moments. The Hamiltonian of this system is given by eq. (6.57) or (6.58). The electromagnetic field has the symmetry of a space group and we assume that there is a corresponding potential with this same symmetry. As symmetry transformations we consider elements of the Shubnikov group $S(4)$. The symmetry group is now a nontrivial magnetic space group G. For an unprimed element $g \in G$ the substitution operator P_g commutes with the Hamiltonian (6.57), for a primed element $g' \in G$ the operator $\theta_0 P_g$ commutes with H. Taking the spin into account, for an unprimed element $g \in G$ with $g = \{R \mid \boldsymbol{t}\}$ the operators $\pm u(R) P_g$ commute with the Hamiltonian (6.58) and for a primed element $g' = \{R \mid \boldsymbol{t}\}' \in G$ the operators $\pm \theta u(R) P_g$ commute with H. In this context, the cases considered in Ch. 5, §2 have as symmetry group a trivial magnetic space group containing the time reversal T.

The structure of the invariance operator groups for the various cases is seen as follows.

1) For H (6.57) with a trivial magnetic symmetry group $G \times C_2$, the operators commuting with H form a group $P_G + \theta_o P_G$.
2) For H (6.57) with a nontrivial magnetic symmetry group G, the unprimed elements form a subgroup $H \subset G$ of index two. When a' is an arbitrary primed element of G one can write

$$G = H + a'H \quad \text{and} \quad P_G = P_H + (\theta_o P_a) P_H \,.$$

In this case the element $(\theta_o P_a)$ does not in general commute with P_h for all $h \in H$.
3) For H (6.58) and trivial magnetic symmetry group $G \times C_2$ the elements $\pm u(R) P_g$ form a (unitary) representation of the double group G^{d}. Therefore, in this case the group of invariance operators is $T_{G^{\mathrm{d}}} + \theta T_{G^{\mathrm{d}}}$. As θ is of order four, this is not a direct product, although θ commutes with all elements T_α with $\alpha \in G^{\mathrm{d}}$: the intersection of $T_{G^{\mathrm{d}}}$ and the group generated by θ is $\pm \mathbb{1}$. The invariance operator group is an extension of C_2 by G^{d}.
4) For H (6.58) and a nontrivial magnetic symmetry group G the operators $\pm u(R) P_g$ and $\pm \theta u(S) P_h$ with $g = \{R \mid \boldsymbol{t}\} \in G$ and $h' = \{S \mid \boldsymbol{v}\}' \in G$ commute with H. When H is the subgroup of unprimed elements of G and $G = H + a'H$ the invariance operator group is

$$T_{H^{\mathrm{d}}} + (\theta T_a) T_{H^{\mathrm{d}}} \,.$$

In all four cases the invariance operator group contains a subgroup of index two consisting of unitary operators. These unitary operators form a representation of the group of unprimed elements in the spinless case, and of its double group in the case with spin. Moreover, the elements of the invariance operator group which are not in this unitary subgroup are antiunitary. Therefore, the invariance operator group can be written as

$$T_G = T_H + T_{a'} T_H \,, \tag{6.79}$$

where T_H is the subgroup of unitary operators and $T_{a'}$ is an arbitrary antiunitary operator of the group. When the symmetry group is a trivial magnetic group, one can choose $a' = T$ and $T_{a'} = \theta_o$ in the spinless case and $T_{a'} = \sigma_2 \theta_o$ in the case with spin. Notice that G in eq. (6.79) denotes the symmetry group of the field in the spinless case and its double group in the case with spin.

As we discussed in Ch. 5, §2 the operators T_G do not form a representation of G, because not all operators are linear. Suppose that $\mathcal{H}_r$ is a subspace of the Hilbert space $\mathcal{H}$ which is invariant under T_G, but which does not contain a proper invariant subspace. Suppose a basis of $\mathcal{H}_r$ is given by $\psi_1, ..., \psi_m$. Then

$$T_h \psi_i = \sum_{j=1}^{m} D(h)_{ji} \psi_j \qquad (h \in H)$$
$$T_{a'h} \psi_i = \sum_{j=1}^{m} D(a'h)_{ji} \psi_j \qquad (h \in H)\,. \tag{6.80}$$

The matrices $D(G)$ satisfy the relations

$$\begin{aligned} D(h_1)D(h_2) &= D(h_1 h_2) \\ D(h_1)D(a'h_2) &= D(h_1 a'h_2) \\ D(a'h_1)D(h_2)^* &= D(a'h_1 h_2) \\ D(a'h_1)D(a'h_2)^* &= D(a'h_1 a'h_2)\,, \end{aligned} \tag{6.81}$$

which means that the matrices $D(G)$ form a *corepresentation of G*. Corepresentations were introduced by Wigner (see Wigner [1959]). A treatment of the corepresentations of magnetic groups is given in Dimmock and Wheeler [1964], Jansen and Boon [1967], and Bradley and Davies [1968]. A brief treatment will be given below.

The space $\mathcal{H}_r$ carries a representation of H via the unitary operators T_H. Now this representation is either irreducible or reducible. Suppose $\mathcal{H}_r$ is an irreducible representation space for H. As the elements $\psi_1, ..., \psi_m$ form a basis, the same is true for $T_{a'}\psi_1, ..., T_{a'}\psi_m$. With respect to the basis $\phi_i = T_{a'}\psi_i$ $(i = 1, ..., m)$ the matrices of the representation are given by

$$T_h \phi_i = T_h T_{a'} \psi_i = T_{a'} T_{a^{-1}ha} \psi_i = \sum_{j=1}^{m} D(a^{-1}ha)^*_{ji} \phi_j\,. \tag{6.82}$$

Hence the space $\mathcal{H}_r$ carries both $D(H)$ and $D(a^{-1}Ha)^*$. Therefore, these representations are equivalent: there is a matrix S such that $D(a^{-1}ha)^* = S^{-1}D(h)S$ for any $h \in H$. Of course the nonsingular matrix S is just the matrix $D(a')$ because $\phi_i = \Sigma_{j=1}^m D(a')_{ji}\psi_j$. This means that in this case $SS^* = D(a')D(a')^* = D(a^2)$ according to eq. (6.81). Therefore, this situation can only occur when $D(a^{-1}Ha)^* = S^{-1}D(H)S$ with $SS^* = D(a^2)$.

The other possibility is that $\mathcal{H}_r$ carries a reducible representation of H. Let $\phi_1, ..., \phi_d$ be a basis of an irreducible component. Then the space spanned by

$\phi_1, ..., \phi_d, T_{a'}\phi_1, ..., T_{a'}\phi_d$ is invariant under G. Consequently it is the space $\mathcal{H}_r$. We have to check if this space does not contain an invariant proper subspace for G. The matrices corresponding to the elements of T_G with respect to $\phi_1, ..., T_{a'}\phi_d$ are

$$D(h) = \begin{pmatrix} \Delta(h) & 0 \\ 0 & \Delta(a^{-1}ha)^* \end{pmatrix}$$

$$D(a'h) = \begin{pmatrix} 0 & \Delta(a'ha') \\ \Delta(h)^* & 0 \end{pmatrix} \qquad (h \in H) \qquad (6.83)$$

where $\Delta(H)$ is the irreducible matrix representation of H carried by $\phi_1, ..., \phi_d$. We stress the fact that, although we can give the antilinear operators $T_{a'h}$ by matrices, one has to be careful in applying the calculation rules for matrices. The operation of $T_{a'h}$ on an arbitrary element ϕ of $\mathcal{H}_r$ is given by

$$T_{a'h}\phi = T_{a'h} \sum_{i=1}^{m} \alpha_i \phi_i = \sum_{i,j} \alpha_i^* D(a'h)_{ji} \phi_j \; .$$

Furthermore, with respect to another basis $\overline{\phi}_i = \Sigma_{j=1}^m S_{ji}\phi_i$ the matrices are given by (cf. eq. (1.4))

$$\overline{D}(h) = S^{-1}D(h)S \quad \text{and} \quad \overline{D}(a'h) = S^{-1}D(a'h)S^* \qquad (\text{any } h \in H) \, . \quad (6.84)$$

When $\Delta(H)$ is not equivalent to $\Delta(a^{-1}Ha)^*$ it follows that the only matrix S which keeps $\overline{D}(H)$ in reduced form is the direct sum $\lambda \mathbb{1}_d + \mu \mathbb{1}_d$. But such a matrix S can not bring $D(a'H)$ into reduced form. Hence $\mathcal{H}_r$ is irreducible if $\Delta(H) \not\sim \Delta(a^{-1}Ha)^*$. On the other hand, it is also possible that $D(H)$ reduces into two equivalent components. Suppose that the basis is chosen in such a way that both $\Delta(H)$ and $\Delta(a^{-1}Ha)^*$ are unitary matrix representations. Then there is a unitary matrix U such that $\Delta(a^{-1}ha)^* = U^{-1}\Delta(h)U$ (for any $h \in H$). It follows that $\Delta(a^{-1}ha) = U\Delta(a^{-2}ha^2)^*U^{-1} = U\Delta(a^{-2})^*\Delta(h)^*[U\Delta(a^{-2})^*]^{-1}$. Then $\Delta(h) = U\Delta(a^{-1}ha)^*U^{-1} = UU^*\Delta(a^{-2})\Delta(h)\Delta(a^{-2})^{-1}U^{*-1}U^{-1}$. From Schur's lemma it follows that $UU^*\Delta(a^{-2}) = \lambda \mathbb{1}$ or $UU^* = \lambda\Delta(a^2)$. Moreover, one has $\Delta(a^2)^* = U^{-1}\Delta(a^2)U$. Then it is easily proved that $\lambda = \pm 1$:

$$UU^* = \pm\Delta(a^2) \, . \qquad (6.85)$$

We now try to reduce the set (6.83). First we can transform these matrices using eq. (6.84) with

$$S = \begin{pmatrix} \mathbb{1}_{\hat{d}} & 0 \\ 0 & U^{-1} \end{pmatrix}$$

into the form

$$\bar{D}(h) = \begin{pmatrix} \Delta(h) & 0 \\ 0 & \Delta(h) \end{pmatrix} \qquad (h \in H) \qquad (6.86)$$

$$\bar{D}(a'h) = \begin{pmatrix} 0 & \Delta(a'ha')U^{*-1} \\ U\Delta(h)^* & 0 \end{pmatrix}.$$

As discussed in Ch. 2, §2.5, a matrix S which leaves $\bar{D}(h)$ in the same form is necessarily

$$S = \begin{pmatrix} \lambda \mathbb{1} & \mu \mathbb{1} \\ \rho \mathbb{1} & \sigma \mathbb{1} \end{pmatrix}.$$

One has to choose S in such a way that $S^{-1}\bar{D}(a'H)S^*$ is in diagonal block form in order to have a reducible set of matrices and a reducible space $\mathcal{H}_r$. It requires some straightforward algebra to prove that this is impossible if $UU^* = -\Delta(a^2)$. However, if $UU^* = \Delta(a^2)$ one can choose λ, μ, ρ, and σ in such a way that

$$S^{-1}\bar{D}(a'h)S = \begin{pmatrix} \Delta(aha^{-1})U & 0 \\ 0 & -\Delta(aha^{-1})U \end{pmatrix}.$$

As we have supposed that $\mathcal{H}_r$ is irreducible under G, this can not happen. Therefore, we have proved

PROPOSITION 6.1. A space $\mathcal{H}_r$ which is invariant under a group T_G of unitary and antiunitary operators does not contain an invariant proper subspace if and only if one has for the group H (of index two and corresponding to the unitary operators) one of the following properties:

1) $\mathcal{H}_r$ carries an irreducible representation $\Delta(H)$,
2) $\mathcal{H}_r$ carries a reducible representation $\Delta(H) \oplus \Delta(a^{-1}Ha)^*$ with nonequivalent components,
3) $\mathcal{H}_r$ carries a reducible representation $\Delta(H) \oplus \Delta(a^{-1}Ha)^*$ with equivalent components and $\Delta(a^{-1}Ha)^* = S^{-1}\Delta(H)S$ with $SS^* = -\Delta(a^2)$.
When the case 1) occurs, one has the relation $\Delta(a^{-1}Ha)^* = S^{-1}\Delta(H)S$ with $SS^* = \Delta(a^2)$.

To distinguish between the 3 cases one has the following criterion: the character χ of $\Delta(H)$ satisfies

$$\sum_{h\in H} \chi[(a'h)^2] = \begin{cases} N = \text{order of } H & \text{for case 1)} \\ 0 & \text{for case 2)} \\ -N & \text{for case 3)}\,. \end{cases} \tag{6.87}$$

Proof:

$$\sum_{h\in H} \chi(a'ha'h) = \sum_{h\in H}\sum_{i=1}^{d} \Delta(a'ha'h)_{ii} = \sum_{h\in H}\sum_{i,k,l} \Delta(a^2)_{ik}\Delta(a^{-1}ha)_{kl}\Delta(h)_{li}\,.$$

This expression is zero if $\Delta(a^{-1}Ha)^* \not\sim \Delta(H)$ because of eq. (1.15). In the case that $\Delta(a^{-1}Ha)^* = S^{-1}\Delta(H)S$ one obtains

$$\begin{aligned}\sum_{h\in H} \chi(a'ha'h) &= \sum_{h\in H}\sum_{i,k,l,m,n} \Delta(a^2)_{ik}(S^{*-1})_{km}\Delta(h)^*_{mn}S^*_{nl}\Delta(h)_{li} \\ &= \frac{N}{d}\sum_{i,k,l} \Delta(a^2)_{ik}(S^{*-1})_{kl}S^*_{il} = \frac{N}{d}\sum_{i,k} \Delta(a^2)_{ik}(S^*S)_{ik} \\ &= \pm\frac{N}{d}\sum_{i,k} \Delta(a^2)_{ik}\Delta(a^{-2})_{ki} = \pm N\,.\end{aligned}$$

Here we have used that S and $\Delta(H)$ are unitary.

A particular case occurs when one can take a' to be the time reversal. This is the situation we investigated in Ch. 5, §2. In this case $a^2 = \mathbb{1}$ and proposition 6.1 together with eq. (6.87) leads to Herrings criterion.

$$\mathcal{H}_r \text{ is } \begin{cases} \text{reducible if } \Delta(H)^* = S^{-1}\Delta(H)S \text{ with } SS^* = +\mathbb{1} \\ \text{irreducible if } \begin{cases} \Delta(H)^* = S^{-1}\Delta(H)S \text{ with } SS^* = -\mathbb{1}, & \text{or} \\ \Delta(H)^* \not\sim \Delta(H)\,. \end{cases} \end{cases} \tag{6.88}$$

The eq. (6.87) simplifies in this case. For the spinless case $T_{a'} = \theta_0$ and

$$\sum_{h\in H} \chi[(a'h)^2] = \sum_{h\in H} \chi(h^2) .$$

For the case with spin one has $T_{a'} = \theta$ and $\theta^2 = -\mathbb{1}$. Then

$$\sum_{h\in H} \chi[(a'h)^2] = \sum_{h\in H} \chi(-h^2) = -\sum_{h\in H} \chi(h^2) .$$

Therefore, in the case that a' is the time reversal, i.e. that the symmetry group of the field is a trivial magnetic group,

$$\sum_{h\in H} \chi(h^2) = \begin{cases} N, \text{ when } \Delta(H)^* \sim \Delta(H), SS^* = +\mathbb{1} \text{ (spinless)}, SS^* = -\mathbb{1} \text{ (with spin)} \\ 0, \text{ when } \Delta(H)^* \not\sim \Delta(H), \\ -N, \text{ when } \Delta(H)^* \sim \Delta(H), SS^* = -\mathbb{1} \text{ (spinless)}, SS^* = +\mathbb{1} \text{ (with spin)}. \end{cases} \tag{6.89}$$

Proposition 6.1 together with eq. (6.87) give a tool to investigate the degeneracy of energy levels in an electromagnetic field with magnetic symmetry.

6.4.4. *Projective co-representations of magnetic space groups*

Gradually we have generalized the types of transformation groups and the kinds of representations we could handle. Now combining the methods of Ch. 4, §2.2, Ch. 5, §§1.4 and 2.3, and Ch. 6, §§3.2 and 4.3, we can study a charged particle with spin 0 or $\frac{1}{2}$ in an electromagnetic potential which has a magnetic space group as invariance group. The only restriction we make is that we only consider fields which have a potential with the same symmetry. E.g. we will not consider uniform fields. When G is the magnetic space group, for each element $g \in G$ there is an operator T_g commuting with the Hamiltonian, (6.57) for spinless particles, (6.58) for spin $\frac{1}{2}$ particles. The operator T_g is given by

$$T_g = \begin{cases} P_g & \text{for spin } 0, g \text{ an unprimed element of } G, \\ \theta_0 P_g & \text{for spin } 0, g \text{ a primed element of } G, \\ u(R)P_g & \text{for spin } \tfrac{1}{2}, g \text{ an unprimed element of } G, \\ \sigma_2\theta_0 u(R)P_g & \text{for spin } \tfrac{1}{2}, g \text{ a primed element of } G. \end{cases}$$

The operators T_g are unitary or antiunitary and satisfy $T_{g_1}T_{g_2} = \omega(g_1,g_2)T_{g_1g_2}$ with

$$\omega(g_1,g_2) = \begin{cases} 1 \text{ for spinless particles,} \\ \omega_s(R_1,R_2) \quad \text{for spin } \tfrac{1}{2}, g_1 \text{ or } g_2 \text{ unprimed,} \\ -\omega_s(R_1,R_2) \quad \text{for spin } \tfrac{1}{2}, g_1 \text{ and } g_2 \text{ primed,} \end{cases} \tag{6.90}$$

where R_i is the homogeneous part of g_i. Such a homomorphism up to a factor into a group of unitary and antiunitary operators is called a *projective co-representation*. Our problem is the determination of the nonequivalent irreducible projective co-representations of G with factor system (6.90). Notice that the subgroup of unprimed translations U is mapped homomorphically on a group of unitary operators. This makes it possible to use the same methods as in the preceding chapters. Suppose that $\mathcal{H}$ carries an irreducible projective co-representation of G with factor system ω. The subduced representation of U can be brought into diagonal form. Hence $\mathcal{H}$ can be decomposed into a direct sum of spaces $\mathcal{H}_{\boldsymbol{k}}$ carrying a unitary representation of U characterized by a vector $\boldsymbol{k}$ in the Brillouin zone. If ψ belongs to $\mathcal{H}_{\boldsymbol{k}}$, the vector $T_g\psi$ with $g = (\{R\,|\,\boldsymbol{t}\}, \epsilon = \pm 1) \in G$ ($\epsilon = 1$ for an unprimed element, $\epsilon = -1$ for a primed element) belongs to $\mathcal{H}_{\boldsymbol{k}'}$ with $\boldsymbol{k}' \equiv \epsilon R\boldsymbol{k}$. The space $\mathcal{H}_{\boldsymbol{k}}$ is invariant under the group of $\boldsymbol{k}$ which is defined in this case by

$$G_{\boldsymbol{k}} = \{g \in G \,|\, \epsilon R\boldsymbol{k} \equiv \boldsymbol{k}\}\,. \tag{6.91}$$

Decompose G into cosets of $G_{\boldsymbol{k}}$: $G = G_{\boldsymbol{k}} + g_2G_{\boldsymbol{k}} + \ldots + g_sG_{\boldsymbol{k}}$ and take a basis $\psi_{11}, \ldots, \psi_{1d}$ of $\mathcal{H}_{\boldsymbol{k}}$. Then $\psi_{\mu i} = T_{g_\mu}\psi_{1i}$ ($\mu = 1, \ldots, s$; $i = 1, \ldots, d$) form a basis for the sd-dimensional representation of G carried by $\mathcal{H}$. The representation of G is expressed in terms of the representation of $G_{\boldsymbol{k}}$ carried by $\mathcal{H}_{\boldsymbol{k}}$ via

$$T_g\psi_{\mu i} = T_gT_{g_\mu}\psi_{1i} = \frac{\omega(g,g_\mu)}{\omega(g_\nu,h)}\,T_{g_\nu}T_h\psi_{1i} \qquad (\text{with } gg_\mu = g_\nu h \in g_\nu G_{\boldsymbol{k}})$$

$$= \begin{cases} \dfrac{\omega(g,g_\mu)}{\omega(g_\nu,h)} \sum\limits_j D_{\boldsymbol{k}}(h)_{ji}\psi_{\nu j} \quad \text{if } g_\nu \text{ is unprimed} \\[2ex] \dfrac{\omega(g,g_\mu)}{\omega(g_\nu,h)} \sum\limits_j D_{\boldsymbol{k}}(h)^*_{ji}\psi_{\nu j} \quad \text{if } g_\nu \text{ is primed}\,. \end{cases} \tag{6.92}$$

In this way one has reduced the problem to the determination of the irreducible co-representations of $G_{\boldsymbol{k}}$ with the property that its restriction to the subgroup U is the direct sum of d representations characterized by $\boldsymbol{k}$. Define on $\mathcal{H}_{\boldsymbol{k}}$ an operator P_R for each element R of the point group $K_{\boldsymbol{k}}$ of $G_{\boldsymbol{k}}$ by

$$P_R = \exp(-i\boldsymbol{k}\cdot\boldsymbol{t}_R)T_{(\{R|\boldsymbol{t}_R\},\epsilon)}\ .$$

It is easily shown that P_R does not depend on the choice of the element $\{R|\boldsymbol{t}_R\}$. Moreover, the operators P_R form a projective co-representation of $K_{\boldsymbol{k}}$ with factor system

$$\omega'(R,R') = \omega(R,R')\exp\left[i(R^{-1}\boldsymbol{k}-\epsilon\boldsymbol{k})\cdot\boldsymbol{t}_{R'}\right]\ . \qquad (6.93)$$

Hence the sd-dimensional co-representation of G is found by induction from a co-representation of $G_{\boldsymbol{k}}$ which in turn is determined by an irreducible projective co-representation of $K_{\boldsymbol{k}}$ with factor system (6.93). It is possible to find all these co-representations of $K_{\boldsymbol{k}}$ in a way similar to that used for projective representations. We will not discuss this method here, since this would lead us to far, but refer to Janssen [1972], where the method of determination of these projective co-representations is treated.

It will be clear that the theory of irreducible representations of space groups is a special case. Then there are no primed elements in G. For a spinless particle $\omega = 1$ in (6.93) which reduces to eq. (4.26), for a spin $\frac{1}{2}$ particle one obtains eq. (5.12). Another special case is that of a trivial magnetic space group $G = G_0 \times J$, where J is generated by the time reversal transformation T. The group U is the ordinary translation subgroup of the nonmagnetic space group G_0. Hence the Brillouin zones of G and G_0 are the same. For a given $\boldsymbol{k}$ in the Brillouin zone, the group of $\boldsymbol{k}$ is $G_{\boldsymbol{k}} = G_{0\boldsymbol{k}} + H_{\boldsymbol{k}}$, where $G_{0\boldsymbol{k}}$ is the group of $\boldsymbol{k}$ for G_0, whereas $H_{\boldsymbol{k}} = \{(\{R|\boldsymbol{t}\},\epsilon=-1)\in G\,|\,R\boldsymbol{k}\equiv-\boldsymbol{k}\} = \{\{R|\boldsymbol{t}\}\in G_0\,|\,R\boldsymbol{k}\equiv-\boldsymbol{k}\}$. One can distinguish two cases: 1) $H_{\boldsymbol{k}}$ is empty, 2) there is an element $g_0 \in G_0$ with $R_0\boldsymbol{k}\equiv-\boldsymbol{k}$, in which case $H_{\boldsymbol{k}} = g_0G_{0\boldsymbol{k}}$, or for the point groups $M_{\boldsymbol{k}} = R_0K_{0\boldsymbol{k}}$. In the first case the dimension of the representation of G is twice that of $G_{0\boldsymbol{k}}$, because $G_{\boldsymbol{k}} = G_{0\boldsymbol{k}}$, but the number of points in the star of $\boldsymbol{k}$ is for G twice that for G_0. In the second case the index of $G_{\boldsymbol{k}}$ in G is equal to the index of $G_{0\boldsymbol{k}}$ in G_0. Then the relation between the representations of G and G_0 corresponding to $\boldsymbol{k}$ is determined by the representation $D_{\boldsymbol{k}}(G_{0\boldsymbol{k}})$. According to §4.3, one has for an ordinary representation (i.e. when $\omega = 1$ in eq. (6.93))

$$\sum_{S\in M_{\boldsymbol{k}}}\chi(S^2) = \begin{cases} n/s & \text{when there is no additional degeneracy} \\ -n/s & \text{when there is additional degeneracy.} \end{cases}$$

Here n is the order of the point group of $G_{0\boldsymbol{k}}$. This is in agreement with eq. (5.22) for a symmorphic group.

To conclude this section we give a recipee to find all nonequivalent irreducible projective co-representations with factor system (6.93) of an arbitrary magnetic space group G. The proof of the validity of the procedure is analogous to that for ordinary space groups and will not be given here.

1) Consider the group U of unprimed translations and construct its Brillouin zone.
2) Divide this Brillouin zone into stars. The star of a vector $\boldsymbol{k}$ is the set $\{\epsilon R\boldsymbol{k} \mid \text{all } (R,\epsilon) \text{ from the point group } K \text{ of } G\}$.
3) Determine a fundamental region in the Brillouin zone in which exactly one vector $\boldsymbol{k}$ from each star is found and take one $\boldsymbol{k}$ in this fundamental region.
4) Determine the group $G_{\boldsymbol{k}}$ (6.91) and its point group $K_{\boldsymbol{k}}$.
5) Find all nonequivalent irreducible projective co-representations of $K_{\boldsymbol{k}}$ with factor system (6.93). Unfortunately they can not yet be found in tabular form as for the unitary projective representations, but they can be found from the ordinary co-representations of a larger group. (See Appendix C.)
6) Induce from each of these projective co-representations a projective co-representation of G via eq. (6.92).

6.5. Other applications

In the preceding sections and chapters we have considered a number of physical systems and their symmetry. The considerations were concerned with the three levels one can distinguish in the application of group theory: the determination of the symmetry, the determination of the invariance operator group and finally the physical consequences. As symmetry transformations, we considered elements of the Shubnikov group $S(4)$, to be combined with permutations of particles for many-particle systems. The invariance operators form an ordinary or projective representation or co-representation of the symmetry group. The third level was treated only briefly. Once one has the invariance operator group and the irreducible representations, one can apply the general theory of Ch. 2 to obtain selection rules, symmetry adapted functions and so on.

Apart from the systems considered here, there are a lot of other problems which can be treated along the same lines. We will just mention some of them.

In Ch. 4, §3 we discussed selection rules for space groups. These can be used to derive selection rules for the interaction of infra-red light with crystals. The interaction is given in the dipole approximation by a matrix element

$\langle f | \boldsymbol{p} | i \rangle$. As the momentum operator $\boldsymbol{p}$ has well defined transformation properties for space group elements, the vanishing of the matrix elements is determined by the rules of Ch. 4, §3. For infrared absorption and Raman scattering on crystals, this was done in Birman [1963] and Chen et al. [1968].

In the theory of second order phase transitions, one considers phases with different symmetry such that the symmetry group of one phase is a subgroup of that of the other phase (see Landau and Lifshitz [1959] Ch. 12). The behaviour around the transition point is partly determined by the symmetry. A review of the application of group theory to ferroelectric phase transitions is given in Birman [1967].

The symmetry transformations we have considered were Shubnikov transformations. One can go further and determine the relativistic four-dimensional symmetry group, which is the subgroup of the inhomogeneous Lorentz group leaving the system invariant. This leads to crystallography in four dimensions. Four-dimensional crystallographic groups were studied by Hermann [1948], Hurley [1951], [1966] and Neubüser [1969], who considered subgroups of $O(4)$, and by Janner and Ascher [1969], [1970] and by Janssen et al. [1969], who studied subgroups of the inhomogeneous Lorentz group. Of course, in relativistic quantum mechanics one has the Dirac equation of the Schrödinger equation.

Apart from some aspects of the theory of lattice vibrations, the symmetries discussed were symmetries of quantum mechanical systems. For classical mechanics the use of group theory is much less developed. Apart from vibrations, group theory is used for the study of macroscopic properties of crystals. E.g. the electromagnetic properties are described by the dielectric and magnetic susceptibility tensors. There is a close relation between the form of these tensors (the number of independent parameters) and the symmetry of the crystal. A discussion of this point is given in Baghavantam [1966], and for magnetic crystals in Birss [1964].

Still other examples can be found in the list of references.

EXERCISES

1.1 Construct the multiplication tables for all groups with order smaller than or equal to six.

1.2 Consider the group of permutations of n elements S_n. Let $O(n)$ be the group of all $n \times n$ orthogonal matrices. Let E_{ij} be a $n \times n$ matrix with 1 on the ij-position (i-th row and j-th column) and everywhere else 0. Define a mapping $\varphi: S_n \to O(n)$ by

$$\varphi(\sigma) = \sum_{i=1}^{n} E_{\sigma(i),i} \qquad (\text{all } \sigma \in S_n) .$$

Show that φ is a homomorphism. What is its kernel? Is φ an epimorphism or a monomorphism?

1.3 Consider the set S of matrices of the form $\begin{pmatrix} 1 & n \\ 0 & 1 \end{pmatrix}$ with integer n. Show: a) the set is not equivalent with a unitary set, b) it is not fully reducible, and c) with the usual matrix multiplication as product rule the set is a group isomorphic to the additive group of integers.

1.4 Consider a complex linear vector space V. A linear function on V is a function f with the property $f(\alpha \boldsymbol{x} + \beta \boldsymbol{y}) = \alpha f(\boldsymbol{x}) + \beta f(\boldsymbol{y})$, with $\boldsymbol{x}, \boldsymbol{y} \in V$ and $\alpha, \beta \in \mathbb{C}$. Show that the set of all linear functions on V form a linear vector space $\widetilde{V}$, called the *dual space*. Show that V and $\widetilde{V}$ are of the same dimension. If $\boldsymbol{e}_1, ..., \boldsymbol{e}_n$ form a basis of V the functions $\widetilde{\boldsymbol{e}}_i$ defined by $\widetilde{\boldsymbol{e}}_i(\boldsymbol{e}_j) = \delta_{ij}$ form a basis of $\widetilde{V}$.

1.5 Let T be a representation of a group G in a linear vector space V. Define a mapping $\widetilde{T}$ of G into the group of nonsingular transformations of $\widetilde{V}$ by $\widetilde{T}_g f(x) = f(T_g^{-1} x)$. Show that $\widetilde{T}$ is a representation of G. Let $D(G)$ be a matrix representation of T_G with respect to the basis $e_1, ..., e_n$. What is the matrix representation of $\widetilde{T}_G$ with respect to the basis $\widetilde{e}_1, ..., \widetilde{e}_n$? The representations T_G and $\widetilde{T}_G$ are called *adjoint or contragredient representations*.

1.6 Let $D(G)$ be a matrix representation of a group G. Show that the mappings $g \to D(g)^*$ and $g \to \widetilde{D}(g^{-1})$ are also matrix representations of G. Show that $D(G)^*$ and $\widetilde{D}(G)$ are irreducible if and only if $D(G)$ is irreducible.

1.7 Consider the representation of S_3 defined in exercise 1.2. Is this representation reducible? What are the invariant subspaces of the representation space?

1.8 Let C_n be the cyclic group of order n and P a projective representation of this group. Show that its factor system ω is associated to the trivial one.

1.9 Define analogous to the regular representation a projective regular representation of a group G with a given factor system ω by $P_g \mathfrak{a} = \omega(g, a)(\mathfrak{g}\mathfrak{a})$. Show that this is a representation with factor system ω. What are its character and irreducible components?
Show that the sum of the squares of the dimensions of the nonequivalent irreducible projective representations with factor system ω is equal to the order of the group (analogue of Burnside's theorem).

2.1 Let $\mathcal{H}$ be the Hilbert space of complex valued functions on the unit sphere. For each rotation g we define a mapping P_g of $\mathcal{H}$ into itself by $P_g f(x,y,z) = f(g^{-1}(x,y,z))$. Show that $g \to P_g$ is a unitary representation of the group of three-dimensional rotations. Consider $f(x,y,z) = x$. What is the subspace of $\mathcal{H}$ generated by all $P_g f$ $(g \in O(3))$?

2.2 Let G be the group generated by A (a 120° rotation around the z-axis) and B (a 180° rotation around the x-axis). Show that this group is isomorphic to S_3. Construct the operations ρ_{ij}^{α} with $\alpha = 1, 2, 3$ and $i,j = 1, ..., d_\alpha$. Construct with these operators basis functions for the irreducible components of the representation $g \to P_g$ defined in exercise 2.1. What is the arbitrariness in the choice of these basis functions?

2.3 What is the symmetry group of the potential

$$V(x,y,z) = \tfrac{1}{2}m(\omega_1^2 x^2 + \omega_2^2 y^2 + \omega_3^2 z^2)?$$

Distinguish the various possibilities for the values of ω_1, ω_2 and ω_3.

2.4 A second rank tensor α belonging to a representation D of a group G is and object transforming as $\alpha_{ij} \to \alpha'_{ij} = \Sigma_{kl} D(g)_{ik} D(g)_{jl} \alpha_{kl}$. [In particular, a second rank tensor is an object transforming under linear transformations R of a real three-dimensional vector space as $\alpha' = R\alpha\widetilde{R}$.] Show that the tensors form a linear vector space carrying the product representation $D(G) \otimes D(G)$. Show that the symmetric tensors $(\alpha = \widetilde{\alpha})$ as well as the antisymmetric tensors $(\alpha = -\widetilde{\alpha})$ form invariant subspaces. Give basis functions for these subspaces. Show that the invariant tensors $(\alpha' = \alpha)$ form a subspace carrying the trivial representation. Give a basis for this subspace, and prove that its dimension is equal to the multiplicity of the trivial representation in the product $D(G) \otimes D(G)$.

2.5 Prove that the operator $\rho^\alpha = (d_\alpha/N)\, \Sigma_{g \in G}\, \chi^*(g)\, T_g$ is a projection operator on the space of functions transforming according to the irreducible representation $D_\alpha(G)$. Prove that

$$\langle \rho^\alpha \psi \,|\, \rho^\beta \psi' \rangle = \delta_{\alpha\beta} \langle \rho^\alpha \psi \,|\, \rho^\alpha \psi' \rangle\,.$$

2.6 Discuss the Stark effect (level splitting of an atomic level in a homogeneous electric field) for the $n = 2$ level of the hydrogen atom from a group theoretical point of view. Notice that for $\boldsymbol{E}$ along the z-axis the interaction Ez transforms as Y_1^0.

2.7 Consider the Abelian group generated by the central inversion $I = -\mathbb{1}$ and time reversal T. According to which representation of this group transform the density operator ρ, the current $\boldsymbol{j}$, the position $\boldsymbol{r}$ and the angular momentum $\boldsymbol{L}$, respectively?

2.8 The momentum $\boldsymbol{p}$ transforms according to an irreducible three-dimensional representation of the symmetry group of the cube O. Construct an invariant tensor operator with the components of $\boldsymbol{p}$.

2.9 Consider a hydrogen atom. The transitions between the various levels by the electric dipole interaction is determined by the matrix elements of $\boldsymbol{A} = e\boldsymbol{r}$ ($\boldsymbol{r}$ is the position operator). Determine the selection rules of this operator between states with given angular momentum and parity.

2.10 Consider a particle in a rectangular box with infinitely hard walls. The potential is given by

$$V(x,y,z) = \begin{cases} 0 \text{ for } |x| \leqslant \frac{1}{2}L_1, |y| \leqslant \frac{1}{2}L_2, |z| \leqslant \frac{1}{2}L_3 \,, \\ \infty \text{ outside} \end{cases}$$

What is the symmetry of this potential for a) L_1, L_2, L_3 all different, b) $L_1 = L_2 \neq L_3$, c) $L_1 = L_2 = L_3$? Consider the lowest energy levels and determine to which representation of the symmetry group they belong.

3.1 Determine all finite subgroups of $SO(2)$, the group of rotations in two dimensions. Which of them are crystallographic? Give for each of the crystallographic ones an invariant lattice. Show that all crystallographic point groups in two dimensions may be obtained by adding the reflection on an arbitrary line to the groups of rotations. Derive the 10 plane geometric crystal classes.

3.2 Show that a lattice Λ is completely determined by its Wigner–Seitz cell. Prove that Λ is invariant under an orthogonal transformation A if and only if the Wigner–Seitz cell is invariant under A.

3.3 Derive the character table for the octahedral group O. Determine the reduction of the 15 Kronecker products of the five irreducible representations. Do the same for the five antisymmetrized Kronecker products $D_\alpha \wedge D_\alpha$ using eq. (3.9).

3.4 Consider an atom in a crystal with one electron outside closed shells. Let $H = H_0 + H_1 + H_2$ be the Hamiltonian. Suppose that the symmetry groups of H_0, H_1, and H_2 are respectively $O(3)$, $m3$, and $3m$ ($3m \subset m3 \subset O(3)$). Determine the splitting of a $l = 2$ level when H_1 is weak compared to H_0, and H_2 is weak compared to H_1.

3.5 The functions x, y, z defined on the unit sphere form a basis for the representation $D^{(1)}$ of $SO(3)$. Construct a basis in this space which reduces the subduced representation of $D_4 \cong 422$.

3.6 Consider a tensor of rank two (cf. exercise 2.4). Show that the space of antisymmetric tensors carries a representation $\Gamma_2 \oplus \Gamma_3$ of $D_3 \cong 32$. What is the representation carried by the space of symmetric tensors? Use this to show that there are no antisymmetric invariant tensors for this group, and that the dimension of the space of symmetric invariant tensors is two. Prove that such a tensor can be brought into the form

$$\alpha = \begin{pmatrix} a & \frac{1}{2}a & 0 \\ \frac{1}{2}a & a & 0 \\ 0 & 0 & b \end{pmatrix} \qquad (a, b \in \mathbb{R}) .$$

3.7 Suppose that an eigenspace of H carries a representation $D(G)$ of the group G. Prove that the space of three-particle states with all 3 particles in this eigenspace carries a representation $D(G) \otimes D(G) \otimes D(G)$. Prove that the completely antisymmetric states form an invariant subspace which carries a representation with character

$$\chi(g) = (\tfrac{1}{6})\,[\chi(g)^3 - 3\chi(g^2)\chi(g) + 2\chi(g^3)] ,$$

where $\chi(G)$ is the character of $D(G)$. Prove that the completely symmetric states carry a representation with character

$$\chi(g) = (\tfrac{1}{6})\,[\chi(g)^3 + \chi(g^3)]\ .$$

3.8 Discuss the splitting of a level with 2 p-electrons under influence of their Coulomb repulsion and an intermediate crystal field of cubic symmetry.

3.9 Show that a second rank tensor which is invariant under the point group 4 is also invariant under $4/mmm$.

3.10 a) Determine a set of two generators for the point group $m3m$. Which point groups are subgroups of this group?
b) Determine a set of generators for $6/mmm$. Which point groups are subgroups of this group?
c) Show that every point group corresponds to a group of integral matrices with respect to a basis $\boldsymbol{e}_1, \boldsymbol{e}_2, \boldsymbol{e}_3$ which is either orthonormal or consists of unit vectors with $\angle(\boldsymbol{e}_1\boldsymbol{e}_2) = 60^\circ$, $\boldsymbol{e}_3 \perp \boldsymbol{e}_1, \boldsymbol{e}_2 \perp \boldsymbol{e}_3$.

4.1 Prove that a function which is invariant under a lattice group U can be expanded in a Fourier series in which only the components belonging to $\boldsymbol{k}$ vectors in the reciprocal lattice Λ^* can be different from zero.

4.2 Let $V(\boldsymbol{r})$ denote a three-dimensional potential. Suppose that the group of translations leaving this potential invariant form a lattice group. Prove that the symmetry group of $V(\boldsymbol{r})$ is a space group.

4.3 Construct a basis for an irreducible representation of the space group Pb, which has a monoclinic lattice group U, a point group m, and a nonprimitive translation $\frac{1}{2}\boldsymbol{a}_3$ associated with the generator of the point group. The representation is characterized by the vector $\boldsymbol{k}$ and the representation Γ_2 of the point group C_2.

4.4 Diffraction of X-rays by a crystal is determined by a matrix element $\langle \boldsymbol{k}' | V | \boldsymbol{k} \rangle$ of the potential $V(\boldsymbol{r})$ between plane waves with wave vectors $\boldsymbol{k}$ and $\boldsymbol{k}'$.

a) Show that from the invariance of $V(\boldsymbol{r})$ under the translation group U follow the Laue conditions $\boldsymbol{k}' \equiv \boldsymbol{k} \ (\mathrm{mod}\ \Lambda^*)$.

b) If $V(\boldsymbol{r})$ is invariant under the space group Pb (see exercise 4.3), one has

$$\langle \boldsymbol{k}' | V | \boldsymbol{k} \rangle = \sum_{\boldsymbol{K} \in \Lambda^*}{}' \, a_{\boldsymbol{K}} \, \delta(\boldsymbol{k}' - \boldsymbol{k} - \boldsymbol{K}) \, ,$$

where the prime means restriction of the summation to the sublattice of all elements $\boldsymbol{K} \in \Lambda^*$ with $\boldsymbol{K} \cdot \boldsymbol{a}_3 = 4\pi \times$ integer. Prove this.

4.5 Consider a space group G with translation subgroup U. We want a representation of G in the space of functions which obey periodic boundary conditions $\psi(\boldsymbol{r}) = \psi(\boldsymbol{r} + N\boldsymbol{a})$ for any $\boldsymbol{r}$ and all $\boldsymbol{a} \in U$. Define the subgroup U^s of all elements $N\boldsymbol{a}$ of U. Prove that the space considered carries a representation of G with $D(\boldsymbol{a}') = \mathbb{1}$ for any $\boldsymbol{a}' \in U^s$. Show that U^s is an invariant subgroup of G. Each representation of the factor group G/U^s gives a representation of G with $D(\boldsymbol{a}') = \mathbb{1}$ for $\boldsymbol{a}' \in U^s$. Prove that we get all the representations with this property in this way.

4.6 Let g be a metric tensor for a lattice with basis $\boldsymbol{a}_1, \boldsymbol{a}_2, \boldsymbol{a}_3$. Show that the metric tensor of the associated basis of the reciprocal lattice is $4\pi^2 g^{-1}$. Show that Λ_1^* and Λ_2^* belong to the same Bravais class if and only if Λ_1 and Λ_2 belong to the same Bravais class. Therefore, there is a one-to-one correspondence between the classes of the direct and the reciprocal lattices. Give this correspondence for the 14 three-dimensional Bravais classes.

4.7 Suppose that G is a symmorphic space group. Each representation $D(G)$ subduces a representation $D(K)$ of the point group K and a representation $D(U)$ of the translation subgroup U. Show that D is completely determined by $D(K)$ and $D(U)$. How are the representations $D(K)$ and $D(U)$ for a representation characterized by a vector in general position in the Brillouin zone?

5.1 Show that the expression (5.3) is a solution of eq. (5.2) for R given by its axis $\boldsymbol{n}$ and its rotation angle φ. Use the fact that any rotation can be written as the product of rotations α, β, γ (the Euler angles) around, respectively, the z-axis, the y-axis and again the z-axis.

5.2 The mapping $1 \rightarrow \mathbb{1}$, $T \rightarrow \lambda \sigma_2 \theta_0$ is a co-representation of the group generated by the time reversal T up to a complex factor. Show that this factor must be of modulus one in order to have an antiunitary operator, and that it is impossible to choose λ in such a way that the mapping becomes an ordinary co-representation.

5.3 Construct the double group of the tetrahedral group T. Choose one from each pair of elements of T^{d} which is mapped on the same element of T. Show that one obtains a set of 2×2 matrices which forms an irreducible projective representation P of T with nontrivial factor system. Choose the matrices corresponding to the generators α and β in such a way that $P(\alpha)^3 = \mathbb{1}$, $P(\beta)^2 = -\mathbb{1}$, $[P(\alpha)P(\beta)]^3 = -\mathbb{1}$.

5.4 Determine the spin representations of the space group $Fd3m$ (the diamond group) for the points $\Gamma(\boldsymbol{k} = 0)$ and $\Delta(k_y = k_z = 0, k_x \neq 0)$ in the Brillouin zone.

5.5 Show that the Pauli matrices and the two-dimensional unit matrix form a projective representation of D_2. What is its factor system? The group of matrices generated by σ_1 and σ_2 is a group of order eight. Discuss the structure of this group (the quaternion group). Is it isomorphic to a crystallographic point group? Derive its character table and discuss the relation with the table for D_2 given in the appendix.

5.6 Consider an electron in a potential $V = V_0 + V_i$, where V_0 is spherically symmetric, V_i has the symmetry of a point group K and both are invariant under time reversal. Show that the levels of V_0 have no additional degeneracy. Discuss the level splitting under the influence of V_i.

5.7 Determine the additional degeneracies for a spin $\frac{1}{2}$ particle in a potential with point group symmetry $\overline{6}m2$, resp. $m3m$.

5.8 Apply eq. (5.22) to determine the additional degeneracies of levels of an electron in a crystal potential with space group $F\overline{4}3m$. Do this for the points Γ, Δ and Σ in the Brillouin zone (fig. 4.6).

APPENDIX A

CHARACTERS OF ORDINARY AND PROJECTIVE REPRESENTATIONS OF CRYSTALLOGRAPHIC POINT GROUPS

In the following tables are given the characters of the nonequivalent irreducible ordinary and projective representations of the 18 three-dimensional abstract point groups. For each isomorphism class of point groups (characters of isomorphic groups are the same) is given the character table for the ordinary representations. Here the character is a class function and giving the character of each class the character table is completely determined by the abstract group.

For projective representations the situation is completely different. First one can divide the representations into similarity classes. The factor systems of representations from one similarity class are associated. One can choose one factor system from each similarity class. Finally one can determine the equivalence classes of irreducible representations with a given factor system. The characters of equivalent representations are the same, but the character is by no means determined by the similarity class, as one can multiply each matrix by an arbitrary phase factor. Therefore, one has to fix a choice of factor systems, one from each similarity class. Once one has done this, the equivalence classes are distinguished by the character. However, the character is here not a class function.

A factor system is determined as follows. Suppose the group K is generated by $\alpha_1, ..., \alpha_\nu$ with defining relations $\Phi_i(\alpha_1, ..., \alpha_\nu) = \epsilon, (i = 1, ..., r)$. When P is a projective representation, one has $\Phi_i(P(\alpha_1), ..., P(\alpha_\nu)) = g_i \mathbb{1}$, where g_i is a phase factor. The elements $g_1, ..., g_r$ are completely determined by P, but also by the factor system ω of P. On the other hand ω is not determined by $g_1, ..., g_r$. However, it can be shown (Janssen [1972]) that i) elements $g_1, ..., g_r$ determine the class of ω, ii) the elements $g_1, ..., g_r$ and a choice of an expression $\alpha = w_\alpha (\alpha_1, ..., \alpha_\nu)$ for any $\alpha \in K$ determines ω. An example is given below. In the tables values of $g_1, ..., g_r$ are given for one factor system from each similarity class. Moreover, expressions $w_\alpha(\alpha_1, ..., \alpha_\nu)$ are given for all elements of the groups. Then for each set $g_1, ..., g_r$, i.e. for each factor system, the characters are given for the nonequivalent irreducible representations. Similar representations P' are found from $P'(\alpha) = u(\alpha)SP(\alpha)S^{-1}$ (with $\alpha \in K$, S a nonsingular matrix and u a phase factor) which has character $\chi'(\alpha) = u(\alpha)\chi(\alpha)$ and factor system determined by $\Phi_i(P'(\alpha_1), ..., P'(\alpha_\nu)) = \Phi_i(u(\alpha_1), ..., u(\alpha_\nu))g_i$.

If one wants to know the nonequivalent irreducible representations of K with given factor system ω, one calculates $g_i = \Phi_i(P(\alpha_1), ..., P(\alpha_\nu))$. It is always possible to find an associated representation such that the elements $g_1, ..., g_r$ are those given in the tables.

Example. For the group D_2 one can take generators α and β with defining relations $\alpha^2 = \beta^2 = (\alpha\beta)^2 = \epsilon$. It has two nonassociated factor systems. The trivial one (ordinary representations) is determined by $P(\alpha)^2 = P(\beta)^2 = [P(\alpha)P(\beta)]^2 = \mathbb{1}$, the nontrivial one by $P(\alpha)^2 = P(\beta)^2 = -[P(\alpha)P(\beta)]^2 = \mathbb{1}$. When the four elements of D_2 are written as ϵ, α, β, and $\alpha\beta$, the nontrivial factor system is given by

$$\omega(\alpha,\alpha) = P(\alpha)^2 = 1$$

$$\omega(\alpha,\beta) = P(\alpha)P(\beta)P(\alpha\beta)^{-1} = 1$$

$$\omega(\beta,\alpha) = P(\beta)P(\alpha)P(\alpha\beta)^{-1} = -1\ , \text{ etc.}$$

For this factor system there is one equivalence class of irreducible representations given by the character $\chi(\epsilon) = 2, \chi(\alpha) = \chi(\beta) = \chi(\alpha\beta) = 0$. For the trivial factor system there are four nonequivalent one-dimensional representations.

Isomorphism class C_1: only ordinary representations

Element	ϵ
Γ_1	1

Isomorphism class C_2: only ordinary representations

Classes	$[\epsilon]$	$[\alpha]$
Γ_1	1	1
Γ_2	1	−1

Isomorphism class C_3: only ordinary representations

Classes	$[\epsilon]$	$[\alpha]$	$[\alpha^2]$
Γ_1	1	1	1
Γ_2	1	ω^2	ω^4
Γ_3	1	ω^4	ω^2

$\omega = \exp(\pi i/3)$

Isomorphism class C_4: only ordinary representations

Classes	$[\epsilon]$	$[\alpha]$	$[\alpha^2]$	$[\alpha^3]$
Γ_1	1	1	1	1
Γ_2	1	i	-1	$-i$
Γ_3	1	-1	1	-1
Γ_4	1	$-i$	-1	i

Isomorphism class C_6: only ordinary representations

Classes	$[\epsilon]$	$[\alpha]$	$[\alpha^2]$	$[\alpha^3]$	$[\alpha^4]$	$[\alpha^5]$
Γ_1	1	1	1	1	1	1
Γ_2	1	ω	ω^2	-1	ω^4	ω^5
Γ_3	1	ω^2	ω^4	1	ω^2	ω^4
Γ_4	1	-1	1	-1	1	-1
Γ_5	1	ω^4	ω^2	1	ω^4	ω^2
Γ_6	1	ω^5	ω^4	-1	ω^2	ω

Isomorphism class D_2: multiplicator $M \cong C_2$
$P(\alpha)^2 = P(\beta)^2 = \mathbb{1}$, $[P(\alpha)P(\beta)]^2 = \lambda \mathbb{1}$

$\lambda = 1$	Classes	$[\epsilon]$	$[\alpha]$	$[\beta]$	$[\alpha\beta]$
	Γ_1	1	1	1	1
	Γ_2	1	-1	1	-1
	Γ_3	1	1	-1	-1
	Γ_4	1	-1	-1	1
$\lambda = -1$	Elements	ϵ	α	β	$\alpha\beta$
	Γ_5'	2	0	0	0

Isomorphism class D_3: only ordinary representations

	Classes	$[\epsilon]$	$[\alpha]$	$[\beta]$
	Γ_1	1	1	1
	Γ_2	1	1	−1
	Γ_3	2	−1	0

Isomorphism class D_4: multiplicator $M \cong C_2$
$P(\alpha)^4 = \lambda \mathbb{1}$, $P(\beta)^2 = [P(\alpha)P(\beta)]^2 = \mathbb{1}$

$\lambda = 1$	Classes	$[\epsilon]$	$[\alpha^2]$	$[\alpha]$	$[\beta]$	$[\alpha\beta]$
	Γ_1	1	1	1	1	1
	Γ_2	1	1	1	−1	−1
	Γ_3	1	1	−1	1	−1
	Γ_4	1	1	−1	−1	1
	Γ_5	2	−2	0	0	0

$\lambda = -1$	Elements	ϵ	α^2	α	α^3	β	$\alpha^2\beta$	$\alpha\beta$	$\alpha^3\beta$
	Γ_6'	2	0	$i\sqrt{2}$	$i\sqrt{2}$	0	0	0	0
	Γ_7'	2	0	$-i\sqrt{2}$	$-i\sqrt{2}$	0	0	0	0

Isomorphism class D_6: multiplicator $M \cong C_2$
$P(\alpha)^6 = P(\beta)^2 = \mathbb{1}$, $[P(\alpha)P(\beta)]^2 = \lambda \mathbb{1}$

$\lambda = 1$	Classes	$[\epsilon]$	$[\alpha^2]$	$[\beta]$	$[\alpha^3]$	$[\alpha]$	$[\alpha\beta]$
	Γ_1	1	1	1	1	1	1
	Γ_2	1	1	−1	1	1	−1
	Γ_3	2	−1	0	2	−1	0
	Γ_4	1	1	1	−1	−1	−1
	Γ_5	1	1	−1	−1	−1	1
	Γ_6	2	−1	0	−2	1	0

$\lambda = -1$	Elements	ϵ	α^2	α^4	β	$\alpha^2\beta$	$\alpha^4\beta$	α^3	α	α^5	$\alpha\beta$	$\alpha^3\beta$	$\alpha^5\beta$
	Γ_7'	2	2	2	0	0	0	0	0	0	0	0	0
	Γ_8'	2	−1	−1	0	0	0	0	$i\sqrt{3}$	$-i\sqrt{3}$	0	0	0
	Γ_9'	2	−1	−1	0	0	0	0	$-i\sqrt{3}$	$i\sqrt{3}$	0	0	0

Isomorphism class T: multiplicator $M \cong C_2$
$P(\alpha)^3 = \mathbb{1}$, $P(\beta)^2 = [P(\alpha)P(\beta)]^3 = \lambda \mathbb{1}$

$\lambda = 1$	Classes	$[\epsilon]$	$[\alpha]$	$[\alpha^2]$	$[\beta]$
	Γ_1	1	1	1	1
	Γ_2	1	ω^2	ω^4	1
	Γ_3	1	ω^4	ω^2	1
	Γ_4	3	0	0	−1

$\lambda=-1$	Elements	1	2	3	4	5	6	7	8	9	10	11	12
	Γ_5'	2	−1	1	1	1	−1	−1	−1	−1	0	0	0
	Γ_6'	2	ω^5	ω^2	ω^2	ω^2	ω^5	ω^5	ω^5	ω^5	0	0	0
	Γ_7'	2	ω	ω^4	ω^4	ω^4	ω	ω	ω	ω	0	0	0

Isomorphism class O: multiplicator $M \cong C_2$
$P(\alpha)^4 = \lambda \mathbb{1}$, $P(\beta)^3 = [P(\alpha)P(\beta)]^2 = \mathbb{1}$

$\lambda = 1$	Classes	$[\epsilon]$	$[\beta]$	$[\alpha^2]$	$[\alpha]$	$[\alpha\beta]$
	Γ_1	1	1	1	1	1
	Γ_2	1	1	1	−1	−1
	Γ_3	2	−1	2	0	0
	Γ_4	3	0	−1	1	−1
	Γ_5	3	0	−1	−1	1

$\lambda=-1$	Elements	1	2	3	4	5	6	7	8	9	10	11	12
	Γ_6'	2	−1	1	−1	−1	−1	1	1	1	0	0	0
	Γ_7'	2	−1	1	−1	−1	−1	1	1	1	0	0	0
	Γ_8'	4	1	−1	1	1	1	−1	−1	−1	0	0	0

	13	14	15	16	17	18	19	20	21	22	23	24
Γ_6'	$i\sqrt{2}$	$i\sqrt{2}$	$-i\sqrt{2}$	$-i\sqrt{2}$	$-i\sqrt{2}$	$-i\sqrt{2}$	0	0	0	0	0	0
Γ_7'	$-i\sqrt{2}$	$-i\sqrt{2}$	$i\sqrt{2}$	$i\sqrt{2}$	$i\sqrt{2}$	$i\sqrt{2}$	0	0	0	0	0	0
Γ_8'	0	0	0	0	0	0	0	0	0	0	0	0

To give the irreducible characters of the projective representations of the direct product groups $K \times C_2$ we use the fact that the Kronecker product of an irreducible representation with factor system ω and an ordinary one-dimensional representation is again an irreducible representation with the same factor system. The Kronecker product representation is similar to the original one, but the representations are in general not equivalent. In the following tables we give nonequivalent, but similar representations as Kronecker products.

Isomorphism class $C_4 \times C_2$: multiplicator $M \cong C_2$
$P(\alpha)^4 = P(\beta)^2 = \mathbb{1}$, $P(\alpha)P(\beta) = \lambda P(\beta)P(\alpha)$

	Elements	ϵ	α	α^2	α^3	β	$\alpha\beta$	$\alpha^2\beta$	$\alpha^3\beta$
$\lambda = 1$	Γ_1	1	1	1	1	1	1	1	1
	Γ_2	1	i	-1	$-i$	1	i	-1	$-i$
		$\Gamma_3 = (\Gamma_2)^2$, $\Gamma_4 = (\Gamma_2)^3$							
	Γ_5	1	1	1	1	-1	-1	-1	-1
		$\Gamma_6 = \Gamma_5 \otimes \Gamma_2$, $\Gamma_7 = \Gamma_5 \otimes (\Gamma_2)^2$, $\Gamma_8 = \Gamma_5 \otimes (\Gamma_2)^3$							
$\lambda = -1$	Γ_9	2	0	2	0	0	0	0	0
		$\Gamma_{10} = \Gamma_9 \otimes \Gamma_2$							

Isomorphism class $C_6 \times C_2$: multiplicator $M \cong C_2$
$P(\alpha)^6 = P(\beta)^2 = \mathbb{1}$, $P(\alpha)P(\beta) = \lambda P(\beta)P(\alpha)$

	Elements	ϵ	α	α^2	α^3	α^4	α^5	β	$\alpha\beta$	$\alpha^2\beta$	$\alpha^3\beta$	$\alpha^4\beta$	$\alpha^5\beta$
$\lambda = 1$	Γ_1	1	1	1	1	1	1	1	1	1	1	1	1
	Γ_2	1	ω	ω^2	-1	ω^4	ω^5	1	ω	ω^2	-1	ω^4	ω^5
		$\Gamma_3 = (\Gamma_2)^2$, $\Gamma_4 = (\Gamma_2)^3$, $\Gamma_5 = (\Gamma_2)^4$, $\Gamma_6 = (\Gamma_2)^5$											
	Γ_7	1	1	1	1	1	1	-1	-1	-1	-1	-1	-1
		$\Gamma_8 = \Gamma_7 \otimes (\Gamma_2)$, $\Gamma_9 = \Gamma_7 \otimes (\Gamma_2)^2$, $\Gamma_{10} = \Gamma_7 \otimes (\Gamma_2)^3$, $\Gamma_{11} = \Gamma_7 \otimes (\Gamma_2)^4$, $\Gamma_{12} = \Gamma_7 \otimes (\Gamma_2)^5$											
$\lambda = -1$	Γ_{13}	2	0	2	0	2	0	0	0	0	0	0	0
		$\Gamma_{14} = \Gamma_{13} \otimes \Gamma_2$, $\Gamma_{15} = \Gamma_{13} \otimes (\Gamma_2)^2$											

Isomorphism class $D_2 \times C_2$: multiplicator $M \cong C_2 \times C_2 \times C_2$
$P(\alpha)^2 = \lambda_1 \mathbb{1}$, $P(\alpha)P(\gamma) = \lambda_2 P(\gamma)P(\alpha)$, $P(\beta)P(\gamma) = \lambda_3 P(\gamma)P(\beta)$
$P(\beta)^2 = P(\gamma)^2 = [P(\alpha)P(\beta)]^2 = \mathbb{1}$

λ_1 λ_2 λ_3		ϵ	α	β	$\alpha\beta$	γ	$\alpha\gamma$	$\beta\gamma$	$\alpha\beta\gamma$
1 1 1	Γ_1	1	1	1	1	1	1	1	1
	Γ_2	1	−1	1	−1	1	−1	1	−1
	Γ_3	1	1	−1	−1	1	1	−1	−1
	Γ_4	1	−1	−1	1	1	−1	−1	1
	Γ_5	1	1	1	1	−1	−1	−1	−1
		$\Gamma_6 = \Gamma_5 \otimes \Gamma_2$, $\Gamma_7 = \Gamma_5 \otimes \Gamma_3$, $\Gamma_8 = \Gamma_5 \otimes \Gamma_4$							
−1 1 1	Γ_9	2	0	0	0	2	0	0	0
		$\Gamma_{10} = \Gamma_9 \otimes \Gamma_5$							
1 −1 1	Γ_{11}	2	0	2	0	0	0	0	0
		$\Gamma_{12} = \Gamma_{11} \otimes \Gamma_3$							
1 −1 1	Γ_{13}	2	$2i$	0	0	0	0	0	0
		$\Gamma_{14} = \Gamma_{13} \otimes \Gamma_2$							
−1 −1 1	Γ_{15}	2	0	0	0	0	0	2	0
		$\Gamma_{16} = \Gamma_{15} \otimes \Gamma_3$							
−1 1 −1	Γ_{17}	2	0	0	0	0	$2i$	0	0
		$\Gamma_{18} = \Gamma_{17} \otimes \Gamma_2$							
1 −1 −1	Γ_{19}	2	0	0	$2i$	0	0	0	0
		$\Gamma_{20} = \Gamma_{19} \otimes \Gamma_2$							
−1 −1 −1	Γ_{21}	2	0	0	0	0	0	0	$2i$
		$\Gamma_{22} = \Gamma_{21} \otimes \Gamma_2$							

Isomorphism class $D_4 \times C_2$: multiplicator $M \cong C_2 \times C_2 \times C_2$
$P(\alpha)^4 = \lambda_1 \mathbb{1}$, $P(\alpha)P(\gamma) = \lambda_2 P(\gamma)P(\alpha)$, $P(\beta)P(\gamma) = \lambda_3 P(\gamma)P(\beta)$
$P(\beta)^2 = P(\gamma)^2 = [P(\alpha)P(\beta)]^2 = \mathbb{1}$

$\lambda_1\ \lambda_2\ \lambda_3$		ϵ	α^2	α	α^3	β	$\alpha^2\beta$	$\alpha\beta$	$\alpha^3\beta$	γ	$\alpha^2\gamma$	$\alpha\gamma$	$\alpha^3\gamma$	$\beta\gamma$	$\alpha^2\beta\gamma$	$\alpha\beta\gamma$	$\alpha^3\beta\gamma$
1 1 1	Γ_1	1	1	1	1	1	1	1	1	1	1	1	1	1	1	1	1
	Γ_2	1	1	1	1	−1	−1	−1	−1	1	1	1	1	−1	−1	−1	−1
	Γ_3	1	1	−1	−1	1	1	−1	−1	1	1	−1	−1	1	1	−1	−1
	Γ_4	1	1	−1	−1	−1	−1	1	1	1	1	−1	−1	−1	−1	1	1
	Γ_5	2	−2	0	0	0	0	0	0	2	−2	0	0	0	0	0	0
	Γ_6	1	1	1	1	1	1	1	1	−1	−1	−1	−1	−1	−1	−1	−1
		$\Gamma_7 = \Gamma_6 \otimes \Gamma_2$, $\Gamma_8 = \Gamma_6 \otimes \Gamma_3$, $\Gamma_9 = \Gamma_6 \otimes \Gamma_4$, $\Gamma_{10} = \Gamma_6 \otimes \Gamma_5$															
−1 1 1	Γ_{11}	2	0	$i\sqrt{2}$	$i\sqrt{2}$	0	0	0	0	2	0	$i\sqrt{2}$	$\sqrt{2}$	0	0	0	0
		$\Gamma_{12} = \Gamma_{11} \otimes \Gamma_3$, $\Gamma_{13} = \Gamma_{11} \otimes \Gamma_6$, $\Gamma_{14} = \Gamma_{11} \otimes \Gamma_8$															
1 −1 1	Γ_{15}	2	2	0	0	2	2	0	0	0	0	0	0	0	0	0	0
		$\Gamma_{16} = \Gamma_{15} \otimes \Gamma_2$															
	Γ_{17}	2	−2	0	0	0	0	0	0	0	0	0	0	2	−2	0	0
		$\Gamma_{18} = \Gamma_{17} \otimes \Gamma_6$															
1 1 −1	Γ_{19}	2	2	2	2	0	0	0	0	0	0	0	0	0	0	0	0
		$\Gamma_{20} = \Gamma_{19} \otimes \Gamma_3$															
	Γ_{21}	2	−2	0	0	0	0	0	0	0	0	$2i$	$-2i$	0	0	0	0
		$\Gamma_{22} = \Gamma_{21} \otimes \Gamma_6$															
−1 −1 1	Γ_{23}	4	0	0	0	0	0	0	0	0	0	0	0	0	0	0	0
−1 1 −1	Γ_{24}	2	0	$i\sqrt{2}$	$i\sqrt{2}$	0	0	0	0	0	$2i$	$\sqrt{2}$	$-\sqrt{2}$	0	0	0	0
		$\Gamma_{25} = \Gamma_{24} \otimes \Gamma_3$, $\Gamma_{26} = \Gamma_{24} \otimes \Gamma_6$, $\Gamma_{27} = \Gamma_{24} \otimes \Gamma_8$															
1 −1 −1	Γ_{28}	2	2	0	0	0	0	2	2	0	0	0	0	0	0	0	0
		$\Gamma_{29} = \Gamma_{28} \otimes \Gamma_2$															
	Γ_{30}	2	−2	0	0	0	0	0	0	0	0	0	0	0	0	−2	−2
		$\Gamma_{31} = \Gamma_{30} \otimes \Gamma_6$															
−1 −1 −1	Γ_{32}	4	0	0	0	0	0	0	0	0	0	0	0	0	0	0	0

Isomorphism class $D_6 \times C_2$: multiplicator $M \cong C_2 \times C_2 \times C_2$
$P(\alpha)^6 = \lambda_1 \mathbb{1}$, $P(\alpha)P(\gamma) = \lambda_2 P(\gamma)P(\alpha)$, $P(\beta)P(\gamma) = \lambda_3 P(\gamma)P(\beta)$, $P(\beta)^2 = P(\gamma)^2 = [P(\alpha)P(\beta)]^2 = \mathbb{1}$

$\lambda_1\ \lambda_2\ \lambda_3$		ϵ	α^2	α^4	β	$\alpha^2\beta$	$\alpha^4\beta$	α^3	α^5	α	$\alpha^3\beta$	$\alpha^5\beta$	$\alpha\beta$	γ	$\alpha^2\gamma$	$\alpha^4\gamma$	$\beta\gamma$	$\alpha^2\beta\gamma$	$\alpha^4\beta\gamma$	$\alpha^3\gamma$	$\alpha^5\gamma$	$\alpha\gamma$	$\alpha^3\beta\gamma$	$\alpha^5\beta\gamma$	$\alpha\beta\gamma$
1 1 1	Γ_1	1	1	1	1	1	1	1	1	1	1	1	1	1	1	1	1	1	1	1	1	1	1	1	1
	Γ_2	1	1	1	−1	−1	−1	1	1	1	−1	−1	−1	1	1	1	−1	−1	−1	1	1	1	−1	−1	−1
	Γ_3	2	−1	−1	0	0	0	2	−1	−1	0	0	0	2	−1	−1	0	0	0	2	−1	−1	0	0	0
	Γ_4	1	1	1	1	1	1	−1	−1	−1	−1	−1	−1	1	1	1	1	1	1	−1	−1	−1	−1	−1	−1
	Γ_7	1	1	1	1	1	1	1	1	1	1	1	1	−1	−1	−1	−1	−1	−1	−1	−1	−1	−1	−1	−1
	$\Gamma_5 = \Gamma_4 \otimes \Gamma_2$, $\Gamma_6 = \Gamma_4 \otimes \Gamma_3$, $\Gamma_8 = \Gamma_7 \otimes \Gamma_2$, $\Gamma_9 = \Gamma_7 \otimes \Gamma_3$, $\Gamma_{10} = \Gamma_7 \otimes \Gamma_4$, $\Gamma_{11} = \Gamma_7 \otimes \Gamma_5$, $\Gamma_{12} = \Gamma_7 \otimes \Gamma_6$																								
−1 1 1	Γ_{13}	2	−2	2	0	0	0	0	0	0	0	0	0	2	−2	2	0	0	0	0	0	0	0	0	0
	Γ_{15}	2	1	−1	0	0	0	0	$\sqrt{3}$	$-\sqrt{3}$	0	0	0	2	1	−1	0	0	0	0	$\sqrt{3}$	$-\sqrt{3}$	0	0	0
	$\Gamma_{14} = \Gamma_{13} \otimes \Gamma_7$, $\Gamma_{16} = \Gamma_{15} \otimes \Gamma_4$, $\Gamma_{17} = \Gamma_{15} \otimes \Gamma_7$, $\Gamma_{18} = \Gamma_{15} \otimes \Gamma_{10}$																								
1 −1 1	Γ_{19}	2	−2	2	2	−2	2	0	0	0	0	0	0	0	0	0	0	0	0	0	0	0	0	0	0
	Γ_{21}	4	2	−2	0	0	0	0	0	0	0	0	0	0	0	0	0	0	0	0	0	0	0	0	0
	$\Gamma_{20} = \Gamma_{19} \otimes \Gamma_2$																								
1 1 −1	Γ_{22}	2	−2	2	0	0	0	$-2i$	$2i$	$2i$	0	0	0	0	0	0	0	0	0	0	0	0	0	0	0
	Γ_{24}	2	1	−1	0	0	0	$2i$	i	i	0	0	0	0	$-i\sqrt{3}$	$-i\sqrt{3}$	0	0	0	0	$\sqrt{3}$	$-\sqrt{3}$	0	0	0
	$\Gamma_{23} = \Gamma_{22} \otimes \Gamma_4$, $\Gamma_{25} = \Gamma_{24} \otimes \Gamma_4$, $\Gamma_{26} = \Gamma_{24} \otimes \Gamma_7$, $\Gamma_{27} = \Gamma_{24} \otimes \Gamma_{10}$																								
−1 −1 1	Γ_{28}	2	−2	2	0	0	0	0	0	0	0	0	0	0	0	2	2	2	0	0	0	0	0	0	0
	Γ_{30}	4	2	−2	0	0	0	0	0	0	0	0	0	0	0	0	0	0	0	0	0	0	0	0	0
	$\Gamma_{29} = \Gamma_{28} \otimes \Gamma_7$																								

−1 1 −1	Γ_{31}	2	−2	2	0	0	0	0	0	0	0	0	0	0	0	0	0	0	0	2	$2i$	$2i$	0	0	0
	Γ_{33}	2	1	−1	0	0	0	0	$\sqrt{3}$	$-\sqrt{3}$	0	0	0	0	$-i\sqrt{3}$	$-i\sqrt{3}$	0	0	0	$2i$	i	i	0	0	0
	$\Gamma_{32} = \Gamma_{31} \otimes \Gamma_7$, $\Gamma_{34} = \Gamma_{33} \otimes \Gamma_4$, $\Gamma_{35} = \Gamma_{33} \otimes \Gamma_7$, $\Gamma_{36} = \Gamma_{33} \otimes \Gamma_{10}$																								
1 −1 −1	Γ_{37}	2	−2	2	0	0	0	0	0	0	$-2i$	$2i$	$2i$	0	0	0	0	0	0	0	0	0	0	0	0
	Γ_{39}	4	2	−2	0	0	0	0	0	0	0	0	0	0	0	0	0	0	0	0	0	0	0	0	0
	$\Gamma_{38} = \Gamma_{37} \otimes \Gamma_4$																								
−1 −1 −1	Γ_{40}	2	−2	2	0	0	0	0	0	0	0	0	0	0	0	0	0	0	0	0	0	0	2	−2	−2
	Γ_{42}	4	2	−2	0	0	0	0	0	0	0	0	0	0	0	0	0	0	0	0	0	0	0	0	0
	$\Gamma_{41} = \Gamma_{40} \otimes \Gamma_7$																								

Isomorphism class $T \times C_2$: multiplicator $M \cong C_2$

$P(\alpha)^3 = P(\gamma)^2 = \mathbb{1}$, $P(\alpha)P(\gamma) = P(\gamma)P(\alpha)$, $P(\beta)P(\gamma) = P(\gamma)P(\beta)$, $P(\beta)^2 = [P(\alpha)P(\beta)]^3 = \lambda\,\mathbb{1}$

		ϵ	α	$\beta\alpha\beta$	$\beta\alpha$	$\alpha\beta$	α^2	$\alpha\beta\alpha$	$\alpha^2\beta$	$\beta\alpha^2$	β	$\alpha\beta\alpha^2$	$\alpha^2\beta\alpha$	γ	α	$\beta\alpha\beta\gamma$	$\beta\alpha\gamma$	$\alpha\beta\gamma$	$\alpha^2\gamma$	$\alpha\beta\alpha\gamma$	$\alpha^2\beta\gamma$	$\beta\alpha^2\gamma$	$\beta\gamma$	$\alpha\beta\alpha^2\gamma$	$\alpha^2\beta\alpha\gamma$
$\lambda = 1$	Γ_1	1	1	1	1	1	1	1	1	1	1	1	1	1	1	1	1	1	1	1	1	1	1	1	1
	Γ_2	1	ω^2	ω^2	ω^2	ω^2	ω^4	ω^4	ω^4	ω^4	1	1	1	1	ω^2	ω^2	ω^2	ω^2	ω^4	ω^4	ω^4	ω^4	1	1	1
	Γ_4	3	0	0	0	0	0	0	0	0	−1	−1	−1	3	0	0	0	0	0	0	0	0	−1	−1	−1
	Γ_5	1	1	1	1	1	1	1	1	1	1	1	1	−1	−1	−1	−1	−1	−1	−1	−1	−1	−1	−1	−1
	$\Gamma_3 = \Gamma_2 \otimes \Gamma_2$, $\Gamma_6 = \Gamma_5 \otimes \Gamma_2$, $\Gamma_7 = \Gamma_5 \otimes \Gamma_2 \otimes \Gamma_2$, $\Gamma_8 = \Gamma_5 \otimes \Gamma_4$,																								
$\lambda = -1$	Γ_9	2	−1	1	1	1	−1	−1	−1	−1	0	0	0	2	−1	1	1	1	−1	−1	−1	−1	0	0	0
	$\Gamma_{10} = \Gamma_9 \otimes \Gamma_2$, $\Gamma_{11} = \Gamma_9 \otimes \Gamma_3$, $\Gamma_{12} = \Gamma_9 \otimes \Gamma_5$, $\Gamma_{13} = \Gamma_9 \otimes \Gamma_6$, $\Gamma_{14} = \Gamma_9 \otimes \Gamma_7$																								

Isomorphism class $O \times C_2$: multiplicator $M \cong C_2 \times C_2$

$P(\alpha)^4 = \lambda_1 \mathbb{1}$, $P(\alpha)P(\gamma) = \lambda_2 P(\gamma)P(\alpha)$, $P(\beta)^3 = [P(\alpha)P(\beta)]^2 = P(\gamma)^2 = \mathbb{1}$, $P(\beta)P(\gamma) = P(\gamma)P(\beta)$

λ_1 λ_2		ϵ	β	$\alpha\beta^2\alpha$	$\alpha^2\beta$	$\beta\alpha^2$	β^2	$\beta\alpha^2\beta$	$\alpha\beta\alpha^3$	$\alpha^2\beta^2$	α^2	$\beta\alpha^2\beta^2$	$\beta^2\alpha^2\beta$	α	α^3	$\alpha^3\beta$	$\beta\alpha^3$	$\beta^2\alpha$	$\alpha\beta^2$	$\alpha^2\beta^2\alpha$	$\beta\alpha$	$\alpha\beta$	$\alpha\beta^2\alpha^2$	$\alpha\beta^2\alpha^2\beta$	$\beta^2\alpha\beta^2$
1 1	Γ_1	1	1	1	1	1	1	1	1	1	1	1	1	1	1	1	1	1	1	1	1	1	1	1	1
	Γ_2	1	1	1	1	1	1	1	1	1	1	1	1	−1	−1	−1	−1	−1	−1	−1	−1	−1	−1	−1	−1
	Γ_3	2	−1	−1	−1	−1	−1	−1	−1	−1	2	2	2	0	0	0	0	0	0	0	0	0	0	0	0
	Γ_4	3	0	0	0	0	0	0	0	0	−1	−1	−1	−1	−1	−1	−1	−1	−1	1	1	1	1	1	1
	Γ_6	1	1	1	1	1	1	1	1	1	1	1	1	1	1	1	1	1	1	1	1	1	1	1	1
	$\Gamma_5 = \Gamma_4 \otimes \Gamma_2$, $\Gamma_7 = \Gamma_6 \otimes \Gamma_2$, $\Gamma_8 = \Gamma_6 \otimes \Gamma_3$, $\Gamma_9 = \Gamma_6 \otimes \Gamma_4$, $\Gamma_{10} = \Gamma_6 \otimes \Gamma_4 \otimes \Gamma_2$																								
−1 1	Γ_{11}	2	−1	1	−1	−1	−1	1	1	1	0	0	0	$i\sqrt{2}$	$i\sqrt{2}$	$-i\sqrt{2}$	$-i\sqrt{2}$	$-i\sqrt{2}$	$-i\sqrt{2}$	0	0	0	0	0	0
	Γ_{15}	4	1	−1	1	1	1	−1	−1	−1	0	0	0	0	0	0	0	0	0	0	0	0	0	0	0
	$\Gamma_{12} = \Gamma_{11} \otimes \Gamma_2$, $\Gamma_{13} = \Gamma_{11} \otimes \Gamma_6$, $\Gamma_{14} = \Gamma_{11} \otimes \Gamma_7$, $\Gamma_{16} = \Gamma_{15} \otimes \Gamma_6$																								
1 −1	Γ_{17}	2	2	2	2	2	2	2	2	2	2	2	2	0	0	0	0	0	0	0	0	0	0	0	0
	Γ_{18}	2	−1	−1	−1	−1	−1	−1	−1	−1	2	2	2	0	0	0	0	0	0	0	0	0	0	0	0
	Γ_{20}	6	0	0	0	0	0	0	0	0	−2	−2	−2	0	0	0	0	0	0	0	0	0	0	0	0
	$\Gamma_{19} = \Gamma_{18} \otimes \Gamma_6$																								
−1 −1	Γ_{21}	4	1	−1	−1	−1	1	1	1	1	0	0	0	0	0	0	0	0	0	0	0	0	0	0	0
	Γ_{23}	4	−2	2	2	2	−2	−2	−2	−2	0	0	0	0	0	0	0	0	0	0	0	0	0	0	0
	$\Gamma_{22} = \Gamma_{21} \otimes \Gamma_6$																								

(continued)

x x x x	γ	$\beta\gamma$	$\alpha\beta^2\alpha\gamma$	$\alpha^2\beta\gamma$	$\beta\alpha^2\gamma$	$\beta^2\gamma$	$\beta\alpha^2\beta\gamma$	$\alpha\beta\alpha^3\gamma$	$\alpha^2\beta^2\gamma$	$\alpha^2\gamma$	$\beta\alpha^2\beta^2\gamma$	$\beta^2\alpha^2\beta\gamma$	$\alpha\gamma$	$\alpha^3\gamma$	$\alpha^3\beta\gamma$	$\beta\alpha^3\gamma$	$\beta^2\alpha\gamma$	$\alpha\beta^2\gamma$	$\alpha^2\beta^2\alpha\gamma$	$\beta\alpha\gamma$	$\alpha\beta\gamma$	$\alpha\beta^2\alpha^2\gamma$	$\alpha\beta^2\alpha^2\beta\gamma$	$\beta^2\alpha\beta^2\gamma$
Γ_1	1	1	1	1	1	1	1	1	1	1	1	1	1	1	1	1	1	1	1	1	1	1	1	1
Γ_2	1	1	1	1	1	1	1	1	1	1	1	1	−1	−1	−1	−1	−1	−1	−1	−1	−1	−1	−1	−1
Γ_3	2	−1	−1	−1	−1	−1	−1	−1	−1	2	2	2	0	0	0	0	0	0	0	0	0	0	0	0
Γ_4	3	0	0	0	0	0	0	0	0	−1	−1	−1	−1	−1	−1	−1	−1	−1	1	1	1	1	1	1
Γ_6	−1	−1	−1	−1	−1	−1	−1	−1	−1	−1	−1	−1	−1	−1	−1	−1	−1	−1	−1	−1	−1	−1	−1	−1
Γ_{11}	2	−1	1	−1	−1	−1	1	1	1	0	0	0	$i\sqrt{2}$	$i\sqrt{2}$	$-i\sqrt{2}$	$-i\sqrt{2}$	$-i\sqrt{2}$	$-i\sqrt{2}$	0	0	0	0	0	0
Γ_{15}	4	1	−1	1	1	1	−1	−1	−1	0	0	0	0	0	0	0	0	0	0	0	0	0	0	0
Γ_{17}	0	0	0	0	0	0	0	0	0	0	0	0	0	0	0	0	0	0	0	0	0	0	0	0
Γ_{18}	0	$-i\sqrt{3}$	$-i\sqrt{3}$	$-i\sqrt{3}$	$-i\sqrt{3}$	$i\sqrt{3}$	$i\sqrt{3}$	$i\sqrt{3}$	$i\sqrt{3}$	0	0	0	0	0	0	0	0	0	0	0	0	0	0	0
Γ_{20}	0	0	0	0	0	0	0	0	0	0	0	0	0	0	0	0	0	0	0	0	0	0	0	0
Γ_{21}	0	$-i\sqrt{3}$	$i\sqrt{3}$	$i\sqrt{3}$	$i\sqrt{3}$	$i\sqrt{3}$	$i\sqrt{3}$	$i\sqrt{3}$	$i\sqrt{3}$	0	0	0	0	0	0	0	0	0	0	0	0	0	0	0
Γ_{23}	0	0	0	0	0	0	0	0	0	0	0	0	0	0	0	0	0	0	0	0	0	0	0	0

APPENDIX B

EXTRA REPRESENTATIONS OF THE CRYSTALLOGRAPHIC DOUBLE GROUPS

In the following we give those extra representations of the double groups which can not, by a suitable choice of phase factor, be transformed into a representation of the (single) group. Moreover, we give only the extra representations of the double groups of rotation groups (crystallographic point groups of the first kind). The double group of a point group of the second kind is isomorphic to the double group of the group of the first kind obtained from the former by multiplication of the elements with determinant -1 by the central inversion. The double group of direct product groups $G \times C_2$ is $G^d \times C_2$, when C_2 is generated by the central inversion I. So we consider only the double groups of the crystallographic point groups 222, 422, 622, 23 and 432.

222^d	$[\epsilon]$	$[-\epsilon]$	$[\pm\alpha]$	$[\pm\beta]$	$[\pm\alpha\beta]$
Γ_5'	2	-2	0	0	0

422^d	$[\epsilon]$	$[-\epsilon]$	$[\pm\alpha^2]$	$[\alpha]$	$[-\alpha]$	$[\pm\beta]$	$[\pm\alpha\beta]$
Γ_6'	2	-2	0	$\sqrt{2}$	$-\sqrt{2}$	0	0
Γ_7'	2	-2	0	$-\sqrt{2}$	$\sqrt{2}$	0	0

622^d	$[\epsilon]$	$[-\epsilon]$	$[\alpha^2]$	$[-\alpha^2]$	$[\pm\beta]$	$[\pm\alpha^3]$	$[\alpha^5]$	$[-\alpha^5]$	$[\pm\alpha^3\beta]$
Γ_8'	2	-2	1	-1	0	0	$\sqrt{3}$	$-\sqrt{3}$	0
Γ_9'	2	-2	1	-1	0	0	$-\sqrt{3}$	$\sqrt{3}$	0
Γ_7'	2	-2	-2	2	0	0	0	0	0

23^d	$[\epsilon]$	$[-\epsilon]$	$[\alpha]$	$[-\alpha]$	$[\alpha^2]$	$[-\alpha^2]$	$[\pm\beta]$
Γ_5'	2	-2	1	-1	1	-1	0
Γ_6'	2	-2	ω	ω^4	ω^2	ω^5	0
Γ_7'	2	-2	ω^5	ω^2	ω^4	ω	0

$\omega = \exp(\pi i/3)$

432^d	$[\epsilon]$	$[-\epsilon]$	$[\beta]$	$[-\beta]$	$[\pm\alpha^2]$	$[\alpha]$	$[-\alpha]$	$[\pm\alpha\beta]$
Γ_6'	2	−2	1	−1	0	$\sqrt{2}$	$-\sqrt{2}$	0
Γ_7'	2	−2	1	−1	0	$-\sqrt{2}$	$\sqrt{2}$	0
Γ_8'	4	−4	−1	1	0	0	0	0

APPENDIX C

PROJECTIVE CO-REPRESENTATIONS OF FINITE GROUPS

Here we will give some definitions and propositions from the theory of projective co-representations. We give only those results we need for Ch. 6, §4.4 and leave out the proofs. A more profound treatment and the proofs can be found in Janssen [1972].

Let G be a group, H a subgroup of G. Either $H = G$ or H is of index two in G. A *projective co-representation* of G with respect to H in a Hilbert space $\mathcal{H}$ is a homomorphism up to a factor from G onto a group of linear and anti-linear operators on $\mathcal{H}$. A *projective unitary/Antiunitary representation* of G with respect to H (a PUA rep of (G,H)) is a mapping P of G into the set of unitary and antiunitary operators on $\mathcal{H}$ such that

$$\begin{aligned}&1)\ P(\alpha)P(\beta) = \omega(\alpha,\beta)P(\alpha\beta) \qquad (\forall\alpha,\beta\in G, |\omega(\alpha,\beta)| = 1)\,,\\ &2)\ P(\alpha) \text{ is unitary if } \alpha\in H, \text{ antiunitary if } \alpha\in G-H\,.\end{aligned} \tag{C1}$$

An *ordinary unitary/antiunitary representation* (UA rep) of (G,H) is a PUA rep with $\omega(\alpha,\beta) = 1$ $(\alpha,\beta\in G)$. A *projective unitary representation* of G is a PUA rep of (G,G). An ordinary unitary representation of G is a UA rep of (G,G).

A PUA rep P of (G,H) in $\mathcal{H}$ is *irreducible* if $\mathcal{H}$ does not contain a proper G-invariant subspace. Two PUA reps P and P' are *equivalent* if there is a unitary operator U such that $P'(\alpha) = UP(\alpha)U^\dagger$ (all $\alpha\in G$). Two PUA reps P and P' are *similar* if there are a unitary operator U and for any $\alpha\in G$ a complex number $u(\alpha)$ of absolute value 1 such that

$$P'(\alpha) = u(\alpha)\,UP(\alpha)U^\dagger \qquad (\text{all } \alpha\in G)\,. \tag{C2}$$

The mapping $\omega: G\times G\to\mathbb{C}$ is called a *factor system*. Because of the associativity of the product of operators one has

$$\omega(\alpha,\beta)\omega(\alpha\beta,\gamma)=\begin{cases}\omega(\alpha,\beta\gamma)\,\omega(\beta,\gamma) & \text{if } \alpha\in H\\ \omega(\alpha,\beta\gamma)\,\omega(\beta,\gamma)^* & \text{if } \alpha\in G-H\,.\end{cases}$$

Now we define the action of $\alpha \in G$ on a complex number of absolute value 1 by

$$\lambda^\alpha=\begin{cases}\lambda & \text{for } \alpha\in H\\ \lambda^* & \text{for } \alpha\in G-H\,.\end{cases} \tag{C3}$$

Then we have for the factor system ω:

$$\omega(\alpha,\beta)\,\omega(\alpha\beta,\gamma)=\omega(\alpha,\beta\gamma)\,\omega(\beta,\gamma)^\alpha\,. \tag{C4}$$

Two equivalent PUA reps have the same factor system. Two similar PUA reps P and P' have factor systems ω and ω' such that

$$\omega'(\alpha,\beta)=\omega(\alpha,\beta)\,u(\alpha)\,u(\beta)^\alpha u(\alpha\beta)^{-1}\,. \tag{C5}$$

Two factor systems ω and ω' satisfying eq. (C5) are called *associated factor systems*.

One can show that for any factor system ω satisfying (C4) there is a PUA rep. When ω and ω' satisfy eq. (C4), then also $(\omega\omega')(\alpha,\beta)=\omega(\alpha,\beta)\,\omega'(\alpha,\beta)$, $\omega_0(\alpha,\beta)=1$ and $(\omega^{-1})(\alpha,\beta)=\omega(\alpha,\beta)^{-1}$ do so. Hence the factor systems form an Abelian group Z. A subgroup B of Z is formed by those factor systems which are associated with the trivial one. They satisfy

$$\omega(\alpha,\beta)=\frac{u(\alpha)\,u(\beta)^\alpha}{u(\alpha\beta)} \qquad (\text{all } \alpha,\beta\in G) \tag{C6}$$

for some mapping u from G into the group of complex numbers of absolute value 1. The factor group Z/B is called the *co-multiplicator* $M(G,H)$. Its elements are the classes of associated factor systems.

When G is a finite group generated by $\alpha_1, ..., \alpha_\nu$ with defining relations $\Phi_i(\alpha_1, ..., \alpha_\nu)=\epsilon$ $(i=1, ..., r)$, and P is a PUA rep, one has

$$\Phi_i(P(\alpha_1), ..., P(\alpha_\nu))=\lambda_i\,\mathbb{1}\,, \tag{C7}$$

where $\lambda_1, ..., \lambda_r$ are complex numbers with $|\lambda_i|=1$. One can prove the following propositions.

PROPOSITION C1. When $P(G)$ is a PUA rep which gives eq. (C7), the elements $\lambda_1, ..., \lambda_r$ determine the association class of the factor system ω of $P(G)$.
PROPOSITION C2. When for any $\alpha \in G$ one chooses a fixed word $\alpha = w_\alpha(\alpha_1, ..., \alpha_\nu)$ such that $P(\alpha) = w_\alpha(P(\alpha_1), ..., P(\alpha_\nu))$, the elements $\lambda_1, ..., \lambda_r$ determine the factor system ω of $P(G)$ completely.
Hence it is useful to give the nonassociated factor systems of G by nonequivalent sets $\{\lambda_i\}$, like we did also for PU reps in Appendix A.

Consider an extension R of the Abelian group A by G (cf. Ch. 4, § 1.8) with factor system $m: G \times G \to A$ and with action of G on A given by

$$\varphi(\alpha)\, a = \begin{cases} a & \text{if } \alpha \in H \\ a^{-1} & \text{if } \alpha \in G - H\,. \end{cases} \qquad (a \in A, \alpha \in G) \tag{C8}$$

The subgroup of R which is mapped by the canonical epimorphism σ on $H \subseteq G$ is denoted by U. Now consider an irreducible UA rep D of (R, U) and a mapping $r: G \to R$ such that $\sigma r(\alpha) = \alpha$ (all $\alpha \in G$). For $a \in A$ one has $D(a)D(r) = D(r)D(a)$ if $r \in U$ and $D(a)^{-1}D(r) = D(r)D(a)$ if $r \in R - U$. Now one has the following generalization of Schur's lemma.
LEMMA. If a unitary operator S satisfies $SU = US$, $AS^{-1} = SA$ for any unitary operator U and any antiunitary operator A from an irreducible UA rep, it is a scalar multiple of the identity operator.
This lemma implies that the unitary operator $D(a)$ satisfies

$$D(a) = \chi(a)\, \mathbb{1}\,.$$

Then we define

$$P(\alpha) = D(r(\alpha)) \qquad (\alpha \in G)\,. \tag{C9}$$

One has consequently the relation

$$P(\alpha)P(\beta) = D(r(\alpha))D(r(\beta)) = D(m(\alpha,\beta))P(\alpha\beta) = \chi(m(\alpha,\beta))P(\alpha\beta)\,, \tag{C10}$$

which means that $P(G)$ is a PUA rep of G with factor system

$$\omega(\alpha,\beta) = \chi(m(\alpha,\beta))\,. \tag{C11}$$

For fixed mapping $r: G \to R$ any irreducible UA rep of (R, U) determines in this way an irreducible PUA rep of (G,H). One can show that every PUA rep of (G,H) can be obtained from the irreducible UA reps of a couple (R,U), where the finite group R is an extension of the co-multiplicator $M(G,H)$ by G. However, we will not investigate this problem, but the problem of the *determination of all irreducible PUA reps of* (G,H) *with given factor system* ω.

We consider again a finite group G generated by $\alpha_1, ..., \alpha_\nu$ with defining relations $\Phi_i(\alpha_1, ..., \alpha_\nu) = \epsilon$ $(i = 1, ..., r)$. A PUA rep of (G,H) with factor system ω determines a set $\lambda_1, ..., \lambda_r$ by eq. (C7). When the factor system is of order d, i.e. when $\omega^d(\alpha,\beta) = 1$ for all $\alpha,\beta \in G$, the numbers $\lambda_1, ..., \lambda_r$ are d-th roots of unity. Suppose that

$$\lambda_i = e_d^{n_i} \qquad (i = 1, ..., r) , \tag{C12}$$

where e_d is a primitive d-th root. Then one can show the following.
PROPOSITION C3. A PUA rep of (G,H) with a set $\lambda_1, ..., \lambda_r$ (determined by eq. (C7)) consisting of d-th roots of unitary can be obtained from a UA rep of (R, U), where R is an extension of the cyclic group of order d by G, generated by $a, \overline{\alpha}_1, ..., \overline{\alpha}_\nu$ with defining relations

$$\begin{aligned} &a^d = e \\ &\Phi_i(\overline{\alpha}_1, ..., \overline{\alpha}_\nu) = a^{n_i} \qquad (i = 1, ..., r;\ n_i \text{ given by C12}) \\ &\overline{\alpha}_i a \overline{\alpha}_i^{-1} = a^{\alpha_i} , \end{aligned} \tag{C13}$$

and where U is $\sigma^{-1}(H)$ for σ the canonical epimorphism $R \to G$.

Example. Suppose that G is the dihedral group D_3, H is its subgroup C_3. The group G is generated by α_1 and α_2 with defining relations $(\alpha_1)^3 = (\alpha_2)^2 = (\alpha_1\alpha_2)^2 = \epsilon$. H is cyclic and generated by α_1. We will determine the irreducible PUA reps of (D_3, C_3) with factor system ω determined by

$$P(\alpha_1)^3 = -P(\alpha_2)^2 = -[P(\alpha_1)P(\alpha_2)]^2 = \mathbb{1}$$

or

$$\lambda_1 = 1 , \quad \lambda_2 = \lambda_3 = -1 . \tag{C14}$$

First we determine the factor system. To do that we have to choose expressions in the generators for all elements of G. One can take: $\epsilon, \alpha_1, \alpha_1^2, \alpha_2, \alpha_1\alpha_2,$

$\alpha_1^2\alpha_2$. As $P(\alpha_1)P(\alpha_2) = P(\alpha_2)P(\alpha_1)^2$, one finds $\omega(\alpha_1,\alpha_1)\,\mathbb{1} = P(\alpha_1)P(\alpha_1) \times P(\epsilon)^{-1} = \mathbb{1}$, $\omega(\alpha_2,\alpha_2)\,\mathbb{1} = P(\alpha_2)P(\alpha_2)P(\epsilon)^{-1} = -\mathbb{1}$, etc. The factor system is the following.

ω	α_1	α_1^2	α_2	$\alpha_1\alpha_2$	$\alpha_1^2\alpha_2$
α_1	1	1	1	1	1
α_1^2	1	1	1	1	1
α_2	1	1	−1	−1	−1
$\alpha_1\alpha_2$	1	1	−1	−1	−1
$\alpha_1^2\alpha_2$	1	1	−1	−1	−1

According to proposition C3 the irreducible PUA reps of G with the given factor system can be obtained from the UA reps of a group of order 12. This group is generated by $a, \bar{\alpha}_1, \bar{\alpha}_2$ with defining relations $a^2 = e, \bar{\alpha}_1^3 = e, \bar{\alpha}_2^2 = a$, $(\bar{\alpha}_1\bar{\alpha}_2)^2 = a, \bar{\alpha}_1 a = a\bar{\alpha}_1, \bar{\alpha}_2 a = a^{-1}\bar{\alpha}_2 = a\bar{\alpha}_2$. The subgroup U is formed by the elements $e, \bar{\alpha}_1, \bar{\alpha}_1^2, a, a\bar{\alpha}_1, a\bar{\alpha}_1^2$ which is a cyclic group of order six generated by $a\bar{\alpha}_1$. This group has 6 nonequivalent irreducible unitary representations. To determine the UA reps of (R,U) one can use eq. (6.87). One has

$$\sum_{u\in U} \chi((\bar{\alpha}_2 u)^2) = 6\chi(a)\,.$$

Using the character table for $U \cong C_6$ (Appendix A) one sees that for $\Gamma_1, \Gamma_3, \Gamma_5$ the irreducible UA reps of R subduce irreducible (one-dimensional) representations of U, whereas for $\Gamma_2, \Gamma_4, \Gamma_6$ one has $6\chi(a) = -6$, which means that in those cases the irreducible UA reps of (R,U) subduce reducible representations of U with two identical components. Hence R has 3 one-dimensional and 3 two-dimensional irreducible UA reps with respect to U. These 6 UA reps determine 6 nonequivalent PUA reps of (G,H). Three of them (those with $\chi(a) = 1$) have trivial factor systems. The other three give PUA reps with the factor system found above. Hence for a factor system with $\lambda_1, \lambda_2, \lambda_3$ given by (C14) there are 3 nonequivalent irreducible PUA reps of (D_3, C_3).

It will be clear that the ordinary projective representations as discussed in Ch. 1 and in Appendix A can be treated in the same way. In this case the action of G on the complex numbers and the action (C8) on A are trivial.

APPENDIX D

THE THREE-DIMENSIONAL SPACE GROUPS

In the following tables the 219 three-dimensional space groups are given by their generating elements. A space group is generated by 3 basis translations and the generators of the point group combined with nonprimitive translations. The space groups are divided into seven systems. For each system we choose a basis of the lattice or a sublattice such that we obtain the metric tensors of table 4.2. In the case of a centered lattice the primitive translations are obtained from our basis by the centering matrix. The primitive translations inside a unit cell of the (sub)lattice are given for each Bravais class. All point group elements and nonprimitive translations are given with respect to the basis of the (sub)lattice.

One obtains all elements of the space group as follows. As the generators of the point group are given with their nonprimitive translations, and each element R of the point group can be written as a product of the generators, the nonprimitive translation belonging to R can be found using the relation

$$\boldsymbol{t}_{RR'} = \boldsymbol{t}_R + R\boldsymbol{t}_{R'} \qquad \text{(up to a primitive translation)}\,.$$

Notice that for each isomorphism class of space groups only one representative is given. In particular, the nonprimitive translations are not uniquely determined. By choosing another origin one obtains the nonprimitive translation $\boldsymbol{t}_{R'} = \boldsymbol{t}_R + (\mathbb{1} - R)\boldsymbol{v}$, where $\boldsymbol{v}$ is an arbitrary vector. Equivalent nonprimitive translations are easily obtained from the tables.

Example. The space group $I2_1 2_1 2_1$ belongs to the body-centered orthorhombic Bravais class. With respect to an orthogonal basis of a primitive orthorhombic lattice the primitive translations are (n_1, n_2, n_3) and $(n_1 + \frac{1}{2}, n_2 + \frac{1}{2}, n_3 + \frac{1}{2})$ with n_1, n_2, n_3 integers. The elements of the point group with their nonprimitive translations are

$$\left\{\begin{pmatrix}-1&0&0\\0&-1&0\\0&0&1\end{pmatrix}\middle|\begin{pmatrix}\frac{1}{2}\\0\\\frac{1}{2}\end{pmatrix}\right\},\quad \left\{\begin{pmatrix}1&0&0\\0&-1&0\\0&0&-1\end{pmatrix}\middle|\begin{pmatrix}\frac{1}{2}\\\frac{1}{2}\\0\end{pmatrix}\right\},\quad \left\{\begin{pmatrix}-1&0&0\\0&1&0\\0&0&-1\end{pmatrix}\middle|\begin{pmatrix}0\\\frac{1}{2}\\\frac{1}{2}\end{pmatrix}\right\}.$$

The nonprimitive translations associated with the generators R_1 and R_2 can also be chosen as $\boldsymbol{t}_{R_1} = (\frac{1}{2}+x, y, \frac{1}{2})$, $\boldsymbol{t}_{R_2} = (\frac{1}{2}, \frac{1}{2}+y, z)$, where x, y, z are arbitrary real numbers. The point group generators are given by numbers indicated in table 4.3.

I Triclinic, primitive
Primitive translations in unit cell: (000)

Arithmetic class	Isomorphism class point group	R_1	t_{R_1}	R_2	t_{R_2}	R_3	t_{R_3}	Symbol	
$P1$	C_1	1	000					$P1$	C_1^1
$P\bar{1}$	C_2	2	000					$P\bar{1}$	C_i^1

II Monoclinic, primitive
Primitive translations in unit cell: (000)

Arithmetic class	Isomorphism class point group	R_1	t_{R_1}	R_2	t_{R_2}	R_3	t_{R_3}	Symbol	
Pm	C_2	7	000					Pm	C_s^1
		7	$0\frac{1}{2}0$					Pb	C_s^2
$P2$	C_2	3	000					$P2$	C_2^1
		3	$00\frac{1}{2}$					$P2_1$	C_2^2
$P2/m$	D_2	3	000	7	000			$P2/m$	C_{2h}^1
		3	$00\frac{1}{2}$	7	$00\frac{1}{2}$			$P2_1/m$	C_{2h}^2
		3	$0\frac{1}{2}0$	7	$0\frac{1}{2}0$			$P2/b$	C_{2h}^4
		3	$0\frac{1}{2}\frac{1}{2}$	7	$0\frac{1}{2}\frac{1}{2}$			$P2_1/b$	C_{2h}^5

III Monoclinic, body-centered
Primitive translations in unit cell: (000), $(\frac{1}{2}\frac{1}{2}\frac{1}{2})$

Arithmetic class	Isomorphism class point group	R_1	t_{R_1}	R_2	t_{R_2}	R_3	t_{R_3}	Symbol	
Im	C_2	7	000					Bm	C_s^3
		7	$0\frac{1}{2}0$					Bb	C_s^4
$I2$	C_2	3	000					$B2$	C_2^3
$I2/m$	D_2	3	000	7	000			$B2/m$	C_{2h}^3
		3	$0\frac{1}{2}0$	7	$0\frac{1}{2}0$			$B2/b$	C_{2h}^6

IV Orthorhombic, primitive
Primitive translations in unit cell: (000)

$P222$	D_2	3	000	4	000			$P222$	D_2^1
		3	$00\frac{1}{2}$	4	000			$P222_1$	D_2^2
		3	000	4	$\frac{1}{2}\frac{1}{2}0$			$P2_12_12$	D_2^3
		3	$\frac{1}{2}0\frac{1}{2}$	4	$\frac{1}{2}\frac{1}{2}0$			$P2_12_12_1$	D_2^4
$P2mm$	D_2	8	000	9	000			$P2mm$	C_{2v}^1
		8	$00\frac{1}{2}$	9	000			$Pmc2_1$	C_{2v}^2
		8	$00\frac{1}{2}$	9	$00\frac{1}{2}$			$Pcc2$	C_{2v}^3
		8	$\frac{1}{2}00$	9	$\frac{1}{2}00$			$Pma2$	C_{2v}^4
		8	$\frac{1}{2}00$	9	$\frac{1}{2}0\frac{1}{2}$			$Pca2_1$	C_{2v}^5
		8	$0\frac{1}{2}\frac{1}{2}$	9	$0\frac{1}{2}\frac{1}{2}$			$Pnc2$	C_{2v}^6
		8	$\frac{1}{2}0\frac{1}{2}$	9	000			$Pmn2_1$	C_{2v}^7
		8	$\frac{1}{2}\frac{1}{2}0$	9	$\frac{1}{2}\frac{1}{2}0$			$Pba2$	C_{2v}^8
		8	$\frac{1}{2}\frac{1}{2}0$	9	$\frac{1}{2}\frac{1}{2}\frac{1}{2}$			$Pna2_1$	C_{2v}^9
		8	$\frac{1}{2}\frac{1}{2}\frac{1}{2}$	9	$\frac{1}{2}\frac{1}{2}\frac{1}{2}$			$Pnn2$	C_{2v}^{10}
$Pmmm$	$D_2 \times C_2$	7	000	8	000	9	000	$Pmmm$	D_{2h}^1
		7	$\frac{1}{2}\frac{1}{2}\frac{1}{2}$	8	$\frac{1}{2}\frac{1}{2}\frac{1}{2}$	9	$\frac{1}{2}\frac{1}{2}\frac{1}{2}$	$Pnnn$	D_{2h}^2
		7	000	8	$00\frac{1}{2}$	9	$00\frac{1}{2}$	$Pccm$	D_{2h}^3
		7	$\frac{1}{2}\frac{1}{2}0$	8	$\frac{1}{2}\frac{1}{2}0$	9	$\frac{1}{2}\frac{1}{2}0$	$Pban$	D_{2h}^4
		7	$\frac{1}{2}00$	8	000	9	$\frac{1}{2}00$	$Pmma$	D_{2h}^5
		7	$\frac{1}{2}00$	8	$\frac{1}{2}\frac{1}{2}\frac{1}{2}$	9	$0\frac{1}{2}\frac{1}{2}$	$Pnna$	D_{2h}^6
		7	$\frac{1}{2}0\frac{1}{2}$	8	$\frac{1}{2}0\frac{1}{2}$	9	000	$Pmna$	D_{2h}^7
		7	$\frac{1}{2}00$	8	$00\frac{1}{2}$	9	$\frac{1}{2}0\frac{1}{2}$	$Pcca$	D_{2h}^8
		7	000	8	$\frac{1}{2}\frac{1}{2}0$	9	$\frac{1}{2}\frac{1}{2}0$	$Pbam$	D_{2h}^9
		7	$\frac{1}{2}\frac{1}{2}0$	8	$0\frac{1}{2}\frac{1}{2}$	9	$\frac{1}{2}0\frac{1}{2}$	$Pccn$	D_{2h}^{10}
		7	$00\frac{1}{2}$	8	$0\frac{1}{2}\frac{1}{2}$	9	$0\frac{1}{2}0$	$Pbcm$	D_{2h}^{11}
		7	000	8	$\frac{1}{2}\frac{1}{2}\frac{1}{2}$	9	$\frac{1}{2}\frac{1}{2}\frac{1}{2}$	$Pnnm$	D_{2h}^{12}
		7	$\frac{1}{2}\frac{1}{2}0$	8	000	9	000	$Pmmn$	D_{2h}^{13}
		7	$\frac{1}{2}\frac{1}{2}\frac{1}{2}$	8	$00\frac{1}{2}$	9	$\frac{1}{2}\frac{1}{2}0$	$Pbcn$	D_{2h}^{14}
		7	$\frac{1}{2}0\frac{1}{2}$	8	$0\frac{1}{2}\frac{1}{2}$	9	$\frac{1}{2}\frac{1}{2}0$	$Pbca$	D_{2h}^{15}
		7	$\frac{1}{2}0\frac{1}{2}$	8	$0\frac{1}{2}0$	9	$\frac{1}{2}\frac{1}{2}\frac{1}{2}$	$Pnma$	D_{2h}^{16}

V Orthorhombic, body-centered
Primitive translations in unit cell: (000), $(\frac{1}{2}\frac{1}{2}\frac{1}{2})$

*I*222	D_2	3	000	4	000			*I*222	D_2^8
		3	$\frac{1}{2}0\frac{1}{2}$	4	$\frac{1}{2}\frac{1}{2}0$			$I2_12_12_1$	D_2^9
*I*2*mm*	D_2	8	000	9	000			*I*2*mm*	C_{2v}^{20}
		8	$00\frac{1}{2}$	9	$00\frac{1}{2}$			*Iba*2	C_{2v}^{21}
		8	$\frac{1}{2}00$	9	$\frac{1}{2}00$			*Ima*2	C_{2v}^{22}
Immm	$D_2 \times C_2$	7	000	8	000	9	000	*Immm*	D_{2h}^{25}
		7	000	8	$00\frac{1}{2}$	9	$00\frac{1}{2}$	*Ibam*	D_{2h}^{26}
		7	$0\frac{1}{2}0$	8	$\frac{1}{2}00$	9	$00\frac{1}{2}$	*Ibca*	D_{2h}^{27}
		7	$0\frac{1}{2}0$	8	$0\frac{1}{2}0$	9	000	*Imma*	D_{2h}^{28}

VI Orthorhombic, side-centered
Primitive translations in unit cell: (000), $(\frac{1}{2}\frac{1}{2}0)$

*A*222	D_2	3	000	4	000			*C*222	D_2^6
		3	$00\frac{1}{2}$	4	000			$C222_1$	D_2^5
*C*2*mm*	D_2	8	000	9	000			*C*2*mm*	C_{2v}^{11}
		8	$00\frac{1}{2}$	9	000			$C2_1mc$	C_{2v}^{12}
		8	$00\frac{1}{2}$	9	$00\frac{1}{2}$			*C*2*cc*	C_{2v}^{13}
*A*2*mm*	D_2	7	000	8	000			*A*2*mm*	C_{2v}^{14}
		7	$0\frac{1}{2}0$	8	$0\frac{1}{2}0$			*A*2*bm*	C_{2v}^{15}
		7	$00\frac{1}{2}$	8	$00\frac{1}{2}$			*A*2*ma*	C_{2v}^{16}
		7	$0\frac{1}{2}\frac{1}{2}$	8	$0\frac{1}{2}\frac{1}{2}$			*A*2*ba*	C_{2v}^{17}
Cmmm	$D_2 \times C_2$	7	000	8	000	9	000	*Cmmm*	D_{2h}^{19}
		7	$00\frac{1}{2}$	8	$00\frac{1}{2}$	9	000	*Cmcm*	D_{2h}^{17}
		7	$0\frac{1}{2}\frac{1}{2}$	8	$0\frac{1}{2}\frac{1}{2}$	9	000	*Cmca*	D_{2h}^{18}
		7	000	8	$00\frac{1}{2}$	9	$00\frac{1}{2}$	*Cccm*	D_{2h}^{20}
		7	$\frac{1}{2}00$	8	$\frac{1}{2}00$	9	000	*Cmma*	D_{2h}^{21}
		7	$0\frac{1}{2}\frac{1}{2}$	8	$0\frac{1}{2}\frac{1}{2}$	9	$0\frac{1}{2}\frac{1}{2}$	*Ccca*	D_{2h}^{22}

VII Orthorhombic, face-centered

Primitive translations in unit cell: (000), $(\frac{1}{2}\frac{1}{2}0)$, $(\frac{1}{2}0\frac{1}{2})$, $(0\frac{1}{2}\frac{1}{2})$

$F222$	D_2	3	000	4	000			$F222$	D_2^7
$F2mm$	D_2	8	000	9	000			$F2mm$	C_{2v}^{18}
		8	$\frac{1}{4}\frac{1}{4}\frac{1}{4}$	9	$\frac{1}{4}\frac{1}{4}\frac{1}{4}$			$Fdd2$	C_{2v}^{19}
$Fmmm$	$D_2 \times C_2$	7	000	8	000	9	000	$Fmmm$	D_{2h}^{23}
		7	$\frac{1}{4}\frac{1}{4}\frac{1}{4}$	8	$\frac{1}{4}\frac{1}{4}\frac{1}{4}$	9	$\frac{1}{4}\frac{1}{4}\frac{1}{4}$	$Fddd$	D_{2h}^{24}

VIII Tetragonal, primitive

Primitive translations in unit cell: (000)

$P4$	C_4	12	000				$P4$	C_4^1
		12	$00\frac{1}{4}$			isomorphic	$P4_1$	C_4^2
		12	$00\frac{3}{4}$				$P4_3$	C_4^4
		12	$00\frac{1}{2}$				$P4_2$	C_4^3
$P4/m$	$C_4 \times C_2$	12	000	7	000		$P4/m$	C_{4h}^1
		12	$00\frac{1}{2}$	7	000		$P4_2/m$	C_{4h}^2
		12	$\frac{1}{2}\frac{1}{2}0$	7	$\frac{1}{2}\frac{1}{2}0$		$P4/n$	C_{4h}^3
		12	$\frac{1}{2}\frac{1}{2}\frac{1}{2}$	7	$\frac{1}{2}\frac{1}{2}\frac{1}{2}$		$P4_2/n$	C_{4h}^4
$P422$	D_4	12	000	5	000		$P422$	D_4^1
		12	$\frac{1}{2}\frac{1}{2}0$	5	000		$P42_12$	D_4^2
		12	$00\frac{1}{4}$	5	$00\frac{3}{4}$	isomorphic	$P4_122$	D_4^3
		12	$00\frac{3}{4}$	5	$00\frac{1}{4}$		$P4_322$	D_4^7
		12	$\frac{1}{2}\frac{1}{2}\frac{1}{4}$	5	000	isomorphic	$P4_12_12$	D_4^4
		12	$\frac{1}{2}\frac{1}{2}\frac{3}{4}$	5	000		$P4_32_12$	D_4^8
		12	$00\frac{1}{2}$	5	$00\frac{1}{2}$		$P4_222$	D_4^5
		12	$\frac{1}{2}\frac{1}{2}\frac{1}{2}$	5	000		$P4_22_12$	D_4^6
$P4mm$	D_4	12	000	8	000		$P4mm$	C_{4v}^1
		12	000	8	$\frac{1}{2}\frac{1}{2}0$		$P4bm$	C_{4v}^2
		12	$00\frac{1}{2}$	8	$00\frac{1}{2}$		$P4_2cm$	C_{4v}^3
		12	$\frac{1}{2}\frac{1}{2}\frac{1}{2}$	8	$\frac{1}{2}\frac{1}{2}\frac{1}{2}$		$P4_2nm$	C_{4v}^4
		12	000	8	$00\frac{1}{2}$		$P4cc$	C_{4v}^5
		12	000	8	$\frac{1}{2}\frac{1}{2}\frac{1}{2}$		$P4nc$	C_{4v}^6

		12	$00\frac{1}{2}$	8	000			$P4_2mc$	C_{4v}^7
		12	$00\frac{1}{2}$	8	$\frac{1}{2}\frac{1}{2}0$			$P4_2bc$	C_{4v}^8
$P\bar{4}$	C_4	13	000					$P\bar{4}$	S_4^1
$P\bar{4}2m$	D_4	13	000	4	000			$P\bar{4}2m$	D_{2d}^1
		13	000	4	$00\frac{1}{2}$			$P\bar{4}2c$	D_{2d}^2
		13	000	4	$\frac{1}{2}\frac{1}{2}0$			$P\bar{4}2_1m$	D_{2d}^3
		13	000	4	$\frac{1}{2}\frac{1}{2}\frac{1}{2}$			$P\bar{4}2_1c$	D_{2d}^4
$P\bar{4}m2$	D_4	13	000	5	000			$P\bar{4}m2$	D_{2d}^5
		13	000	5	$00\frac{1}{2}$			$P\bar{4}c2$	D_{2d}^6
		13	000	5	$\frac{1}{2}\frac{1}{2}0$			$P\bar{4}b2$	D_{2d}^7
		13	000	5	$\frac{1}{2}\frac{1}{2}\frac{1}{2}$			$P\bar{4}n2$	D_{2d}^8
$P4/mmm$	$D_4 \times C_2$	12	000	8	000	7	000	$P4/mmm$	D_{4h}^1
		12	000	8	$00\frac{1}{2}$	7	000	$P4/mcc$	D_{4h}^2
		12	000	8	$\frac{1}{2}\frac{1}{2}0$	7	$\frac{1}{2}\frac{1}{2}0$	$P4/nbm$	D_{4h}^3
		12	000	8	$\frac{1}{2}\frac{1}{2}\frac{1}{2}$	7	$\frac{1}{2}\frac{1}{2}\frac{1}{2}$	$P4/nnc$	D_{4h}^4
		12	000	8	$\frac{1}{2}\frac{1}{2}0$	7	$\frac{1}{2}\frac{1}{2}0$	$P4/mbm$	D_{4h}^5
		12	000	8	$\frac{1}{2}\frac{1}{2}\frac{1}{2}$	7	$\frac{1}{2}\frac{1}{2}\frac{1}{2}$	$P4/mnc$	D_{4h}^6
		12	$\frac{1}{2}\frac{1}{2}0$	8	000	7	$\frac{1}{2}\frac{1}{2}0$	$P4/nmm$	D_{4h}^7
		12	$\frac{1}{2}\frac{1}{2}0$	8	$00\frac{1}{2}$	7	$\frac{1}{2}\frac{1}{2}0$	$P4/ncc$	D_{4h}^8
		12	$00\frac{1}{2}$	8	000	7	000	$P4_2/mmc$	D_{4h}^9
		12	$00\frac{1}{2}$	8	$00\frac{1}{2}$	7	000	$P4_2/mcm$	D_{4h}^{10}
		12	$\frac{1}{2}\frac{1}{2}\frac{1}{2}$	8	$\frac{1}{2}\frac{1}{2}0$	7	$\frac{1}{2}\frac{1}{2}\frac{1}{2}$	$P4_2/nbc$	D_{4h}^{11}
		12	$\frac{1}{2}\frac{1}{2}\frac{1}{2}$	8	$\frac{1}{2}\frac{1}{2}\frac{1}{2}$	7	$\frac{1}{2}\frac{1}{2}\frac{1}{2}$	$P4_2/nnm$	D_{4h}^{12}
		12	$00\frac{1}{2}$	8	$\frac{1}{2}\frac{1}{2}0$	7	$\frac{1}{2}\frac{1}{2}\frac{1}{2}$	$P4_2/mbc$	D_{4h}^{13}
		12	$\frac{1}{2}\frac{1}{2}\frac{1}{2}$	8	$\frac{1}{2}\frac{1}{2}\frac{1}{2}$	7	000	$P4_2/mnm$	D_{4h}^{14}
		12	$\frac{1}{2}\frac{1}{2}\frac{1}{2}$	8	000	7	$\frac{1}{2}\frac{1}{2}\frac{1}{2}$	$P4_2/nmc$	D_{4h}^{15}
		12	$\frac{1}{2}\frac{1}{2}\frac{1}{2}$	8	$00\frac{1}{2}$	7	$\frac{1}{2}\frac{1}{2}\frac{1}{2}$	$P4_2/ncm$	D_{4h}^{16}

IX Tetragonal, body-centered

Primitive translations in unit cell: (000), $(\frac{1}{2}\frac{1}{2}\frac{1}{2})$

$I4$	C_4	12	000					$I4$	C_4^5
		12	$0\frac{1}{2}\frac{1}{4}$					$I4_1$	C_4^6
$I4/m$	$C_4 \times C_2$	12	000	7	000			$I4/m$	C_{4h}^5
		12	$0\frac{1}{2}\frac{1}{4}$	7	$0\frac{1}{2}\frac{1}{4}$			$I4_1/a$	C_{4h}^6
$I422$	D_4	12	000	15	000			$I422$	D_4^9
		12	$0\frac{1}{2}\frac{1}{4}$	15	000			$I4_1 22$	D_4^{10}
$I4mm$	D_4	12	000	8	000			$I4mm$	C_{4v}^9
		12	000	8	$\frac{1}{2}\frac{1}{2}0$			$I4cm$	C_{4v}^{10}
		12	$0\frac{1}{2}\frac{1}{4}$	8	000			$I4_1md$	C_{4v}^{11}
		12	$0\frac{1}{2}\frac{1}{4}$	8	$00\frac{1}{2}$			$I4_1cd$	C_{4v}^{12}
$I\bar{4}$	C_4	13	000					$I\bar{4}$	S_4^2
$I\bar{4}m2$	D_4	13	000	5	000			$I\bar{4}m2$	D_{2d}^9
		13	000	5	$00\frac{1}{2}$			$I4c2$	D_{2d}^{10}
$I\bar{4}2m$	D_4	13	000	4	000			$I42m$	D_{2d}^{11}
		13	000	4	$0\frac{1}{2}\frac{1}{4}$			$I42d$	D_{2d}^{12}
$I4/mmm$	$D_4 \times C_2$	12	000	8	000	7	000	$I4/mmm$	D_{4h}^{17}
		12	000	8	$00\frac{1}{2}$	7	000	$I4/mcm$	D_{4h}^{18}
		12	$0\frac{1}{2}\frac{1}{4}$	8	000	7	$0\frac{1}{2}\frac{1}{4}$	$I4_1/amd$	D_{4h}^{19}
		12	$0\frac{1}{2}\frac{1}{4}$	8	$00\frac{1}{2}$	7	$0\frac{1}{2}\frac{1}{4}$	$I4_1/acd$	D_{4h}^{20}

X Trigonal, rhombohedral

Primitive translations in unit cell of hexagonal lattice

(000), $(\frac{1}{3}\frac{1}{3}\frac{1}{3})$, $(\frac{2}{3}\frac{2}{3}\frac{2}{3})$

$R3$	C_3	14	000			$R3$	C_3^4
$R32$	D_3	14	000	6	000	$R32$	D_3^7
$R\bar{3}$	C_6	15	000			$R\bar{3}$	C_{3i}^2
$R3m$	D_3	14	000	10	000	$R3m$	C_{3v}^5
		14	000	10	$00\frac{1}{2}$	$R3c$	C_{3v}^6
$R3\overline{m}$	D_6	15	000	10	000	$R\bar{3}m$	D_{3d}^5
		15	000	10	$00\frac{1}{2}$	$R\bar{3}c$	D_{3d}^6

XI Hexagonal, primitive

Primitive translations in unit cell: (000)

$P3$	C_3	14	000				$P3$	C_3^1
		14	$00\frac{1}{3}$			isomorphic	$P3_1$	C_3^2
		14	$00\frac{2}{3}$				$P3_2$	C_3^3
$P312$	D_3	14	000	5	000		$P312$	D_3^1
		14	$00\frac{1}{3}$	5	000	isomorphic	$P3_112$	D_3^3
		14	$00\frac{2}{3}$	5	000		$P3_212$	D_3^5
$P321$	D_3	14	000	6	000		$P321$	D_3^2
		14	$00\frac{1}{3}$	6	$00\frac{2}{3}$	isomorphic	$P3_121$	D_3^4
		14	$00\frac{2}{3}$	6	$00\frac{1}{3}$		$P3_221$	D_3^6
$P\bar{3}$	C_6	15	000				$P\bar{3}$	C_{3i}^1
$P31m$	D_3	14	000	11	000		$P31m$	C_{3v}^2
		14	000	11	$00\frac{1}{2}$		$P31c$	C_{3v}^4
$P3m1$	D_3	14	000	10	000		$P3m1$	C_{3v}^1
		14	000	10	$00\frac{1}{2}$		$P3c1$	C_{3v}^3
$P\bar{3}1m$	D_6	15	000	11	000		$P\bar{3}1m$	D_{3d}^1
		15	000	11	$00\frac{1}{2}$		$P\bar{3}1c$	D_{3d}^2
$P\bar{3}1m$	D_6	15	000	10	000		$P\bar{3}m1$	D_{3d}^3
		15	000	10	$00\frac{1}{2}$		$P\bar{3}c1$	D_{3d}^4
$P6$	C_6	16	000				$P6$	C_6^1
		16	$00\frac{1}{6}$			isomorphic	$P6_1$	C_6^2
		16	$00\frac{5}{6}$				$P6_5$	C_6^3
		16	$00\frac{1}{3}$			isomorphic	$P6_2$	C_6^4
		16	$00\frac{2}{3}$				$P6_4$	C_6^5
		16	$00\frac{1}{2}$				$P6_3$	C_6^6
$P6mm$	D_6	16	000	10	000		$P6mm$	C_{6v}^1
		16	000	10	$00\frac{1}{2}$		$P6cc$	C_{6v}^2
		16	$00\frac{1}{2}$	10	$00\frac{1}{2}$		$P6_3cm$	C_{6v}^3
		16	$00\frac{1}{2}$	10	000		$P6_3mc$	C_{6v}^4

$P622$	D_6	16	000	5	000				$P622$	D_6^1
		16	$00\frac{1}{6}$	5	$00\frac{1}{6}$			isomorphic	$P6_122$	D_6^2
		16	$00\frac{5}{6}$	5	$00\frac{5}{6}$				$P6_522$	D_6^3
		16	$00\frac{1}{3}$	5	$00\frac{1}{3}$			isomorphic	$P6_222$	D_6^4
		16	$00\frac{2}{3}$	5	$00\frac{2}{3}$				$P6_422$	D_6^5
		16	$00\frac{1}{2}$	5	$00\frac{1}{2}$				$P6_322$	D_6^6
$P6/m$	$C_6 \times C_2$	16	000	7	000				$P6/m$	C_{6h}^1
		16	$00\frac{1}{2}$	7	$00\frac{1}{2}$				$P6_3/m$	C_{6h}^2
$P\bar{6}$	C_6	17	000						$P\bar{6}$	C_{3h}^1
$P\bar{6}m2$	D_6	17	000	5	000				$P\bar{6}m2$	D_{3h}^1
		17	$00\frac{1}{2}$	5	000				$P\bar{6}c2$	D_{3h}^2
$P\bar{6}2m$	D_6	17	000	6	000				$P\bar{6}2m$	D_{3h}^3
		17	$00\frac{1}{2}$	6	000				$P\bar{6}2c$	D_{3h}^4
$P6/mmm$	$D_6 \times C_2$	16	000	10	000	7	000		$P6/mmm$	D_{6h}^1
		16	000	10	$00\frac{1}{2}$	7	000		$P6/mcc$	D_{6h}^2
		16	$00\frac{1}{2}$	10	$00\frac{1}{2}$	7	$00\frac{1}{2}$		$P6_3/mcm$	D_{6h}^3
		16	$00\frac{1}{2}$	10	000	7	$00\frac{1}{2}$		$P6_3/mmc$	D_{6h}^4

XII Cubic, primitive
Primitive translations in unit cell: (000)

$P23$	T	18	000	3	000				$P23$	T^1
		18	000	3	$\frac{1}{2}0\frac{1}{2}$				$P2_13$	T^4
$Pm3$	$T \times C_2$	18	000	3	000	2	000		$Pm3$	T_h^1
		18	000	3	000	2	$\frac{1}{2}\frac{1}{2}\frac{1}{2}$		$Pn3$	T_h^2
		18	000	3	$\frac{1}{2}0\frac{1}{2}$	2	000		$Pa3$	T_h^6
$P432$	O	12	000	19	000				$P432$	O^1
		12	$\frac{1}{2}\frac{1}{2}\frac{1}{2}$	19	000				$P4_232$	O^2
		12	$\frac{3}{4}\frac{1}{4}\frac{3}{4}$	19	000			isomorphic	$P4_332$	O^6
		12	$\frac{1}{4}\frac{3}{4}\frac{1}{4}$	19	000				$P4_132$	O^7
$P\bar{4}3m$	O	13	000	19	000				$P\bar{4}3m$	T_d^1
		13	$\frac{1}{2}\frac{1}{2}\frac{1}{2}$	19	000				$P\bar{4}3n$	T_d^4

$Pm3m$	$O \times C_2$	12	000	19	000	2	000	$Pm3m$	O_h^1
		12	000	19	000	2	$\frac{1}{2}\frac{1}{2}\frac{1}{2}$	$Pn3n$	O_h^2
		12	$\frac{1}{2}\frac{1}{2}\frac{1}{2}$	19	000	2	000	$Pm3n$	O_h^3
		12	$\frac{1}{2}\frac{1}{2}\frac{1}{2}$	19	000	2	$\frac{1}{2}\frac{1}{2}\frac{1}{2}$	$Pn3m$	O_h^4

XIII Cubic, body-centered

Primitive translations in unit cell: (000), $(\frac{1}{2}\frac{1}{2}\frac{1}{2})$

$I23$	T	18	000	3	000			$I23$	T^3
		18	000	3	$\frac{1}{2}0\frac{1}{2}$			$I2_13$	T^5
$Im3$	$T \times C_2$	18	000	3	000	2	000	$Im3$	T_h^5
		18	000	3	$0\frac{1}{2}0$	2	000	$Ia3$	T_h^7
$I432$	O	12	000	19	000			$I432$	O^5
		12	$\frac{3}{4}\frac{1}{4}\frac{3}{4}$	19	000			$I4_132$	O^8
$I\bar{4}3m$	O	13	000	19	000			$I\bar{4}3m$	T_d^3
		13	$\frac{3}{4}\frac{1}{4}\frac{3}{4}$	19	000			$I\bar{4}3d$	T_d^6
$Im3m$	$O \times C_2$	12	000	19	000	2	000	$Im3m$	O_h^9
		12	$\frac{3}{4}\frac{1}{4}\frac{3}{4}$	19	000	2	000	$Ia3d$	O_h^{10}

XIV Cubic, face-centered

Primitive translations in unit cell: (000), $(\frac{1}{2}\frac{1}{2}0)$, $(\frac{1}{2}0\frac{1}{2})$, $(0\frac{1}{2}\frac{1}{2})$

$F23$	T	18	000	3	000			$F23$	T^2
$Fm3$	$T \times C_2$	18	000	3	000	2	000	$Fm3$	T_h^3
		18	000	3	000	2	$\frac{1}{4}\frac{1}{4}\frac{1}{4}$	$Fd3$	T_h^4
$F432$	O	12	000	19	000			$F432$	O^3
		12	$\frac{1}{4}\frac{1}{4}\frac{1}{4}$	19	000			$F4_132$	O^4
$F\bar{4}3m$	O	13	000	19	000			$F\bar{4}3m$	T_d^2
		13	$00\frac{1}{2}$	19	000			$F\bar{4}3c$	T_d^5
$Fm3$	$O \times C_2$	12	000	19	000	2	000	$Fm3m$	O_h^5
		12	$00\frac{1}{2}$	19	000	2	000	$Fm3c$	O_h^6
		12	$\frac{1}{4}\frac{1}{4}\frac{1}{4}$	19	000	2	$\frac{1}{4}\frac{1}{4}\frac{1}{4}$	$Fd3m$	O_h^7
		12	$\frac{1}{4}\frac{1}{4}\frac{1}{4}$	19	000	2	$\frac{1}{4}\frac{1}{4}\frac{3}{4}$	$Fd3c$	O_h^8

BIBLIOGRAPHY

In the following list we give the references cited in the text as well as some other papers treating solid state problems by group theoretical methods, but which were not discussed in the preceding pages.

In the list appear a number of books on related subjects which can be useful for consultation, further reading or a deeper understanding. A book which has much in common with the present one is Streitwolf [1967]. Other books especially concerned with the symmetry of the solid state are Knox and Gold [1967], which includes many reprints of important papers, and the reprint volume Meijer [1964]. For the mathematical treatment of crystallographic point groups and space groups we recommend Burckhardt [1966]. The representation theory of these groups is treated in Bradley and Cracknell [1971]. In Buerger [1963] the space groups and their elements are discussed more from the point of view of a crystallographer.

Altmann, S.L. and C.J. Bradley, 1965: Lattice harmonics II, Rev. Mod. Phys. **37**, 33–45.

Altmann, S.L. and A.P. Cracknell, 1965: Lattice harmonics I, Rev. Mod. Phys. **37**, 19–32.

Ascher, E., 1966a: Role of some maximal subgroups in continuous phase transitions, Phys. Letters **20**, 352–357.

Ascher, E., 1966b: Some properties of spontaneous currents, Helv. Phys. Acta **39**, 40–48.

Ascher, E., 1967: Symmetry changes in continuous phase transitions, a simplified theory applied to V_3Si, Chem. Phys. Letters **1**, 69–72.

Ascher, E. and A. Janner, 1965a: Subgroups of black–white point groups, Acta Cryst. **18**, 325–330.

Ascher, E. and A. Janner, 1965b: Algebraic aspects of crystallography I: space groups as extensions, Helv. Phys. Acta **38**, 551–572.

Ascher, E. and A. Janner, 1968: Algebraic aspects of crystallography II: nonprimitive translations in space groups, Comm. Math. Phys. **11**, 138–167.

Aviram, A. and J. Zak, 1968: Matrix elements in systems with nonunitary symmetry, J. Math. Phys. **9**, 2138–45.

Backhouse, N.B., 1970, 1971: Projective representations of space groups II and III, Qu. J. Math. Oxford (2) **21**, 277–295; **22**, 277–290.

Backhouse, N.B. and C.J. Bradley, 1970: Projective representations of space groups I, Qu. J. Math. Oxford (2) **21**, 203–222.

Bell, D.G., 1954: Group theory and crystal lattices, Rev. Mod. Phys. **26**, 311.
Belov, N.V., N.N. Neronova and T.S. Smirnova, 1957: Soviet Phys. Cryst. **2**, 311.
Bethe, H., 1929: Termaufspaltung in Kristallen, Ann. Phys. **3**, 133 (English edition by Consultants Bureau, New York).
Bhagavantam, S., 1966: Crystal symmetry and physical properties, Academic Press, New York.
Bieberbach, L., 1911: Über die Bewegungsgruppen der Euklidischen Raüme, Math. Ann. **70**, 297–336.
Bieberbach, L., 1912: II Die Gruppen mit einem endlichen Fundamentalbereich, Math. Ann. **70**, 400–412.
Birman, J.L., 1962: Space group selection rules, Phys. Rev. **127**, 1093.
Birman, J.L., 1963: Theory of infrared and Raman processes: selection rules in diamond and zincblende, Phys. Rev. **131**, 1489–96.
Birman, J.L., 1966: Full-group and subgroup methods in crystal physics, Phys. Rev. **150**, 771–782.
Birman, J.L., 1967: Contribution to "Ferroelectricity" (Weller ed.), Elsevier.
Birss, R.R., 1964: Symmetry and magnetism, North-Holland, Amsterdam.
Boerner, H., 1967: Representations of groups, North-Holland, Amsterdam.
Boon, M., 1964: The dependence of the anisotropic Knight shift on crystal symmetry, Physica **30**, 1326–40.
Born, M. and K. Huang, 1954: Dynamical theory of crystal lattices, Oxford University Press, Oxford.
Bouckaert, L., R. Smoluchowski and E. Wigner, 1936: Theory of the Brillouin zone and symmetry properties of wave functions in crystals, Phys. Rev. **50**, 58–67.
Bradley, C.J., 1966: Space groups and selection rules, J. Math. Phys. **7**, 1145–52.
Bradley, C.J. and A.P. Cracknell, 1966: Corepresentations of magnetic point groups, Prog. Theor. Phys. **36**, 648–9.
Bradley, C.J. and A.P. Cracknell, 1970: Some comments on the theory of space group representations, J. Phys. C3, 610–618.
Bradley, C.J. and A.P. Cracknell, 1971: Mathematical theory of symmetry in solids, Oxford University Press, Oxford.
Bradley, C.J. and B.L. Davies, 1968: Magnetic groups and their corepresentations, Rev. Mod. Phys. **40**, 359–379.
Bradley, C.J., D.E. Wallis and A.P. Cracknell, 1970: Some comments on the theory of space group representations II, J. Phys. C3, 619–626.
Brinkman, W. and R.J. Elliott, 1966: Space group theory for spin waves, J. Appl. Phys. **37**, 1457–59.
Brinkman, W., 1967: Magnetic symmetry and spin waves, J. Appl. Phys. **38**, 939–943.
Brott, C., 1966: Program zur Bestimmung absolut irreduzibler Charaktere und Darstellungen endlicher Gruppen, Diplomarbeit, University of Kiel.
Brown, E., 1964: Bloch electrons in a uniform magnetic field, Phys. Rev. **133** A, 1038.
Brown, E., 1968: Aspects of group theory in electron dynamics, Solid State Physics (ed. Seitz and Turnbull) Academic Press, New York, **22**, 313–408.
Buerger, M.J., 1963: Elementary crystallography, John Wiley, New York.
Burckhardt, J.J., 1966: Die Bewegungsgruppen der Kristallographie (2nd ed.) Birkhaüser, Basel.
Chen Li-Ching, R. Berenson and J.L. Birman, 1968: Space group selection rules: *Fm3m*, Phys. Rev. **170**, 639–648.

Clifford, A.H., 1937: Representations induced in an invariant subgroup, Ann. Math. **38**, 533–550.

Coleman, A.J., 1968: Induced and subduced representations, in: "Group theory and its applications" (M. Loebl ed.) Academic Press, New York.

Cornwell, J.F., 1969: Group theory and electronic energy bands in solids, North-Holland, Amsterdam.

Cornwell, J.F., 1971: Origin dependence of the symmetry labelling of electron and phonon states in crystals, Phys. stat. sol. **43**, 763–767.

Cracknell, A.P., 1968: Crystal field theory and the Shubnikov point groups, Adv. Phys. **17**, 367–420.

Cracknell, A.P., 1969: Symmetry adapted functions for double point groups I, Non-cubic groups, Proc. Cambridge Phil. Soc. **65**, 567–578.

Dimmock, J.O., 1963: Use of symmetry in the determination of magnetic structures, Phys. Rev. **130**, 1337–44.

Dimmock, J.O. and R.G. Wheeler, 1964: Symmetry properties of magnetic crystals, in: "Mathematics of physics and chemistry II" (Margenau and Murphy eds.) D. Van Nostrand, Princeton.

Doering, W., 1959: Die Strahldarstellungen der kristallographischen Gruppen, Z. Naturforsch. **14a**, 343–350.

Elliott, R.J. and R. Loudon, 1960: Group theory of scattering processes in crystals, J. Phys. Chem. Solids **15**, 146.

Fast, G. and T. Janssen, 1971: Determination of *n*-dimensional space groups by means of an electronic computer, J. Comp. Phys. **7**, 1–11.

Fedorow, E. von, 1892: Zusammenstellung der kristallographischen Resultate, Z. Krist. **20**.

Fischbeck, H.J., 1963: Theory of a Bloch electron in a magnetic field, Phys. stat. sol. **3**, 1082, 2399.

Fischbeck, H.J., 1968: General theory of the Bloch electron in a magnetic field, Phys. stat. sol. **30**, 779–789.

Fischbeck, H.J, 1970: Theory of Bloch electrons in a magnetic field, Phys. stat. sol. **38**, 11–62.

Flodmark, S., 1968: Group theory in solid state physics, in: "Group theory and its applications" (M. Loebl ed.) Academic Press, New York.

Flodmark, S. and E. Blokker, 1967: A computer program for calculation of irreducible representations of finite groups, Int. J. Quantum Chem. **1s**, 703–711.

Frei, V., 1967: On energy bands of crystals with lowered point symmetry, Phys. stat. sol. **22**, 381–390.

Glück, M., Y. Gur and J. Zak, 1967: Double representations of space groups, J. Math. Phys. **8**, 787–790.

Griffith, J.S., 1961: The theory of transition-metal ions, Cambridge University Press, Cambridge.

Hall, M., 1959: The theory of groups. Mac Millan, New York.

Hamermesh, M., 1962: Group theory, Addison Wesley, Reading.

Harper, P.G., 1955: General motion of conduction electrons in a uniform magnetic field, Proc. Phys. Soc. (London) **A68**, 879–892.

Harter, W.G., 1969: Algebraic theory of ray representations of finite groups, J. Math. Phys. **10**, 739–752.

Heesch, H., 1929: Z. Krist. **71**, 95.

Heine, V., 1960: Group theory in quantum mechanics, Pergamon Press, New York.

Henry, N.F.M. and K. Lonsdale, 1965: International Tables for X-ray crystallography, I symmetry groups, Kynoch Press, Birmingham.

Herman, F., 1959: Lattice vibrational spectrum of germanium, J. Phys. Chem. Sol. **8**, 405–18, 421–2.

Hermann, C., 1948: Kristallographie in Raüme beliebiger Dimensionszahl, Acta Cryst. **2**, 139–145.

Herring, C., 1937a: Effect of time reversal symmetry on energy bands of crystals, Phys. Rev. **52**, 361–365.

Herring, C., 1937b: Accidental degeneracy in the energy bands of crystals, Phys. Rev. **52**, 365–373.

Herzfeld, C. and P. Meijer, 1961: Group theory and crystal field theory, in: "Solid state physics" (Seitz and Turnbull eds) Vol. 12.

Hurley, A.C., 1951: Finite rotation groups and crystal classes in 4 dimensions, Proc. Cambridge Phil. Soc. **47**, 650–661.

Hurley, A.C., 1966a: Ray representations of point groups and irreducible representations of space groups and double space groups, Phil. Trans. Roy. Soc. A **260**, 1–36.

Hurley, A.C., 1966b: Finite rotation groups and crystal classes in four dimensions, in: "Quantum theory of atoms, molecules and the solid state" (Löwdin ed.) Academic Press, New York.

Janner, A., 1966: On Bravais classes of magnetic lattices, Helv. Phys. Acta **39**, 665–682.

Janner, A. and E. Ascher, 1970a: Space-time symmetries of crystal diffraction, Physica **46**, 162–164.

Janner, A. and E. Ascher, 1970b: Relativistic symmetry groups of uniform electromagnetic fields, Physica **48**, 425–446.

Janner, A. and T. Janssen, 1971: Electromagnetic compensating gauge transformations, Physica **53**, 1–27.

Jansen, L. and M. Boon, 1967: The theory of finite groups, North-Holland, Amsterdam.

Janssen, T., 1969: Crystallographic groups in space and time III, Physica **42**, 71–92.

Janssen, T., 1972: On projective unitary/antiunitary representations of finite groups, J. Math. Phys., **13**, 342–51.

Kitz, A., 1965: Über die irreduzible Darstellungen der Raumgruppen und die Strahlvorstellungen der kristallographischen Punktgruppen, Phys. stat. sol. **8**, 813–829.

Klauder, L.T. and J.G. Gay, 1968: Note on Zak's method for constructing representations of space groups, J. Math. Phys. **9**, 1408–09.

Knox, R. and A. Gold, 1967: Symmetry properties in the solid state, Benjamin, New York.

Koptzik, V.A., 1966: Shubnikovskie Gruppi (in Russian) Izdatelstwo Moskowskowo Universiteta, Moscow.

Koster, G.K., 1957: Space groups and their representations, in: "Solid state physics" (Seitz and Turnbull eds.) Vol. 5.

Koster, G., J. Dimmock, R. Wheeler and H. Statz, 1963: Properties of the 32 point groups, MIT-press, Cambridge (Mass.).

Kovalev, V., 1961: Irreducible representations of space groups (in Russian) Izdatelstwo Ak. Nauk SSSR, Kiev.

Landau, L. and E. Lifshitz, 1959: Statistical physics, Pergamon, London.

Lax, M., 1965: Subgroup techniques in crystal and molecular physics, Phys. Rev. **138**, A 793–802.

Lax, M. and J.J. Hopfield, 1961: Selection rules connecting different points in the Brillouin zone, Phys. Rev. **124**, 115.

Litvin, D.B. and J. Zak, 1968: Clebsch–Gordan coefficients for space groups, J. Math. Phys. **9**, 212–221.

Lomont, J.S., 1959: Applications of finite groups, Academic Press, New York.

Löwdin, P.O., 1967: Group algebra, convolution algebra and applications to quantum mechanics, Rev. Mod. Phys. **39**, 259–287.

Luehrmann, A.W., 1968: Crystal symmetry of plane-wave-like functions, I symmorphic groups, Adv. Phys. **17**, 1–78.

Lyubarski, G., 1962: Anwendungen der Gruppentheorie in der Physik, VEB Deutsche Verlag, Berlin.

Mac Kay, J., 1968: A method for computing the character table of a finite group, in: "Computers in mathematical research" (Churchhouse ed.) North-Holland, Amsterdam.

Mackey, G.W., 1955: Lectures given at the University of Chicago.

Mackey, G.W., 1958: Unitary representations of group extensions, Acta Math. **99**, 265–311.

Maradudin, A.A. and S.H. Vosko, 1968: Symmetry properties of the normal vibrations of a crystal, Rev. Mod. Phys. **40**, 1–37.

Maradudin, A.A., E.W. Montroll and G. Weiss, 1963: Theory of lattice dynamics in the harmonic approximation, Academic Press, New York.

Messiah, A., 1961: Quantum mechanics, North-Holland, Amsterdam.

Meijer, P.H., 1964: Group theory and solid state physics, I A selection of papers, Gordon and Breach, New York.

Miller, S. and W.F. Love, 1967: Tables of irreducible representations of space groups and co-representations of magnetic space groups, Pruett Press, Boulder.

Murthy, M.V., 1966: Ray representations of finite nonunitary groups, J. Math. Phys. **7**, 853–857.

Nussbaum, A., 1966: Crystal symmetry, group theory and band structure calculations, in: "Solid state physics" (Seitz and Turnbull eds.) Academic Press.

Opechowski, W., 1940: Sur les groupes cristallographiques doubles, Physica **7**, 552–562.

Opechowski, W. and R. Guccione, 1965: Magnetic symmetry, in: "Magnetism" (Rado and Suhl eds.) Vol. IIa, Academic Press.

Opechowski, W. and W.G. Tam, 1969: Invariance groups of the Schrödinger equation for the case of a uniform magnetic field, Physica **42**, 529–556.

Overhof, H. and U. Roessler, 1968: Magnetic space groups and their irreducible representations, Phys. stat. sol. **26**, 461–468.

Puff, H., 1970: Contribution to the theory of cubic harmonics, Phys. stat. sol. **41**, 11–22.

Raghavacharyulu, I.V.V., 1961a: Representations of space groups, Can. J. Phys. **39**, 830–839.

Raghavacharyulu, I.V.V., 1961b: Normal vibrations in crystals, Can. J. Phys. **39**, 1704–20.

Rudra, P., 1965a: On projective representations of finite groups, J. Math. Phys. **6**, 1273–77.

Rudra, P., 1965b: On irreducible representations of space groups, J. Math. Phys. **6**, 1278–82.

Rudra, P., 1966: On irreducible tensor operators for finite groups, J. Math. Phys. **7**, 935–937.

Schoenflies, A., 1891: Krystallsysteme und Krystallstructur, Leipzig.

Schur, I., 1904: Über die Darstellungen der endlichen Gruppen durch gebrochene lineare Substitutionen, J. reine u. angew. Math. **127**, 20–50.

Schur, I., 1907: Untersuchungen über die Darstellung der endlichen Gruppen durch gebrochene lineare Substitutionen, J. reine u. angew. Math. **132**, 85–137.

Seitz, F.: A matrix-algebraic development of the crystallographic groups, Z. Krist. I **88**, 433 (1934); II **90**, 289 (1935); III **91**, 336 (1935); IV **94**, 100 (1936).

Shubnikov, A.V., N.V. Belov et al., 1964: Coloured symmetry (Holser ed.) MacMillan, New York.

Streitwolf, H.W., 1967: Gruppentheorie in der Festkörperphysik, Geest und Portig, Leipzig.

Tam, W.G., 1969: Invariance group of the Schrödinger equation for the case of a uniform magnetic field II, Physica **42**, 557–564.

Tinkham, M., 1964: Group theory and quantum mechanics, Mac Graw Hill, New York.

Waeber, W., 1969: Lattice vibrations of gallium, J. Phys. C **2**, 882.

Warren, J.L., 1968: Further considerations on the symmetry properties of the normal vibrations of a crystal, Rev. Mod. Phys. **40**, 38–76.

Wigner, E., 1932: The operation of time reversal in quantum mechanics, Göttinger Nachrichten 546–559.

Wigner, E., 1939: On unitary representations of the inhomogeneous Lorentz group, Ann. Math. **40**, 149–204.

Wigner, E., 1959: Group theory, Academic Press, New York.

Wintgen, G., 1941: Zur Darstellungstheorie der Raumgruppen, Math. Ann. **118**, 195–215.

Zak, J., 1960: Method to obtain the character tables of nonsymmorphic space groups, J. Math. Phys. **1**, 165–171.

Zak, J., 1962: Selection rules for integrals of Bloch functions, J. Math. Phys. **3**, 1278–79.

Zak, J., 1964a: Magnetic translation group, Phys. Rev. **134**, A 1602–11.

Zak, J., 1964b: Group-theoretical consideration of Landau level broadening in crystals, Phys. Rev. **136**, A 776–780.

Zak, J., 1969: The irreducible representations of space groups, Benjamin, New York.

Zassenhaus, H., 1947: Über einen Algorithmus zur Bestimmung der Raumgruppen, Comm. Math. Helv. **21**, 117–141.

INDEX